Dreamscapes of Modernity

Dreamscapes of Modernity

Sociotechnical Imaginaries and the Fabrication of Power

EDITED BY SHEILA JASANOFF
AND SANG-HYUN KIM

The University of Chicago Press
Chicago and London

Sheila Jasanoff is the Pforzheimer Professor of Science and Technology Studies at the Harvard Kennedy School. **Sang-Hyun Kim** is associate professor at the Research Institute of Comparative History and Culture at Hanyang University in Korea.

The University of Chicago Press, Chicago 60637
The University of Chicago Press, Ltd., London
© 2015 by The University of Chicago
All rights reserved. Published 2015.
Printed in the United States of America

24 23 22 21 20 19 18 17 16 15   1 2 3 4 5

ISBN-13: 978-0-226-27649-6 (cloth)
ISBN-13: 978-0-226-27652-6 (paper)
ISBN-13: 978-0-226-27666-3 (e-book)
DOI: 10.7208/chicago/9780226276663.001.0001

Library of Congress Cataloging-in-Publication Data

Dreamscapes of modernity : sociotechnical imaginaries and the fabrication of power / edited by Sheila Jasanoff and Sang-Hyun Kim.
    pages cm
  Includes bibliographical references and index.
    ISBN 978-0-226-27649-6 (cloth : alk. paper) — ISBN 978-0-226-27652-6 (pbk. : alk. paper) — ISBN 978-0-226-27666-3 (e-book) 1. Science—Social aspects. 2. Technological innovations—Social aspects. I. Jasanoff, Sheila. II. Kim, Sang-Hyun, 1967–
    Q175.5.D74 2015
    303.48'3—dc23

                        2014050176

⊚ This paper meets the requirements of ANSI/NISO Z39.48-1992 (Permanence of Paper).

CONTENTS

CONTENTS

ONE

Future Imperfect: Science, Technology, and the Imaginations of Modernity

SHEILA JASANOFF

Technological innovation often follows on the heels of science fiction, lagging authorial imagination by decades or longer. One hundred fifty years passed between the youthful Mary Shelley's fantastic story of a vengeful creature brought to life by Dr. Frankenstein and the production of new life forms in twentieth-century biological laboratories (Shelley 2008 [1818]). Jules Verne's Nautilus, piloted by Captain Nemo, took to the ocean depths well before real submarines went on such long or distant voyages (Verne 1887). At the dawn of the Progressive Era, the American socialist Edward Bellamy (1889) foresaw an economy fueled by rapid communication, credit cards, and in-home delivery of goods; a hundred years on, those imagined revolutions have become routine. Aldous Huxley (1932) fantasized about an assembly line of artificial human reproduction to serve state purposes twenty years before the unraveling of the structure of DNA, which in turn paved the way for the currently forbidden cloning of human beings. Arthur C. Clarke (1968) created the scheming, lip-reading computer Hal thirty years before IBM programmers developed Deep Blue to beat chess master Gary Kasparov at his own game. And interplanetary travel was in the minds of such writers as H.G. Wells, Fred Wilcox, and Fred Hoyle appreciably before Neil Armstrong stepped onto the moon with his "giant leap for mankind."

Belying the label "science fiction," however, works in this genre are also fabulations of social worlds, both utopic and dystopic. Shelley's lab-generated monster turns murderous because he is excluded from society by his abnormal birth and hence is denied the blessings of companionship and social life enjoyed by his creator. Jules Verne's Nemo, a dispossessed Indian prince driven by hatred of the British colonialists who exploited his land and destroyed his family, seeks freedom and scientific enlightenment in the ocean depths. Biopower runs amok in Aldous Huxley's imagined world, overwhelming human dignity and autonomy in the name of collective needs under authoritarian rule. Equally concerned with the interplay of social and material innovation, but reversing the emotional gears, Edward Bellamy's look backward from an imagined 2000 offers, first, an optimistic account of a new social order and only secondarily a foray into technological unknowns. And as a dystopic counterpoint, George Orwell's (1949) *Nineteen Eighty-Four* presents a world of totalitarian thought control overseen by a technologically advanced, all-seeing, all-knowing, 24/7 surveillance state—whose real-life counterpart Edward Snowden, the whistle-blowing, twenty-first-century American contractor, famously revealed in the US National Security Agency.

Oddly, though, many nonfictional accounts of how technology develops still treat the material apart from the social, as if the design of tools and machines, cars and computers, pharmaceutical drugs and nuclear weapons were not in constant interplay with the social arrangements that inspire and sustain their production. In popular discourse the word "technology" tends to be equated with machine or invention, something solid, engineered, black boxed, and these days most likely an instrument of electronic communication. Yet cars as we know them would never have taken to the roads without the myriad social roles, institutions, and practices spawned by modernity: scientists, engineers, and designers; patents and trademarks; autoworkers and big corporations; regulators; dealers and distributors; advertising companies; and users, from commuters to racers, who ultimately gave cars their utility, appeal, and meaning. Similar observations can be made about contraceptives, computers, cell phones, and countless other artifacts that serve our needs while, to varying degrees, arousing our desires. Technological objects, in other words, are thoroughly enmeshed in society, as integral components of social order; one does not need fictive or futuristic stories to recognize this truth.

Bringing social thickness and complexity back into the appreciation of technological systems has been a central aim of the field of science and technology studies (STS). Historians and social analysts of technology have worked in tandem to remind us that there can be no machines without humans to make them and powerful institutions to decide which technologies are worth our investment (Winner 1986). This literature resists the temptation to construe technology as deterministic. STS scholars tend to bristle at the evolutionary economist's language of strict path dependence (David 1985; Arthur 1994). STS accounts recognize that history matters, as indeed it must, but reject the notion of rigid lock-ins in favor of a more open sense of agency and contingency in society's charting of technological possibilities. Many aspects of the presenting face of technological systems are socially constructed (Bijker et al. 1987). The stamp of conscious or unconscious human choice and user preference marks the design of objects, their weighting of risks and benefits, and the behaviors they encourage, exclude, or seek to regulate (Callon 1987; Jasanoff 2006).

Less frequently encountered in the STS literature, however, are conceptual frameworks that situate technologies within the integrated material, moral, and social landscapes that science fiction offers up in such abundance. To be sure, the normative dimensions of science and technology do not fall wholly outside the scope of STS analysis. STS scholarship acknowledges that science and technology do not unidirectionally shape our values and norms. Rather, and symmetrically, our sense of how we ought to organize and govern ourselves profoundly influences what we make of nature, society, and the "real world." The idiom of coproduction explicitly foregrounds this two-way dynamic:

> Briefly stated, co-production is shorthand for the proposition that the ways in which we know and represent the world (both nature and society) are inseparable from the ways in which we choose to live in it. Knowledge and its material embodiments are at once products of social work and constitutive of forms of social life; society cannot function without knowledge any more than knowledge can exist without appropriate social supports. Scientific knowledge, in particular, is not a transcendent mirror of reality. It both embeds and is embedded in social practices, identities, norms, conventions, discourses, instruments, and institutions—in short, in all the building blocks of what we term the social. The same can be said even more forcefully of technology. (Jasanoff 2004a, 2–3)

For all its analytic potential, however, the notion of coproduction does more to advance the Weberian project of *Verstehen* (understanding subjectively how things fit together) than the scientific goal of *Erklären* (explaining objectively how things come to be as they are). It lacks the specificity that might allow us to elucidate certain persistent problems and difficulties of the modern technoscientific world. Left unaccounted for by the bare idiom of coproduction are some of the biggest "why" questions of history—why upheavals sometimes seem to come from nowhere and why attempts to remake the world sometimes fail despite much concerted effort and expenditure of resources. Puzzles also include cross-national and cross-cultural divergences in technological development that lack obvious grounding in natural, economic, or social disparities. It is important to understand in a time of globalization why different moral valences attach to new scientific ideas and technological inventions throughout the world and why differences persist in what we might call the constitutional position of science and technology in the political order (Jasanoff 2012b; Dennis; Miller, this volume).

The idea of sociotechnical imaginaries confronts some of these challenges head on. Our starting point is the definition Sang-Hyun Kim and I offered in an earlier study of US and South Korean responses to nuclear power: national sociotechnical imaginaries are "collectively imagined forms of social life and social order reflected in the design and fulfillment of nation-specific scientific and/or technological projects" (Jasanoff and Kim 2009, 120). This definition, as we show in this volume, needs to be refined and extended in order to do justice to the myriad ways in which scientific and technological visions enter into the assemblages of materiality, meaning, and morality that constitute robust forms of social life. Sociotechnical imaginaries, as elaborated in the following chapters, are not limited to nation-states as implied in our original formulation but can be articulated and propagated by other organized groups, such as corporations, social movements, and professional societies. Though collectively held, sociotechnical imaginaries can originate in the visions of single individuals or small collectives, gaining traction through blatant exercises of power or sustained acts of coalition building. Only when the originator's "vanguard vision" (Hilgartner 2015) comes to be communally adopted, however, does it rise to the status of an imaginary. Multiple imaginaries can coexist within a society in tension or in a productive dialectical relationship. It often falls to legislatures, courts, the media, or other institutions of power to elevate some imagined futures above others, according them a dominant position for policy purposes. Imaginaries, moreover, encode not only visions of what is attainable through science and technology but also of how life ought, or ought not, to be lived; in this respect they express a society's shared understandings of good and evil.

Taking these complexities into account, we redefine sociotechnical imaginaries in this book as collectively held, institutionally stabilized, and publicly performed visions of desirable futures, animated by shared understandings of forms of social life and social order attainable through, and supportive of, advances in science and technology. This definition privileges the word "desirable" because efforts to build new sociotechnical futures are typically grounded in positive visions of social progress. It goes without saying that imaginations of desirable and desired futures correlate, tacitly or explicitly, with the obverse—shared fears of harms that might be incurred through invention and innovation, or of course the failure to innovate. The interplay between positive and negative imaginings—between utopia and dystopia—is a connecting theme throughout this volume.

In this chapter, I lay out the theoretical precursors that inform our work on sociotechnical imaginaries and outline the major methodological approaches by which we make the term analytically tractable. Imaginaries are securely established in interpretive social theory as a term of art referring to collective beliefs about how society functions. Yet, as I show below, little has been done to link that notion to modernity's grand aspirations and adventures with science and technology. This absence is all the more perplexing because the performative dimensions of a society's self-reproduction—the enactment and reenactment of its imaginaries—so heavily depend on experiment and demonstration, practices that are intimately linked to science and technology (Ezrahi 1990; Hilgartner 2000; Jasanoff 2012b). In contrast to social theory in general, STS theorizing affirms the centrality of science and technology in the making and stabilizing of collectives, although STS has paid relatively less attention to the aspirational and normative dimensions of social order captured by the notion of imaginaries.

Sociotechnical imaginaries as illustrated by the contributors to this collection occupy the blank space between two important literatures, the construction of imaginaries in political and cultural theory and of sociotechnical systems in STS (e.g., Bijker 1997; Bijker et al. 1987). The concept helps explain a number of otherwise troublesome problems: why do technological trajectories diverge across polities and periods; what makes some sociotechnical arrangements more durable than others; how do facts and technologies transcend and reconstruct time and space; and what roles do science and technology play in connecting the individual's subjective self-understanding to a shared social or moral order? The chapter then addresses the practical questions that arise in working with this theoretical concept: when does it make sense to invoke sociotechnical imaginaries and what methods and sources are most appropriate for identifying these constructs and their constitutive elements? Lastly, the chapter lays out a map of the major thematic connections among the empirical case studies that follow.

Imagination as a Social Practice

Modern societies prize imagination as an attribute of the creative individual. It is the faculty that allows the extraordinary person to see beyond the limits of constraining reality and to make or do things that are out of the ordinary. We rightly celebrate the seer, the visionary, the transformative political thinker. But imagination also operates at an intersubjective level, uniting members of a social community in shared perceptions of futures that should or should not be realized. Prior efforts to theorize the collective imagination constitute a fundamentally important strand in the genealogy of sociotechnical imaginaries.

More than a century after the seminal writings of Durkheim and Weber, we take for granted that vibrant societies share common narratives of who they are, where they have come from, and where they are headed. These stories are reflected in rituals of giving and receiving, producing and consuming, birth, marriage, and death. Uncovering these tacit ordering rules even in foreign and distant cultures was the project of anthropology from its colonial origins. Thus, the great structural-functionalist Evans-Pritchard (1937), who helped import Durkheim into anthropology (Kuklick 1992), attributed allegations of witchcraft among the Zande of Central Africa to a logic of averting the chaos of ignorance. Witchcraft on Evans-Pritchard's reading supported order by assigning otherwise inexplicable events to discernible social causes. His student Mary Douglas adopted a similar analytic stance in disentangling beliefs about pollution in premodern societies, eventually extending her ideas to relations between social structures and contemporary perceptions of risk in her work on cultural theory (Douglas 1966; Douglas and Wildavsky 1980). These studies blurred the lines between real and imagined realities, showing how observed facts of nature are refracted through collective desires for logic and order, producing authoritative representations of how the world works—as well as how it should work. In the language of STS, all these works can be seen as broadly illustrative of the phenomenon of coproduction (Jasanoff 2004a).

Early ethnographers did not fail to see that political systems make up a particular kind of imagined reality whose rules are amenable to anthropological investigation. Evans-Pritchard and Meyer Fortes, for example, edited a collection of essays on political systems in sub-Saharan Africa for the International African Institute (Fortes and Evans-Pritchard 1940). Notably, however, this kind of analysis was rarely directed toward modern societies; instead, realist accounts of states predominated in political theory, and little analytic room was left for such nebulous, hard to quantify factors as social imaginations. In his classic work *Imagined Communities*, Benedict Anderson sliced through the divide between ethnography and political science with his now famous definition of a nation as "an imagined political community—

and imagined as both inherently limited and sovereign" (Anderson 1991 [1983], 6). Nationalism, on his reading, is a construct of minds that may never encounter each other in reality but nevertheless are tied together through shared practices of narrating, recollecting, and forgetting. Not only did Anderson's move provide a powerful explanation for what unifies something so heterogeneous and spatially dispersed as a nation, it also validated the cultural, historical, and comparative investigation of the psychosocial attributes of political collectives.

Following Anderson's lead, Charles Taylor (2004) expanded the analysis of collective imaginations to address grand patterns of historical and political thought. How, Taylor asks in the opening pages of *Modern Social Imaginaries*, did modernity come about, with its distinctive complex of new practices and institutions, new ways of living, and new forms of malaise? His explanation can be summed up in two words: imaginaries changed. But how does Taylor define an imaginary, let alone one that looks distinctively modern and social? Here is his answer: "By social imaginary, I mean something much broader and deeper than the intellectual schemes people may entertain when they think about reality in a disengaged mode. I am thinking, rather, of the ways people imagine their social existence, how they fit together with others, how things go on between them and their fellows, the expectations that are normally met, and the deeper normative notions and images that underlie these expectations" (Taylor 2004, 23).

We recognize here, as in Anderson's invocation of the imagination, an anthropological vision that rejects the idea of politics as consisting simply of purposive, rational action. Taylor looks instead to a society's moral practices, those tacit rules for "how things go on between them and their fellows" that make up the foundations of social order. An imaginary in Taylor's scheme of things involves not only common understandings and practices based on a sense of what is real but also "a widely shared sense of legitimacy" about how to order lives in relation to those realities. In STS terms again, this is an incipiently coproductionist perspective that bridges, without explicitly saying so, the epistemic and the normative, the objective and the subjective. But Taylor's imaginaries do not have a space for the material aspects of order.

Social imaginaries in Anderson's and Taylor's analyses can hold very big things together, such as nationhood or modernity. But imaginaries can also operate at substantially smaller scales. Indeed, Arjun Appadurai, whose much admired 1990 essay on globalization and diasporas influenced thought far outside his field, uses the concept of imaginaries to dissolve the notion of a universal, homogeneous modernity. For Appadurai, globalization consists of disjointed flows or "scapes"—of people, technology, money, electronic communications, and ideas—each constituted by the overlapping but not necessarily coherent practices of the people engaging in them:

> No longer mere fantasy (opium for the masses whose real work is elsewhere), no longer simple escape (from a world defined principally by more concrete purposes and structures), no longer elite pastime (thus not relevant to the lives of ordinary people), and no longer mere contemplation (irrelevant for new forms of desire and subjectivity), the imagination has become an organized field of social practices, a form of work (both in the sense of labor and of culturally organized practice) and a form of negotiation between sites of agency ("individuals") and globally defined fields of possibility. (Appadurai 2002, 50)

It is the turn from a purely mentalist notion of the imagination as fantasy to imagination as organized work and practices that puts Appadurai on a continuum with Anderson and Taylor. As we see below, and indeed throughout this volume, this way of thinking about the imagination is also consistent with current trends in science and technology studies, although STS scholars are likely to find troubling Appadurai's implication that "scapes" flow independently of one another in their complex global circulation.

A startling, almost inexplicable omission from all of these classic accounts of social imaginaries is a detailed investigation of modernity's two most salient forces: science and technology. Anderson's imagined communities were bound together by the medium of newsprint, but technologies of communication as such play little or no role in his storytelling, except perhaps via the inclusion of museums and maps (along with the census) in the book's expanded second edition. In three passing mentions, almost as afterthoughts, Taylor in *Modern Social Imaginaries* subsumes science and technology into the aggregated institutional changes that mark the emergence of modernity. But he pays little attention to their instrumental or transformative role, even in relation to the "multiple modernities"—"different ways of erecting and animating the institutional forms that are becoming inescapable" (Taylor 2004, 195)—which he takes to be emblematic of the contemporary condition. Appadurai sees flows of technology as part of the disjointed and multiple nature of current realities, but he too fails to engage with the seminal role of knowledge and its materializations in generating and anchoring imaginaries of social order. These are not accidental gaps but, as the leading STS scholar Bruno Latour has insistently argued, a systematic obscuration in the imagination of the social sciences themselves.

For example, in the famous 1971 debate between Noam Chomsky and Michel Foucault on Dutch television, neither giant of twentieth-century social thought, both deeply attuned to the history and politics of science, paid much attention to material inventiveness, or the *"grille"* of technology, in their accounts of human nature, power, and justice.[1] Bridging this gap in the analysis, indeed in the apprehension, of modernity is a central purpose of this introductory essay and this entire collection.

Curiously, too, performance as a social practice gets short shrift in much of the theorizing on imaginaries, even though theatricality has been part of the machinery of statecraft and rulership from the earliest times. Machiavelli, writing in exile in 1513 and addressing his work to Lorenzo di Piero de Medici, grandson of Lorenzo "the Magnificent," called attention to the importance of spectacle to a ruler's reputation: "Nothing gives a prince more prestige than undertaking great enterprises and setting a splendid example" (Machiavelli 1977 [1513], 65). He noted too the delicate balance that politics must strike between displays of greatness and of familiarity, both essential to the prince's public standing: "He should also at fitting times of the year, entertain his people with festivals and spectacles. And because every city is divided into professional guilds and family groupings, he should be inward with these people, and attend their gatherings from time to time, giving evidence of his humanity and munificence, yet avoiding any compromise to his dignity, for that must be preserved at all costs" (Machiavelli 1977 [1513], 65–66).

Localized in time and place, Machiavelli's prescriptions nonetheless resonated far beyond his immediate circumstances. The cult status of successful European monarchs from Louis XIV of France to Elizabeth I of England, dubbed Gloriana by her subjects, bears witness (Strong 1984, 1987). In Britain, the union of state building with monarchical pomp and pageantry persisted down the ages, through Queen Victoria's acclaimed Diamond Jubilee celebration in 1897, at the high-water mark of the British Empire (Morris 2003), down to the rain-drenched but feel-good Thames flotilla, fittingly led by a royal barge named Gloriana, that provided visual distraction for an economically depressed British nation at Queen Elizabeth II's Diamond Jubilee in the summer of 2012.

That same summer's Olympic Games in London, however, provided spectacles relying less on royal history and more on Britain's artistic and cultural heritage, liberally spiced with high-tech fantasy. Epitomizing that postmodern synthesis was a hugely popular video of Queen Elizabeth herself making a mock parachute landing in the Olympic stadium accompanied by Daniel Craig, the latest incarnation of James Bond in Ian Fleming's perennially popular franchise. The Bond stories showcase not only Britain's mechanical inventiveness but Britishness writ large[2]: the shots of the helicopter daringly skimming under the Thames bridges and the parachute descent itself conjured up Britain's heroic World War II history, when mastery of the air proved essential for the nation's defense. The video—which soon attracted more than a million viewers on YouTube—blended together memory, technology, the monarchy, and popular culture in a performance designed to play to every register in Britain's happiest imaginations of itself. It reinforced nationhood on many levels at once, and it did so in part by appealing to what we call sociotechnical imaginaries.

Performance, Visibility, and Instrumentalism

Bringing performance back into the landscape of political theory helps reposition science and technology as key sites for the constitution of modern social imaginaries. Performances of statehood in modernity are increasingly tied to demonstrations and to public proofs employing scientific and technological instruments; equally, however, acts of popular resistance, from terrorist attacks to Wikileaks, draw on the same repertoires of technoscientific imagination and instrumental action. That histories of science and technology are interwoven with political histories is not in itself a novel claim; in particular, it will not raise eyebrows among social scientists familiar with STS. Yet the mechanics of the interconnections between technoscientific and political practice have not been articulated in detail or systematically. A few landmark works serve as milestones for explorers, but the map of the highways and byways that link science, technology, and state-making lacks its Mercator or even its Ptolemy. Particularly empty of theoretical guidelines is the domain that connects creativity and innovation in science, and even more technology, with the production of power, social order, and a communal sense of justice.

A promising starting point is the notion of "technoscientific imaginaries" developed by George Marcus (1995) and his colleagues in the anthropology of science and technology. At first blush, this term seems to perform the very same bridging that we, too, seek to accomplish in this volume. Yet, while Marcus notes in his editorial introduction that technoscientific imaginaries might have encompassed the "reflective, visionary thoughts of scientists," this is not the direction his essay collection pursues. Instead, in a move more consistent with disciplinary anthropology than STS, Marcus and his colleagues "were much more interested in the imaginaries of scientists tied more closely to their current positionings, practices, and ambiguous locations in which the varied kinds of science they do are possible at all" (Marcus 1995, 4). As in all work on imaginaries, the focus in the resulting, highly individual accounts is on futures and future possibilities, but the context of the imagination is the scientific workplace, and imagination's aims and achievements are tied to forms of scientific production. Our ambition in this book is spatially and temporally larger and more symmetrical. It is to investigate how, through the imaginative work of varied social actors, science and technology become enmeshed in performing and producing diverse visions of the collective good, at expanding scales of governance from communities to nation-states to the planet. This is why we choose the term "sociotechnical" (not technoscientific) to characterize our elaboration of imaginaries.

For this purpose, a more congenial point of departure is *Leviathan and the Air Pump*, the classic account by Steven Shapin and Simon Schaffer (1985) of the conflicts between Robert Boyle and Thomas Hobbes in Restoration England. The book does not use the term "imaginaries," but it is at its heart a story of competing, coproduced imaginations of natural and social orders. Boyle and Hobbes, Shapin and Schaffer argue, were fighting for the same dyad of causes: how to establish truth and how to achieve authority in a time of immense epistemic as well as political upheaval. Their study of the controversy suggests that what was at stake in that revolutionary moment was not simply the legitimacy of scientific experiment, although Boyle the scientist and Hobbes the political philosopher[3] conflicted in their views of whether seeing an experiment could be a valid basis for believing its findings. Implicated as well in these two men's quarrels was the emergence of a democratic public sphere in which authority would depend on experimentally verifiable truths, observable in principle by everyone, rather than on declarations from an inaccessible central authority such as the monarch. In short, the rise of the experimental method—which depends on transparency, a common language for speaking about matters of fact, and the assent of witnesses who are not necessarily in the room with the experimentalist—simultaneously laid the foundations for the political movement toward modern democracy.[4] Experiments, in this telling, were important performative occasions, requiring carefully orchestrated meetings of minds and eyes to build consensus around what was being shown and seen.

The political scientist Yaron Ezrahi carried forward these suggestive connections between epistemic and political performance in his *Descent of Icarus* (Ezrahi 1990). According to Ezrahi, the shift of viewpoints introduced by experimental science eventually permeated political culture, allowing subjects who had previously functioned as mere consumers of the state's displays of authority to become skeptical witnesses of its claims. Democratization entailed in effect the conversion of the "celebratory" eye of the passive subject into the "attestive" gaze of the modern citizen, able to question and evaluate the factual assertions of those in power. We are reminded here of Immanuel Kant's famous description of *Mündigkeit* as "the human being's emergence from his self-incurred minority" (Kant 1996, 11). This is the state of maturity attained by enlightened humans when they learn to think for themselves without leaning on others for guidance. Importantly, however, Kantian enlightenment is tied to an inward capacity to reason on one's own, similar to attaining adulthood or independence, whereas the transformation that Ezrahi posits, following Shapin and Schaffer, relates more to the capacity to apprehend natural facts for what they are, in short, to trust the empirical evidence of one's senses.

Ezrahi's democratic theory reopens a space for political performance, a space in which technology, in addition to science, finds an explicit role. In his political universe, the democratic state is sensitive to a continual need to prove itself to witnessing citizens. This ongoing demand for accountability can most easily be met through public demonstrations of power and efficacy, leading to increasingly instrumental uses of technology. In an evocative passage, Ezrahi calls attention to the ritual that goes on at the Kennedy Space Center when American citizens from all parts of the country are given a tour of the premises, to observe how their state's contributions helped create the marvels on display: "Perhaps the most important artifact is the body of a Saturn 5, a gigantic space leviathan whose carcass lies wide open in a didactic gesture toward curious taxpayers always eager to be informed" (Ezrahi 1990, 42). From an author of Ezrahi's erudition, the use of "leviathan" here is no accident: Saturn 5 is a material manifestation of the American federation, and the tour guide's enthusiastic efforts to tie the machine's components back to the visitors' home states is nothing less than a performance of nationalism to train, and retain, the loyalty of citizens.

Performance becomes yet more centrally the lens through which Ezrahi looks at politics in a later book, *Imagined Democracies* (Ezrahi 2012). Here his focus is on the "necessary fictions" that societies adopt when they perform democracy. Those fictions make democracy work, despite all the hidden backstage machinery that also makes democracy as we experience it a matter of artifice, illusion, and pretense. Technologies seen in this light operate as performative scripts that combine values and interests, materializing and making tangible the invisible components of social imaginaries. Such performances in turn embed technological systems into the "masonry of political world-making" (Ezrahi, personal communication).

Almost inevitably, Ezrahi's historical and imaginative sweep comes at the expense of specificity. *Descent of Icarus* tends to merge all of European culture into a single monolithic formation marked by ambivalence toward technology, as opposed to the instrumental enthusiasm that Ezrahi attributes to the United States. Yet the evolution of engineering and of technological systems, as well as the status and power of engineering in society, followed distinctive paths in Britain, France, and Germany. These national trajectories included institutionalized differences in educational systems for science and engineering, in the role of these fields in elite formation, and in political culture, via what I have called civic epistemologies (Jasanoff 2005). Twentieth-century history might have looked quite different if all of Europe had held uniformly skeptical views toward technology. Science and technology continue to play diverse legitimating functions in the world's newer democracies, corresponding to differences in the nature and status of expertise and in cultural expectations about evidence and proof in the public sphere. Simplifying these subtle variations into binaries, such as esthetic Europe versus utilitarian United States, misses the finer threads that help define the place of science in the distinctive political and constitutional cultures—and imaginaries—of sovereign nations and their polities.

Michel Foucault's assessment of the power of inspection in his elaboration of Jeremy Bentham's idea of the panopticon (Foucault 1979, 195–228) contrasts markedly with the emancipatory role that Ezrahi ascribes to the transparency of science's experimental regime. Ezrahi takes his cues from Jefferson, Paine, and Priestly, all men of the Enlightenment, and perhaps more Kantian than Foucauldian in their commitment to reason. He observes, "once it is the government itself which becomes an object of increasing observation, inspection as a technique of control is transformed into a democratic instrument for holding authority publicly accountable" (Ezrahi 1990, 116). This, however, fails to take on board the constructedness of seeing in all its complexity. The viewer after all construes what she sees; in turn, the viewer's capacity for observation is socially trained in ways that delimit what she can perceive. The state, too, commands innumerable devices that occlude vision and limit transparency, such as large databases, weapons programs, and laws of official secrecy. Sight, to borrow a term from Foucault's repertoire, operates within the *grille* of historical conditioning (Chomsky and Foucault 2006), with "its choices and exclusions" determining what can be seen and what passes unnoticed (consider, for example, the critique of courtroom witnessing in Jasanoff 1998).[5]

Appropriately, in an era dominated by the mass media, the filmmaker Akira Kurosawa provided a memorable challenge to the very possibility of being all seeing. *Rashomon,* Kurosawa's midcentury masterpiece, dramatized how the same "reality" is perceived in radically different ways depending on the position, perspective, and indeed imagination of the observer (Kurosawa 1950). Yet, perversely, vision still remains the great naturalizer. What we "see" in familiar surroundings looks right, epistemically as well as normatively. So the socially conditioned eye can take for granted that all-male orchestras or all-black passengers on the backseats of buses, or even scenes of filth and abject poverty simply represent the rightful order of things. And, as Foucault preeminently observed, when bodies are well disciplined to live inside those orders, what looks natural from on high may not be so different from what looks natural from below. The same collective imaginary may condition and constrain the sense of justice that binds a community.[6] Other ways of seeing and reasoning—ways that would make injustice palpable— may not enter anyone's imagination, even in democratic societies, and hence may never give rise to organized criticism or opposition, let alone to revolutions that could hold power accountable, or at the extreme overthrow it.

To understand order and, its obverse, disorder in contemporary societies, we need an encompassing theoretical framework that draws together our scientifically and culturally conditioned perceptions of reality, our capacity to create new collectives through technological as well as social means, and the changes in expectation that arise when science and technology interact with individual self-awareness and the sense of being well ruled. The idiom of coproduction offers such a framework: it is symmetrically concerned with mutual emergences in how one thinks the world is and what one determines it ought to be (Jasanoff 2004a). Work in the coproductionist vein sensitizes us to the ways in which elements of human subjectivity and agency get bound up with technoscientific advances through adjustments in identities, institutions, and discourses that accompany new representations of things. It offers an entry point into the means by which *is* and *ought* remain fitted together while our awareness of the world and what to make of it both move. Less explicitly, the idiom of coproduction also allows us to consider how time and space are involved in the formation, or reformation, of conceptual, material, and social orders, thereby helping to explicate such pervasive shifts in consciousness as the Reformation, the Enlightenment, decolonization, globalization, racial and sexual emancipation, and modernity itself (Jasanoff 2010; Jasanoff and Martello 2004). More needs to be done, however, as this volume shows, to clarify why, at significant forks in the road, societies opt for particular directions of choice and change over others and why those choices gain stability or, at times, fail to do so.

The Flatness of Networks

One influential way of accounting for both stability and instability is actor network theory (ANT), an STS framework developed by French sociologists Michel Callon and Bruno Latour at the Ecole des Mines in Paris in the 1980s. ANT offers a systematic mode of inquiry into the connections between humans and the animate and inanimate features of the environments they make and inhabit. In other words, it offers a conceptual foundation for examining the nature of the "sociotechnical." In this respect, ANT is a significant strand in the genealogy that sociotechnical imaginaries draw upon, but the two concepts also decisively part company in their treatment of power and normativity.

ANT grew out of a felt need to bring human relations with nonhumans, and with materiality more generally, back into sociology. ANT thus seeks to avoid preconceived analytic boundaries between the components that hold social systems together. All are seen as hybrids composed of heterogeneous elements: people, objects, nonhuman entities, organizations, and texts are taken as interactive participants in the networks that make up the structures of modernity. To correct for the humanistic bias of classical sociology (see Latour 1988, 35–40), Callon and Latour put forward the notion of *actants*, nonhuman agents that mediate among humans and help mold their collectives. This allowed the authors to pursue what they termed a symmetrical approach to society and nature. Callon (1986) famously insisted on using the same terminology to account for modes of resistance and engagement that occur among scientists, fishermen, and scallops when a form of scallop cultivation was imported from Japan to France's St. Brieuc Bay. Latour's provocative history of pasteurization represented microbes as powerful agents, not only channeling Louis Pasteur's efforts to come to grips with them in the laboratory but eventually extending their force outward to transform farming, medicine, markets, and society. In his signature polemical style Latour pronounced, "There are not only 'social' relations. Relations between man and man. Society is not made up just of men, for everywhere microbes intervene" (Latour 1988, 35). Not only microbes but hosts of inanimate objects, such as maps, legal reports, speed bumps, and door locks share space with humans in Latour's ordering of social relations, forming a kind of dark matter of society. To this array, Callon (1998) and his followers have added the instruments that make modern economies function, such as the infamous credit derivatives that were blamed for the worldwide economic collapse in 2008.

These moves are enormously appealing because they dissolve binaries that seem intolerably rigid in complex modern societies: nature-culture, science-society, subject-object, human-nonhuman. More important for the social sciences, ANT's vision of networked societies encourages greater attentiveness to forms of distributed agency and action—and hence of dispersed causality—that disciplinary training tends to simplify or dismiss. The political theorist Timothy Mitchell, who embraced the ANT approach in writing about Egypt's political modernization, placed on a par the military invasion of the country by British forces from the north and the biological invasion by the malarial mosquito from the south. Normal history, Mitchell suggested, errs in giving voice only to humans when narrating such periods of nation building. In reality, things happen in mixed-up ways. It takes power, as Foucault and other historians of the human sciences have long seen, to create demarcations and simplifications in a world of hybridity: "indeed producing the effect of neatly separate realms of reason and the real world, ideas and their objects, the human and the nonhuman, was how power was coming to work in Egypt, and in the twentieth century in general" (Mitchell 2002, 52). It follows for Mitchell that the traditional social sciences uncritically replicate modernity's established forms of power by paying homage to the very intellectual binaries and categorical separations that are the characteristic outputs of modernity (see in this connection Latour 1993).

Truthfulness in the social sciences today, most would agree, demands simultaneous attention to more forms of agency, more pathways of change, and more narratives of causation than single disciplines are wont to provide. In this respect, ANT and the new investigations of materiality (a trend some call "speculative realism") in STS perform a valuable function. They urge us not to take any aspect of the world for granted as natural or given, and hence foreclosed to investigation, even those that seem to hold still and do nothing; but instead to look around at all the compass points from which forces originate to make up reality as we see it. Such analysis in the round should be mindful of all the devices—not only law or policy or culture or armed might—with which power seeks to achieve its ends. Yet this hugely appealing celebration of mixtures, hybrids, and complexity suffers from its own fecundity. It is too distributive, too promiscuous in attributing cause and agency. As even friendly critics have observed (e.g., Mitchell 2002, 52–3; Farias and Bender 2010, 305), it risks a kind of moral nihilism, making all actions and agents seem equally empowered, or disempowered, and therefore equally responsible, or irresponsible, for the networks within which they function. Network-based accounts seem in this respect to play into and reinforce what Ulrich Beck (1998) has called modernity's "organized irresponsibility."

The preoccupation with hybridity also risks establishing a troubling normative equivalence between nonhuman and human agents. Gifted writers can make anything speak, in the sense that their stories give voice to that thing and captivates readers with the subversive pleasure of hearing from entities usually held to be mute. Animals talked and even frogs demanded a ruler in Aesop's popular fables, and life forms seamlessly transmuted into one another in Ovid's fantastic narratives. In our day, when science has liberated nature from such enchantments, giving voice and agency to things can be seen as a form of rebellion, an enterprise of reenchanting. One need only look at the explosion of interest in cyborgs and interspecies ethnography (Haraway 1991, 2003) or the transhuman and the posthuman (Hayles 1999). Yet, it is still humans and their collectives who can imagine a world—or a continent as Helen Tilley (2011) argues in her work on Africa and William Storey's essay on Cecil Rhodes explores in this volume—that is governable by science and technology and emptied of mosquitoes. Only humans can devise the strategies of disciplining and targeted eradication that may accomplish such wonders. Maybe the mosquito can speak, or be ventriloquized by an exceptional storyteller. But can the mosquito imagine? In this book, we argue that imagination, a crucial reservoir of power and action, lodges in the hearts and minds of human agents and institutions, although imagination's skilled implementation requires putting in play the intricate networks whose construction has been the stuff of so much STS analysis.

If networks diffuse responsibility, they can also depoliticize power by making its actions opaque or invisible. Here again, a cardinal virtue of network analysis, namely, its utility in explaining how big formations cohere, calls for a confrontation with critical political theory. Illustrative for these purposes is Bruno Latour's influential essay "Drawing Things Together," in which he argued that the diffusion of scientific ideas can be attributed to two linked phenomena: the production of inscriptions that simplify and "flatten" the world, making "immutable mobiles"; and their subsequent distribution by "centers of calculation" that enable these representations to draw together actors and actions far outside the initial loci of production (Latour 1990).

Latour's most vivid example of the mobility of inscriptions is remarkable for its elision of power:

La Pérouse travels through the Pacific for Louis XVI with the explicit mission of bringing *back* a better map. One day, landing on what he calls Sakhalin, he meets with Chinese and tries to learn from them whether Sakhalin is an island or a peninsula. To his great surprise the Chinese understand geography quite well. An older man stands up and draws a map of his island on the sand with the scale and the details needed by La Pérouse. Another, who is younger, sees that the rising tide will soon erase the map and picks up one of La Pérouse's notebooks to draw the map again with a pencil . . .

What are the differences between the savage geography and the civilized one? (Latour 1990, 24)

A cartographic mission is undertaken at the behest of a king, with explicit aims of advancing science, expanding trade, and establishing French outposts in the Pacific. It is hard to miss the colonizing undertones, albeit in this case both monarch and minion came to violent, untimely ends: La Pérouse's entire expedition mysteriously disappeared in 1788, a year before the revolution that consumed his royal patron's life. Latour's language, however, preserves the hierarchy of center and periphery, contrasting the "savage geography" of the older Chinese man's sand drawing with the "civilized" geography of La Pérouse's scientific team.[7] Latour himself is far too knowing to buy into such easy binaries as "savage" and "civilized": "There is no need to bring a prescientific mind into the picture," he goes on to say. Nevertheless, the relationships he describes appear natural, part of the order of things, and well in line with France's famed *mission civilisatrice*. There is no acknowledgment here of the turbulent histories through which centers of calculation obtain the resources to draw things together or the force and violence often required to make representations circulate. Raw power has little overt place in actor-network narratives, which tend not to disrupt science's own self-presentation as gentlemanly, civilized, and civilizing (by contrast, see Visvanathan 1997; Scott 1998; Jasanoff 2004a, 26–27). Disrupting this flatness, revealing the topographies of power, is one aim of work on sociotechnical imaginaries.

For a sharply contrasting vision of science's mobility, we can turn to Donna Haraway's spirited deconstruction and reconstruction of a site of scientific representation, the African Hall of New York's American Museum of Natural History—a place in which time, space, and power intertwine altogether less innocently (Haraway 1989). Haraway's pathbreaking essay on the dioramas produced by the hunter, photographer, scientific taxidermist, and naturalist Carl Akeley offers a riposte against the formal symmetries between human and nonhuman that form the backbone of ANT. For Haraway (1989, 55), "Sciences are woven of social relations throughout their tissues. The concept of social relations must include the entire complex of interactions among people; objects, including books, buildings, and rocks; and animals." Seen in this light, the dioramas she interprets are first and foremost "meaning machines." Like all machines they freeze social relations, reinforcing an impression of predestination that analysts should seek to dissolve. This Haraway does with gusto, bringing to life the extravagant male egos of Akeley and his friend President Theodore Roosevelt, their intrusive, omnipresent, violent cameras, the dead but immortalized animals, and the silenced voices and bodies of Africans and women, including Akeley's irreverent first wife Delia, who were altogether alive and active during the adventures that secured the trophies for the museum. No longer mute representations of truth to nature, the dioramas become in Haraway's telling raced and gendered objects created to give that era's anxious white American males the illusion that nature is still there to be fought and conquered in trials of male vitality. If Haraway's African exhibit functions as a center of calculation in Latour's terms, then we see that its very construction is a project of politics—it is a site (in my terms) of coproduction. The exhibit reflects and reinforces a specific, historically situated, American sociotechnical imaginary in which nature and manliness are simultaneously defended against threats from urbanization. The science of natural history thus ends up speaking truths subservient to the power of a specific cultural imagination.

Sociotechnical Imaginaries

Sociotechnical imaginaries occupy the theoretically undeveloped space between the idealistic collective imaginations identified by social and political theorists and the hybrid but politically neutered networks or assemblages with which STS scholars often describe reality. Our definition pulls together the normativity of the imagination with the materiality of networks: sociotechnical imaginaries thus are "collectively held and performed visions of desirable futures" (or of resistance against the undesirable), and they are also "animated by shared understandings of forms of social life and social order attainable through, and supportive of, advances in science and technology." Unlike mere ideas and fashions, sociotechnical imaginaries are collective, durable, capable of being performed; yet they are also temporally situated and culturally particular. Moreover, as captured by the adjective "sociotechnical," these imaginaries are at once products of and instruments of the coproduction of science, technology, and society in modernity.

We have located sociotechnical imaginaries in genealogies that refer specifically to theories of national and social identity in political theory and to hybrid collectives in STS. The performative dimension of sociotechnical imaginaries, however, also relates this term to concepts more closely linked to instrumental political action—in other words, to policy as well as politics. Differences between theoretical concepts are notoriously hard to pin down by definition alone, and refining them in practice entails considerable elaboration and boundary work by communities of scholars over time. Nevertheless, it is worth making a few quick observations about the relationship between sociotechnical imaginaries and related ideas in studies of public policy. Possibly closest in spirit is the concept of a *master narrative*. Like an imaginary, a master narrative—such as "American exceptionalism"—offers a rationale for a society's long evolutionary course while also committing that society to keep performing the imagined lines of the story. But a master narrative implies a more monolithic and unchangeable vision, closely bound to a singular retelling of national and cultural history, and not necessarily welcoming of invention or prescriptive of new goals to be achieved.

Discourse shares with imaginaries the properties of being collective and systemic (e.g., Hajer 1995), but it usually focuses on language and is less directly associated with action and performance or with materialization through technology. Political ideologies are perhaps more obviously tied to power and social structure than are sociotechnical imaginaries, but ideology is generally seen as entrenched and immovable. Ideology also lacks the imagination's properties of reaching and striving toward possible futures, and ideology has not typically been analyzed as being encoded in material technologies. Policy itself refers to formal or tacit programs of action, not to the underlying rationale or justification that may be provided by sociotechnical imaginaries. Similarly, a plan conveys the intentionality of sociotechnical imaginaries, but it usually refers to near-term futures with specific, designated goals (e.g., a plan to build a weapon or a highway) and is usually a product of formal institutional authority rather than a shared cultural property. In the same vein, a project usually involves a single, targeted, technological endpoint, such as the Apollo moon landing, the "cure for cancer," the sequencing of the human genome, or the mapping of the brain; such projects may themselves reflect animating sociotechnical imaginaries. Finally, unlike sociotechnical imaginaries, which can be articulated and advocated for from below the seats of power, public reason tends to be shaped by institutionalized relations between citizens and the political authorities who govern them (Jasanoff 2012b).

These contrasts carry us forward to some degree, but a more user-friendly way to operationalize sociotechnical imaginaries is to ask what explanatory work the concept enables. Here we lay out four ways in which sociotechnical imaginaries help overcome some limitations of earlier work in both STS and political theory. The first problem for which the concept provides answers is that of difference, in particular the unexpected divergence of sociotechnical outcomes across political regimes, even among liberal democracies that share fundamental aspirations and commitments. If, as political scientists have traditionally argued, exogenous events drive political agendas (Kingdon 2010), then after momentous happenings convergence rather than divergence ought to be the rule. One might expect, for example, that nuclear disasters like Chernobyl and Fukushima would generate equivalent fear and revulsion around the globe; the hacking of climate scientists' e-mails at the University of East Anglia in 2009 would provoke similar skepticism and distrust toward climate science everywhere; and policy framings for new bio- or nanotechnologies would converge over time from Washington to Brussels and New Delhi to Tokyo. Similarly, revolutionary discoveries such as the decoding of DNA should drive social values toward similar patterns of rejection or acceptance of engineered forms of life. Yet the reverse is often true (Jasanoff 2005, 2011a; Jasanoff and Kim 2009, 2013). Discrepancies persist in responses to new and emerging technologies and technological disasters, suggesting that even earth-shattering events are absorbed and integrated into preexisting imaginaries in ways that forestall globally homogeneous meaning making and policy formulation.

A second problem is time and its companion, change. Past and future connect in a complex dialectic that is widely acknowledged. The past is prologue, but it is also a site of memory excavated and reinterpreted in the light of a society's understanding of the present and its hopes for what lies ahead. As Alberto Melucci (1996, 12) put it, while "the future is born of the past, it is equally true that the past is also continuously shaped by the future." But why do people's expectations of how things fit together (in Charles Taylor's formulation), and how they ought to fit, remain stable for long durations, so that we can speak of extended eras such as modernity? And when sweeping change happens—the Arab Spring, for example, or decolonization, or the fall of the Iron Curtain—where does the impetus come from and how in turn does it take hold? Clearly some account of relative embedding, or rootedness, is needed for us to understand both durability and change; I return to this point in the final chapter. The popular though disconcertingly flat metaphor of networks, whose durability depends on the thickness of horizontal linkages and the density of nodal connectivities, does scant justice either to the historical *longue durée* or to moments when things catastrophically fall apart.

The materiality of technoscience, as this collection demonstrates, is surely implicated in the stability and instability of social arrangements, but just as important are the belief systems out of which those materialities emerge and which give them value and meaning. A better balance needs to be struck between the theoretical poles of abstract idealism and deterministic materialism. By turning to sociotechnical imaginaries, we can engage directly with the ways in which people's hopes and desires for the future—their sense of self and their passion for how things ought to be—get bound up with the hard stuff of past achievements, whether the material infrastructures of roads, power plants, and the security state or the normative infrastructures of constitutional principles, juridical practices, and public reason (Jasanoff 2012b). Technological systems serve on this view a doubly deictic function, pointing back at past cultural achievements and ahead to promising and attainable futures, or to futures to be shunned and avoided.

A third problem, especially familiar to critical geographers, is space. Space and social order are coproduced in part through the spread of ideas and practices—and indeed ideologies—across times and territories. Views and practices originating with individuals or small groups acquire governing force across much wider domains, both physical and temporal. In the concluding chapter I refer to this phenomenon as extension. For STS scholars, it is tempting to put science and technology at the heart of such stories because science is modernity's ultimate traveler, its findings accepted everywhere as universal. Latour (1988), as noted earlier, used actor network theory to great effect in his account of the spread of pasteurization, a case of one man and one scientific idea that—with the help of microbes—took over a country and eventually the world. But are scientists and engineers, and the materialities they harness, really prime movers in building the grand architectures of states and markets, or of empires, let alone the myriad lesser constellations of meaningful spaces nested within those encompassing structures? And does the world really come as unconfigured and available to be reorganized, with only nodal frictions and struggles for power, as some STS accounts of the spread of scientific and technological networks suggest? ANT stories, as we have already observed, risk flattening—even sanitizing—the circulation of knowledge in a world of persistent inequality and dominance.

Sociotechnical imaginaries tackle head-on, and more symmetrically, the complex topographies of power and morality as they intersect with the forces of science and technology. As the term itself suggests, the concept allows for spatial imaginations to preexist and channel the spread of science and technology, instead of only vice versa, as when a Cecil Rhodes, with his dreams of conquest, bestrides Africa like a colossus (Storey, this volume); or when

President Lyndon B. Johnson's press secretary, George Reedy, is persuaded of the "poetry" of the space program on a deserted hilltop outside Austin, Texas (Jasanoff 2004b, 40); or when an institution such as the World Health Organization mobilizes expert technical resources around common fears of a global pandemic (Lakoff; Miller, this volume). Then, too, by allowing for competition among different visions of futures, the framework of sociotechnical imaginaries restores some of the indeterminacy of history and avoids the determinism built into grand narratives of scientific progress, such as pasteurization. From an imaginaries perspective, moreover, space and scale are linked in a normative coupling that cannot as easily be captured by the metaphor of networks. For imaginaries not only help to reconfigure actors' sense of the possible spaces of action but also their sense of the rightness of action, at scales ranging from locality to nation (Barker; Chen; Kim; Moon, this volume) to continent (Smith; Storey, this volume) and to the planet itself (Lakoff; Miller, this volume).

A fourth and final problem that the concept of sociotechnical imaginaries helps tease apart is the relationship between collective formations and individual identity. From Foucault's observations about the capillary effects of power on human bodies to Anderson's characterization of nationhood as a product of communal imaginations to Pierre Bourdieu's (1990) sociological analysis of the individual's habitus as a historically conditioned subjective state—as well as in volumes of work on feminism, critical race theory, and subaltern studies—the relationship between the ideas of rulers and the self-understanding of subjects has long been the stuff of social theory. Accounts of subject formation, with their focus on humans as psychosocial beings, bring to light features of making collectives that tend to get backgrounded in impersonal studies of institutions, as well as in the behaviorist-leaning microsociologies of technoscientific practices favored by many STS scholars. Yet joining a collective does matter to the actors who join it; and those who form and manage collectives are often intensely (if unconsciously) aware of the need to control the emotive registers of adherence and belonging. I am reminded of my own "naturalization" as an American citizen in Ithaca, New York, in 1987, when the presiding judge told us to think of that day, October 22, as our "personal Independence Day"[8]; or of the tug at the heartstrings when newly minted Harvard PhDs are welcomed each year into "the ancient and universal company of scholars." By stressing the roles of memory, language, and performance—in short, by keeping the focus on human actors and their collectively enacted hopes and expectations—the essays in this volume seek to remedy some of the shortcomings of accounts that reduce human agents to mere cogs in machines (see especially Barker; .

Dennis; Hurlbut; Moon; Storey, this volume) or represent them as agents defined chiefly through their struggles with the material elements of the heterogeneous sociotechnical networks they happen to be caught in.

Reflections on Method

A theoretical term is worth little unless it fits into the circumstances of the world, casting light on corners that need illumination. Such terms need to be operationalized, and for that purpose method is indispensable. How can we recognize when something as abstract yet durable as an imaginary is in play and what are its constitutive components? How can we confidently identify a sociotechnical imaginary and be sure that it is not mere rhetorical flourish, institutional ideology, or fleeting policy preference? These questions receive detailed treatment in the following chapters and in the conclusion, but some broad outlines can be sketched here.[9]

As an analytic concept, "sociotechnical imaginary" cuts through the binary of structure and agency: it combines some of the subjective and psychological dimensions of agency with the structured hardness of technological systems, policy styles, organizational behaviors, and political cultures. The methods best suited to studying sociotechnical imaginaries therefore are the methods of interpretive research and analysis that probe the nature of structure-agency relationships through inquiries into meaning making. Although few of these methods are specific to the analysis of sociotechnical imaginaries, they can be applied in ways that are especially attuned to this concept: by attending to the means by which imaginaries frame and represent alternative futures, link past and future times, enable or restrict actions in space, and naturalize ways of thinking about possible worlds.

Perhaps the most indispensable method for studying sociotechnical imaginaries is comparison. Comparing across social and political structures not only helps to identify the content and contours of sociotechnical imaginaries but also avoids the intellectual trap of taking as universal epistemic and ethical assumptions that turn out, on investigation, to be situated and particular. Cross-national comparisons have proved especially useful in revealing the ingrained normative commitments that distinguish political communities, such as their ways of knowing and reasoning (Burri, this volume). These are rarely discernible from inside the safe havens of nation-states, where so much of political culture is accepted as part of the natural order of things; only by adopting the comparatist's estranging gaze does one perceive the artifices of one's own reasoning (Jasanoff 2005, 2011b, 2012b). But comparison need not be limited in kind or scale to nation-states alone.

Comparisons can be conducted across policy sectors or over time to illuminate the distributed character of the practices that hold imaginaries in place (see especially Felt; Hurlbut; Kim, this volume). Then, too, actors in these stories themselves compare, shaping their personal visions in accordance with imagined elsewheres and elsewhens, and those comparisons in turn get woven into social meaning making (Barker; Bowman; Dennis; Storey, this volume). The challenge for analysts is to conduct their own comparisons with epistemic charity and due respect for difference: not to apply universal yardsticks for measuring advances toward, or deviance from, allegedly transcendental ideals but instead to reveal, and destabilize if we are so inclined, the naturalized logics of functioning, self-contained, and self-replicating social and political systems.

Imaginaries by definition are group achievements—for example, of nations (Hecht 1998), ethnic or linguistic communities, social movements (Epstein 1996), or biosocial formations such as carriers of genetic traits for disease (Rabinow 1992; Parthasarathy 2007). Biographies of individuals are therefore not the most obvious route into uncovering the origins of imaginaries, although as several of the following chapters illustrate (Barker; Bowman; Moon; Storey, this volume), individual dreams and aspirations take hold and acquire collective force only when key actors mobilize the resources for making their visions durable. The literature on social movements has engaged with this interplay of subjective identity and action with the possibilities for intersubjectivity created by social norms. Melucci (1996, 33), for example, notes that identity is "both our ability to recognize ourselves and the possibility of being recognized by others." In tracing how individual visions sometimes rise to the status of collectively held objectives, the imaginaries framework urges us to note not only the material instruments that reformers are able to accumulate but also their uses of symbolic and cultural resources, such as images, texts, memories, metaphors, and language itself.

The languages of power, especially the official discourses of the state, have provided fertile ground for social theorists, but once again the coalescence of the collective imagination with scientific and technological production offers particular stream beds along which to direct the flow of such analysis. Some are exemplified in this volume (Burri; Felt; Miller, for example); others can be found in work not explicitly invoking the concept of sociotechnical imaginaries. Policy discourses and processes of issue framing and agenda setting offer one commonly recognized starting point (Gusfield 1981; Schon and Rein 1994; Hajer 1995). Adding to that corpus, we can ask how actors with authority to shape the public imagination construct stories of progress in their programmatic statements and how they blend into these their expectations of science and technology (Barker; Bowman; Dennis; Moon, this volume). Those questions, in turn, can be turned toward specific types of technopolitical order. One may ask, for instance, how narratives define the public good with respect to biotechnology (e.g., Chen; Hurlbut; Kim; Smith, this volume) or how they delimit, control, or contain risk in projects aimed at furthering goods such as energy provision (Jasanoff and Kim 2009, 2013). And, recognizing the potential of imaginaries to configure shared understandings of space and time, one can trace in policy discourse the creation of new geopolitical boundaries (Jasanoff and Martello 2004; Lakoff; Miller, this volume) or references to past achievements in promises (or fears) of future developments (Dennis; Felt; Hurlbut, this volume).

Practices matter in the analysis of sociotechnical imaginaries as they do in all attempts to make sense of the nature of collective life. An imaginary is neither cause nor effect in a conventional sense but rather a continually rearticulated awareness of order in social life (Jasanoff 2012b) and a resulting commitment to that order's coherence and continuity. A sure guide to finding such regularity is to look at how social actors and institutions respond when confronted by events that might disrupt order. Law then emerges as an especially fruitful site in which to examine imaginaries in practice. Legal disputes are in their very nature moments of contestation between disparate understandings of the good; and in the modern world these attach with great regularity to questions about science and technology. Should trees have standing (Stone 1974), should a regulatory agency be entitled to treat nuclear wastes as posing no risk (Jasanoff and Kim 2009), should living organisms be treated as property (Jasanoff 2012a; see also Sunder Rajan 2006)—these and countless other questions of equal or lesser significance have perplexed American legal thought in the past half century, requiring judges to issue rulings that often reproduce dominant sociotechnical imaginaries.

Legal decision making enjoys a special status in American political culture, because US courts are so heavily implicated in solving public problems. But legal practices are equally important to the construction of sociotechnical imaginaries in countries where power is differently allocated among the major branches of government. The practices of the Indian Supreme Court, for example, present striking similarities and differences in comparison with its American counterpart. The Indian high court has been as intimately involved in resolving disputes of a deeply political character on a range of issues involving science and technology, such as environmental protection, rights to life and health, and intellectual property. Yet, because Indian citizens may bring a so-called writ petition asking for direct adjudication of claims of fundamental rights, the Indian Supreme Court has arguably been more open to the imaginations of the poor, or those normally seen as outcasts and outsiders of society, than the US Supreme Court, which is empowered to hear only well-formed cases and controversies. Even among Western industrial nations, comparisons of legal practices may prove extremely productive as a method of identifying and characterizing sociotechnical imaginaries (Jasanoff 2011b).

Policy documents, no less than judicial opinions, can be mined for insights into the framing of desirable futures (or, as Dennis argues, for the "monsters" that policy seeks to keep at bay) as well as for specific verbal tropes and analogies that help identify the elements of the imaginary (e.g., Burri; Bowman; Chen; Felt; Hurlbut; Kim; Miller this volume). Imaginaries, moreover, are not exclusively the property of state actors. National sociotechnical imaginaries may permeate into popular culture, finding expression in the mass media and in nonofficial genres such as advertising (e.g., Felt, this volume) or the popular writings of prominent individuals (Barker; Moon; Storey, this volume). Multinational corporations increasingly act upon imagined understandings of how the world is and ought to be, playing upon the perceived hopes and fears of their customers and clients and thereby propagating notions of technological progress and benefit that cut across geopolitical boundaries (Smith, this volume). Coalitions between corporate interests and the media, through advertising and outright control, are increasingly likely to play a pivotal role in making and unmaking global sociotechnical imaginaries.

Conclusion

The essays in this book deal with questions that are central to any examination of political and social order. What makes a given social system—a nation or polity or movement or community—not only cohere (Benedict Anderson's primary problem) but also be capable of absorbing and coming to terms with its own internal tensions and contradictions? How do the practices of collective imagination resolve conflict and produce consensus? Do powerful imaginaries mainly constrain and exclude action from below, or—like Foucault's *grilles* which train perception and channel action but are themselves open to reshaping—can imaginaries be transformative, as vehicles for reenvisioning and recalibrating human futures? In the latter case, how do new mind-sets break free from older, culturally stereotyped ways of knowing that keep dominant hierarchies in place and alternative imaginations from flowering? How more particularly do the omnipresent agents, instruments, and processes of science and technology—woefully neglected in so much social analysis—help mediate among competing expectations; and to what extent are institutions of power equipped to detect and correct for their own unexamined presuppositions when pursuing or implementing grand visions of progress?

In this chapter and those that follow, we provide methodological point-ers for how to begin addressing such questions. We take comparison, in all its forms, to be a foundational technique, recognizing that comparison can operate across all kinds of organizational variables: political across na-tions and actors; historical through time; geographic in relation to space; economic across sectors; and cultural between groups and societies. His-torical research in our view is essential to the exploration of imaginaries: it is only by following ideas through time that one gains a feel for what is fixed and what is changeable in social self-understandings as well as the reasons why. We have indicated, too, how many of the classical methods for studying social meaning making can be adapted and put to use in the frame-work of sociotechnical imaginaries. For example, the languages, metaphors, and symbols of official political talk can be mined for framings of risk and benefit, attitudes toward regulation and the market, and visions of techno-logically mediated progress or failure and backsliding.

The organization of the volume as a whole traces the basic dynamics of sociotechnical imaginaries: how they stop being personal or actor-centric "vanguard visions" (Hilgartner 2015) and how they instead become collec-tively held reference points and anchors for future projects. Scales matter on this account, as when the power to imagine moves from single "inspired" individuals or small collectives to communities and their leaders to nation-states and supranational global agencies. But the following essays should be seen more as braided together through overlapping themes than as divided into discrete blocks of sectoral or scalar analysis. Thus, historical origins matter centrally in chapters 2–6, modes of imperialism in chapters 2–4, Asian sociotechnical imaginaries in chapters 7–10, new technologies in chapters 9–12, and globalization in chapters 11–14. Paired juxtapositions explore more specific themes, such as memory making in the chapters by Felt and Hurlbut, Indonesian imaginaries of resistance in the chapters by Moon and Barker, and global imaginaries of risk and security in the chapters by Miller and Lakoff.

The book's conclusion offers a more structured and sequential review of the chapters. Here we build on the content of the individual contributions to show how reforming visions are translated into imaginaries through in-terlinked phases of origination, embedding, resistance, and extension. In recapitulating the stories told in each chapter, the conclusion reinforces the work done throughout the volume to dissolve hard-and-fast binaries: between descriptive and normative, structure and agency, material and mental, local and translocal. Imaginaries operate as both glue and solvent, able— when widely disseminated and effectively performed—to preserve continuity across the sharpest ruptures of innovation or, in reverse, to upend firm worlds and make them anew.

Regardless of the methods by which, and the sites in which, they are studied, sociotechnical imaginaries allow us to explore more thoroughly and understand more completely some of the most basic elements of human welfare. These include, most centrally, questions about the stability, durability, and coherence of social arrangements, all the more pressing in the postmodern condition, which has sensitized us to the contingency (Hacking 1999), fluidity, and chaos that often lap at the margins of achieved order. Squarely located in the space of science and technology studies, sociotechnical imaginaries at the same time break disciplinary boundaries, borrowing or building on theories and methods from anthropology, history, sociology, critical legal studies, and political and cultural theory. In this way, the framework avoids the analytic blinders that Latour, Mitchell, and others have rightly cautioned against. Most exhilarating, though, as this volume attests, is the fertile hybridity of the term itself. It offers unfettered entry into the coproduced realities of the known, the made, the remembered, and the desired worlds in which we live and which we have power to refashion through our creative, collective imaginings.

Notes

1. Creativity as a feature of human nature figured prominently in the debate, with Chomsky asserting that he means the everyday linguistic creativity of the child and Foucault insisting that individual creativity is enmeshed in prior, collectively formed regimes of truth within which minds unfold themselves. The word "grille" was used by Foucault in a meaning quite similar to Taylor's social imaginary. A full transcript of the debate can be found online at http://www.chomsky.info/debates/1971xxxx.htm.

2. In 2012, the Bond film series enjoyed its own Golden Jubilee as "the longest running and (adjusting for inflation) most lucrative franchise in cinema." See "From Britain with Love."

3. These designations, as Shapin and Schaffer show, are themselves anachronistic. Hobbes was also a natural philosopher in his day, and his views on the nature of truth were deeply entangled with his political thought. Boyle for his part engaged with Hobbes on matters of political authority. Both men were concerned with the question of ultimate authority to resolve disputes over factual claims.

4. *Leviathan and the Air Pump* has canonical status in STS with respect to its claims about the origins of English experimental science. There is, of course, a large literature on

Thomas Hobbes in political philosophy, but it has not generally engaged with STS on a deeper level. A fairly recent book by Stephen Finn (2006), a political philosopher by training, affords an interesting glimpse into the noncommensurability of disciplines. Finn, following another political philosopher, William Lynch, rejects what Lynch refers to as the "holistic sociology of knowledge thesis" advocated by the authors of *Leviathan and the Air Pump* (Finn 2006, 17). Only Finn repeatedly, and erroneously, refers to the book's second author as "Simon Shapiro."

5. Ezrahi (2012, 288) makes a partial bow to these aspects of what he terms "perceptual indeterminism"and the subjectivity of visual experience in his later work on the imaginaries of democracy.

6. This was a major point of disagreement between Chomsky and Foucault in their 1971 debate. Chomsky, while disavowing any commitment to an idealized notion of justice, argued for the possibility of a gradual evolution toward greater justice ("we must act as sensitive and responsible human beings in that position to imagine and move towards the creation of a better society and also a better system of justice"). Foucault denied any such pragmatic evolution toward betterment without falling victim to the entrenched practices of power within which any notion of justice is embedded. See Chomsky-Foucault debate, *supra,* note 1.

7. Given where La Pérouse sailed, the elderly Chinese was more likely to have been one of the Ainu, the indigenous inhabitants of Japan.

8. We were each handed a red rose and a little rolled-up American flag, a potent symbol of nationhood, in addition to our naturalization certificates.

9. As an accompaniment to this book, the Program on Science, Technology and Society at the Harvard Kennedy School has also created a web-based research platform, which provides additional information on which kinds of primary sources provide researchable insights into sociotechnical imaginaries. See http://sts.hks.harvard.edu /research/platforms/imaginaries/.

References

Anderson, Benedict. 1991 [1983]. *Imagined Communities*. Revised and expanded second edition. London: Verso.

Appadurai, Arjun. 2002. "Disjuncture and Difference in the Global Cultural Economy." Pp. 46–64 in *The Anthropology of Globalization: A Reader,* edited by Jonathan Xavier Inda and Renato Rosaldo. Oxford: Blackwell. An earlier version was published in 1990 in *Public Culture* Spring 2(2):1–24.

Arthur, W. Brian. 1994. *Increasing Returns and Path Dependence in the Economy*. Ann Arbor: University of Michigan Press.

Beck, Ulrich. 1998. "Politics of Risk Society." Pp. 9–22 in *The Politics of Risk Society,* edited by Jane Franklin. Cambridge: Polity Press.

Bellamy, Edward. 1889. *Looking Backward, 2000–1887*. Boston: Houghton Mifflin.

Bijker, Wiebe. 1997. *Of Bicycles, Bakelites, and Bulbs: Toward a Theory of Sociotechnical Change*. Cambridge, MA: MIT Press.

Bijker, Wiebe, Thomas Hughes, and Trevor Pinch, eds. 1987. *The Social Construction of Technological Systems: New Directions in the Sociology and History of Technology*. Cambridge, MA: MIT Press.

Bourdieu, Pierre. 1990. *The Logic of Practice*. Stanford, CA: Stanford University Press.

Callon, Michel. 1986. "Some Elements of a Sociology of Translation: Domestication of the

Scallops and the Fishermen of St. Brieuc Bay." Pp.196–223 in *Power, Action and Belief: A New Sociology of Knowledge?* edited by John Law. London: Routledge.

———. 1987. "Society in the Making." Pp. 83–103 in *The Social Construction of Technological Systems.*

———. 1998. *Laws of the Markets.* Oxford: Blackwell, 1998.

Chomsky, Noam, and Michel Foucault. 2006. *The Chomsky-Foucault Debates.* New York: New Press.

Clarke, Arthur C. 1968. *2001: A Space Odyssey.* London: Penguin.

David, Paul A. 1985. "Clio and the Economics of QWERTY." *American Economic Review* 75 (2): 332–7.

Douglas, Mary. 1966. *Purity and Danger: An Analysis of Concepts of Pollution and Taboo.* London: Routledge and Kegan Paul.

Douglas, Mary, and Aaron Wildavsky. 1982. *Risk and Culture: An Essay on the Selection of Technological and Environmental Dangers.* Berkeley: University of California Press.

Epstein, Steven. 1996. *Impure Science: AIDS, Activism, and the Politics of Knowledge.* Berkeley: University of California Press.

Evans-Pritchard, Edward E. 1937. *Witchcraft, Oracles and Magic among the Azande.* London: Faber and Faber.

Ezrahi, Yaron. 1990. *The Descent of Icarus: Science and the Transformation of Contemporary Democracy.* Cambridge, MA: Harvard University Press.

———. 2012. *Imagined Democracies: Necessary Political Fictions.* New York: Cambridge University Press.

Farias, Ignacio, and Thomas Bender. 2010. *Urban Assemblages: How Actor-Network Theory Changes Urban Studies.* London: Routledge.

Finn, Stephen J. 2006. *Thomas Hobbes and the Politics of Natural Philosophy.* New York: Continuum.

Fortes, Meyer, and E.E. Evans-Pritchard. 1940. *African Political Systems.* Oxford: Oxford University Press.

Foucault, Michel. 1979. *Discipline and Punish.* New York: Vintage.

"From Britain with Love." *Economist,* July 7, 2012.

Gusfield, Joseph. 1981. *The Culture of Public Problems: Drinking-Driving and the Symbolic Order.* Chicago: University of Chicago Press.

Hacking, Ian. 1999. *The Social Construction of What?* Cambridge, MA: Harvard University Press.

Hajer, Maarten. 1995. *The Politics of Environmental Discourse.* Oxford: Oxford University Press.

Haraway, Donna. 1989. *Primate Visions.* New York: Routledge.

———. 1991. *Simians, Cyborgs and Women: The Reinvention of Nature.* London: Routledge.

———. 2003. *The Companion Species Manifesto: Dogs, People, and Significant Otherness.* Chicago: Prickly Paradigm Press.

Hayles, N. Katherine. 1999. *How We Became Posthuman.* Chicago: University of Chicago Press.

Hecht, Gabrielle. 1998. *The Radiance of France: Nuclear Power and National Identity after World War II.* Cambridge, MA: MIT Press.

Hilgartner, Stephen. 2015. "Capturing the Imaginary: Vanguards, Visions, and the Synthetic Biology Revolution." Chapter 3 in *Science andDemocracy: Knowledge as Wealth and Power in the Biosciences and Beyond,* edited by Stephen Hilgartner, Clark Miller, and Rob Hagendijk. Abingdon, Oxon: Routledge.

——— 2000. *Science on Stage: Expert Advice as Public Drama.* Stanford, CA: Stanford University Press.

Huxley, Aldous. 1932. *Brave New World*. New York: Harper and Brothers.

Jasanoff, Sheila. 1998. "The Eye of Everyman: Witnessing DNA in the Simpson Trial." *Social Studies of Science* 28 (5–6): 713–40.

———, ed. 2004a. *States of Knowledge: The Co-Production of Science and Social Order*. London: Routledge.

———. 2004b. "Heaven and Earth: The Politics of Environmental Images." Pp. 31–54 in *Earthly Politics*, edited by Sheila Jasanoff and Marybeth Long Martello (Cambridge, MA: MIT Press).

———. 2005. *Designs on Nature: Science and Democracy in Europe and the United States*. Princeton, NJ: Princeton University Press.

———. 2006. "Technology as a Site and Object of Politics." Pp. 745–63 in *Oxford Handbook of Contextual Political Analysis*, edited by Charles Tilly and Robert Goodin. Oxford: Oxford University Press.

———. 2010. "A New Climate for Society." *Theory, Culture and Society* 27 (2–3):233–53.

———. 2011a. "Cosmopolitan Knowledge: Climate Science and Global Civic Epistemology." Pp. 129–43 in *The Oxford Handbook of Climate Change and Society*, edited by John S. Dryzek, Richard B. Norgaard, and David Schlosberg. Oxford: Oxford University Press.

———, ed. 2011b. *Reframing Rights: Bioconstitutionalism in the Genetic Age*. Cambridge, MA: MIT Press.

———. 2012a. "Taking Life: Private Rights in Public Nature." Pp. 155–83 in *Lively Capital: Biotechnologies, Ethics, and Governance in Global Markets*, edited by Kaushik Sunder Rajan. Durham: Duke University Press.

———. 2012b. *Science and Public Reason*. Abingdon, Oxon: Routledge-Earthscan.

Jasanoff, Sheila, and Sang-Hyun Kim. 2009. "Containing the Atom: Sociotechnical Imaginaries and Nuclear Regulation in the U.S. and South Korea." *Minerva* 47 (2):119–46.

———. 2013. "Sociotechnical Imaginaries and National Energy Policies." *Science as Culture* 22(2):189–96.

Jasanoff, Sheila, and Marybeth Long Martello, eds. 2004. *Earthly Politics: Local and Global in Environmental Governance*. Cambridge, MA: MIT Press.

Kant, Immanuel. 1996 [1784]. "An Answer to the Question: What Is Enlightenment?" (*Beantwortung der Frage: Was Ist Aufklärung?*). Pp. 11–22 in *Practical Philosophy*, translated and edited by Mary J. Gregor. Cambridge: Cambridge University Press.

Kingdon, John. 2010. *Agendas, Alternatives, and Public Policies 2nd edition*. New York: Longman.

Kuklick, Henrika. 1992. *The Savage Within: The Social History of British Anthropology, 1885–1945*. Cambridge: Cambridge University Press.

Kurosawa, Akira. 1950. *Rashomon*. Daiei Film Co., Ltd.

Latour, Bruno. 1988. *The Pasteurization of France*. Cambridge, MA: Harvard University Press.

———. 1990. "Drawing Things Together." Pp. 19–68 in *Representation in Scientific Practice*, edited by Michael Lynch and Steve Woolgar. Cambridge, MA: MIT Press.

———. 1993. *We Have Never Been Modern*. Cambridge, MA: Harvard University Press.

Machiavelli, Niccolò. 1977 [1513]. *The Prince: A New Translation, Backgrounds, Interpretations, Peripherica*. Translated and edited by Robert M. Adams. New York: Norton.

Marcus, George E., ed. 1995. *Technoscientific Imaginaries: Conversations, Profiles, and Memoirs*. Chicago: University of Chicago Press.

Melucci, Alberto. 1996. *The Playing Self: Person and Meaning in the Planetary Society*. Cambridge: Cambridge University Press.

Mitchell, Timothy. 2002. *Rule of Experts: Egypt, Techno-Politics, Modernity.* Berkeley: University of California Press.

Morris, Jan. 2003. *Heaven's Command: An Imperial Progress.* London: Faber and Faber.

Orwell, George. 1949. *Nineteen Eighty-Four.* London: Martin Secker and Warburg.

Parthasarathy, Shobita. 2007. *Building Genetic Medicine: Breast Cancer, Technology, and the Comparative Politics of Health Care.* Cambridge, MA: MIT Press.

Rabinow, Paul. 1992. "Artificiality and Enlightenment: From Sociobiology to Biosociality." Pp. 234–52 in *Zone 6: Incorporations,* edited by Jonathan Crary and Sanford Kwinter. Cambridge, MA: MIT Press.

Schon, Donald A., and Martin Rein. 1994. *Frame Reflection: Toward the Resolution of Intractable Policy Controversies.* New York: Basic Books.

Scott, James C. 1998. *Seeing Like a State: How Certain Schemes to Improve the Human Condition Have Failed.* New Haven: Yale University Press.

Shapin, Steven, and Simon Schaffer. 1985. *Leviathan and the AirPump: Hobbes, Boyle, and the Experimental Life.* Princeton, NJ: Princeton University Press.

Shelley, Mary Wollstonecraft. 2008 [1818]. *Frankenstein: or, The Modern Prometheus.* Edited by M.K. Joseph. Oxford: Oxford University Press.

Stone, Christopher D. 1974. *Should Trees Have Standing?* Los Altos, CA: William Kaufmann.

Strong, Roy C. 1984. *Art and Power: Renaissance Festivals 1450–1650.* Melton, Woodbridge: Boydell Press.

———. 1987. *Gloriana: The Portraits of Queen Elizabeth I.* London: Random House, 1987.

Sunder Rajan, Kaushik. 2006. *Biocapital: The Constitution of Postgenomic Life.* Durham: Duke University Press.

Taylor, Charles. 2004. *Modern Social Imaginaries.* Durham, NC: Duke University Press.

Tilley, Helen. 2011. *Africa as a Living Laboratory: Empire, Development, and the Problem of Scientific Knowledge, 1870–1950.* Chicago: University of Chicago Press.

Verne, Jules. 1887 [1870]. *Twenty Thousand Leagues under the Sea.* New York: Butler Brothers.

Visvanathan, Shiv. 1997. *A Carnival for Science.* Delhi: Oxford University Press.

Winner, Langdon. 1986. "Do Artifacts Have Politics?" Pp. 19–39 in *The Whale and the Reactor.* Chicago: University of Chicago Press.

TWO

Cecil Rhodes and the Making of a Sociotechnical Imaginary for South Africa

WILLIAM KELLEHER STOREY

Cecil Rhodes was no ordinary businessman and politician. On arriving in southern Africa as a teenager in 1871, his initial vision for the region reflected a boyish romanticism. Coming to maturity over the course of the next three decades, he played a key role in fashioning a sociotechnical imaginary for southern Africa by influencing the development of major economic and political institutions. He built the DeBeers Consolidated Mining Company into the business that still, to this day, produces and sells most of the world's diamonds. He also played an important role in gold-mining companies, agricultural modernization, and the extension of railroads and telegraph lines. Another of Rhodes's businesses, the British South Africa Company, founded the settler colony of Rhodesia, today known as the independent country of Zimbabwe. He was a sitting member of the Cape Colony's parliament from 1880 until his death in 1902. He even served a long, turbulent term as the Cape's prime minister from 1890 to 1895, during which time he put in place significant policies related to economic and social development.

As a leader in business and politics, Rhodes did more than any other individual to set South Africa on the path of industrial modernization and racial segregation. Previous Rhodes biographers have focused on his involvement in politics, paying little attention to geology, mining, telegraphs, railroads, and farming in his life story. Yet Rhodes was a visionary leader in business and politics who promoted advanced mine engineering while at the same time pressing for monopoly capitalism and racial discrimination, the sociotechnical imaginary that emerged in late nineteenth-century South Africa.

The concept of sociotechnical imaginaries helps us to articulate his racist political vision with the forces of production that he set in motion. Rhodes's policies, as they were extended from his personal sociotechnical vision into a full-blown collective imaginary, produced wrenching social and environmental changes. Written and visual sources from Rhodes's time show us how he developed his vision, pulled it together from disparate elements, and then made it legible and attractive to key backers. As his vision came to be shared with others and became part of a widely held imaginary, fellow industrialists and politicians gained power to keep an eye on all workers, black and white, unskilled and skilled, while ordering the spaces of their world, through impositions ranging from the designs of the Arts and Crafts Movement to prison-like compounds for migrant workers. The increasingly segregated South Africa that emerged between Rhodes's arrival in 1871 and his death in 1902 illustrates the ways in which individual ideas, industrial revolutions, and political transformations can intersect to form a durable sociotechnical imaginary. This essay will explore the ways in which Cecil Rhodes envisioned and enacted a coproduced material and social order in late nineteenth-century South Africa that laid the foundation for the modern, racist state that emerged in the twentieth century.

Cecil Rhodes was born in Bishop's Stortford, Hertfordshire, in 1853, the fifth son of an Anglican vicar who went on, with his second wife, to have nine sons and two daughters. At the age of seventeen, Cecil Rhodes emigrated to Natal, South Africa, to farm with one of his older brothers. The seventeen-year-old Rhodes was at first quite taken with the landscape of Natal. His early letters home reveal the sense of awe and wonder, staples of English romanticism, that he felt when gazing at the Natal landscape. Viewing the Drakensberg on March 17, 1871, he wrote, "It fills you with a sort of awe to get right into the heart," and "I believe that to see the sun rise in Natal from the top of the Drakensberg, is one of the finest views in the world" (Rhodes House Library)

Rhodes and his brother abandoned farming in 1871, looking for opportunities at the new diamond diggings in Kimberley, along with tens of thousands of other young men. Rhodes set out on a month-long oxcart journey to join his brother at Kimberley, four hundred miles across the Drakensberg Mountains and the semiarid Karoo. Upon arriving at Colesberg Kopje, the Kimberley Mine, he described what he saw in a letter to his mother.

> It is like an immense number of ant-heaps covered with black ants, as thick as can be, the latter represented by human beings; when you understand that there are about 600 claims on the kopje and each claim is generally split into 4, and on each bit there are about 6 blacks and whites working, it gives a total of about ten thousand working every day on a piece of ground 180 yards by 220 . . . The carting on the kopje is done chiefly by mules, as they are so very hardy, and have so few diseases. There are constantly mules, carts and all

going head over heels into the mines below as there are no rails or anything on either side of the roads, nothing but one great broad chasm below . . . On each side of every road there is now a continuous chasm from top to bottom of the kopje varying in depth from 30 to 60 ft. (B. Williams, 1926, 27–28)

Descriptions of plunges into dangerous chasms are reminiscent of romantic notions displaced from natural to industrial landscapes. As Rosalind Williams writes in *Notes on the Underground,* Rhodes's English contemporaries associated the great mines of England and their dangers with the sublime. Sublimity became strongly associated with the awesome discovery of deep, geological time, as well as with what she calls the "aesthetic discovery of industrial technology" (R. Williams 2008, 88–89).

Rhodes saw the mines as an educated, middle-class, mid-nineteenth-century Englishman would, as a theater of the sublime but also as a place to turn a profit. In an early letter to his family, Rhodes envisioned exactly what would happen to the mine.

There are reefs all round these diamond mines, inside which the diamonds are found. The reef is the usual soil of the country, round, red sand just at the top and then a black and white stony shale below. Inside the reef is the diamond-iferous soil. It works just like Stilton cheese, and is as like the composition of Stilton cheese as anything I can compare it to. . . . They have been able to find no bottom yet, and keep on finding steadily at 70 ft. You will understand how enormously rich it is, when I say that a good claim would certainly average a diamond to every load of stuff that was sorted—a load being about 50 buckets . . . Some day I expect to see the kopje one big basin where once there was a large hill (B. Williams, 1926, 27–28).

On the surface, Rhodes's vision of the place as so much Stilton cheese seems a relic of his middle-class, English background. Yet from a geological perspective Rhodes's metaphor was essentially correct. Thanks in large part to Rhodes's own efforts, Colesberg Kopje did turn into a basin, one of the largest dug holes on Earth (see fig. 2.1).

Typically early biographers portrayed him as a visionary. This accords with the reminiscences of his fellow diggers (some of whom became influential in their own right), many of whom recall young Rhodes, tall and thin, dressed simply, sitting quietly at the edge of the mine, thinking about it, or standing or sitting, quietly by himself, on a street corner, or outside a tent. He was not, however, either a loner or a mere dreamer. It was by sharing tents and meals with other young, gentleman miners that Rhodes formed

2.1. DeBeers Mine, Kimberley, ca. 1872. This figure illustrates the early working of one of the two biggest open-pit diamond mines, the other one being the Kimberley Central Mine on the other side of town. Note that various small claims have gone to different levels of the mine. Miners are using a system of winches and ropes to haul out the earth before sorting it nearby. Photo courtesy of the McGregor Museum, Kimberley. MMKP 4447.

his political vision and began to extend it into a sociotechnical imaginary. Mainly he did this by forming friendships with men who served in government and also by forming fledgling business partnerships. Rhodes broke bread with John X. Merriman, also the son of an opinionated clergyman, who had been a member of the Cape parliament since 1869. When parliament was out of session, Merriman was attempting to earn money as a diamond buyer. Later, he became one of the most influential South African politicians of the late nineteenth century. Rhodes would also meet John Blades Currey, who would soon be deeply involved in Cape and Kimberley politics, initially as a magistrate. Rhodes met his future business partner, an athlete from Harrow and Cambridge named Charles Dunell Rudd, who would later help Rhodes to found Rhodesia. Rhodes became friends with Sidney Shippard, an Oxford graduate and a future colonial administrator in southern Africa, who served as attorney general of Griqualand West from 1873 to 1877. Rhodes also "messed" with Jacob Barry, who would become an influential attorney and judge in the Eastern Cape.

The young men from England were themselves involved in the physical labor of digging and sifting, although Rhodes and his friends also hired African laborers to help them work their claims. Biographer Robert Rotberg points out that even in the early stages of his career, as illustrated in his letters home, Rhodes had a strong command of technical details, be they related to engineering, finance, or politics (1988, 72). Rhodes dug for diamonds, bought up claims, and worked as a partner in mining, pumping, and marketing operations.

Unlike most of the men who were involved at the mines, Rhodes pursued new riches as well as older forms of prestige. In 1873 he enrolled at Oriel College, Oxford. Soon after he returned to Kimberley—he was able to fulfill most Oxford requirements between 1876 and 1878—finally earning his degree in 1881. Academically, Rhodes was a mediocrity. He rarely attended lectures and earned a basic or "pass" degree. However, he was successful in other ways. He joined and subsequently led several social clubs, leading most biographers to surmise that he attended Oxford in order to gain the polish and social connections that would advance his dual careers in business and politics. By the time he finished his degree, he was one of the most important business leaders in South Africa, with a large stake in one of the biggest Kimberley mines, the De Beers mine, plus positions in other local mines.

During Rhodes's early years on the diamond fields, the mines boomed, but starting around 1874 several factors caused many small-scale miners to give up. As diggers dug deeper for diamonds, it became more costly to get the stones out of the ground—this at a time when the world economy was going through a downturn. And locally, diggers were used to working in what they called the "yellow ground" near the surface. It was somewhat loose and easily worked with picks and shovels. By 1874 and 1875, diggers were reaching the next stratum, what they called the "blue ground," which was harder. It needed to be dug out, carted, spread out, and weathered on fields, called "floors," for upwards of a year before it was amenable to crushing and sifting. This made it more costly to handle. The deeper stratum of rock, more difficult to work, also raised questions in the minds of many miners about whether or not the mines would still yield plentiful diamonds.

Despite the seeming constraints of geology, Rhodes and his partners kept faith in the diamond mines. Rhodes's knowledge of geology is likely to have helped his highly speculative business a great deal. The buying and selling of claims involved credit, and that in turn involved the backing of bankers and shareholders in London and Paris who had to believe in the mines' future productivity. Rhodes had a knack for cooperation, which helped him to gain knowledge of the mines as diamond mining got more capital intensive. Even so, it is hard to trace his actual knowledge of geology. He did regularly visit the Kimberley Public Library, now renamed the Africana Library, which is next door to the Kimberley Club, where Rhodes and the directors of De-Beers resided when in town. There visitors may even see Rhodes's favorite chair. The library catalogue indicates that Rhodes and other patrons could have consulted twenty-two volumes pertaining to the geology of diamonds. Further works would have been available to them on visits to Cape Town, where the South African Library also collected volumes about geology and mining. Rhodes was also probably aware of debates about the geology of diamonds in the *Cape Monthly Magazine*, a journal that circulated among the educated classes. The magazine and the gentleman scientists it served are strongly associated with the emergence of Cape liberalism, the paradoxical ideology that sought to extend universal freedoms while making pragmatic concessions to the European dominance of Africans that held out rights for white people. Liberalism and science were associated with the emergence of a distinct white South African identity (Dubow 2006). Finally, the manager of Rhodes's mines in Kimberley, Gardner Williams, wrote extensively about diamond geology in his 1905 book, *The Diamond Mines of South Africa*; given the close collaboration between Williams and Rhodes, this constitutes further evidence of Rhodes's knowledge.

Rhodes aimed for dominance in parliamentary politics as well as at the mines. To him, business and politics were key components of his vision for unifying and giving purpose to the British Empire. He made an early statement of his wishes in 1877, in an early will that historians call "The Confession of Faith." In that document, Rhodes wrote the following words: "The idea gleaming and dancing before ones eyes like a will-of-the-wisp at last frames itself into a plan. Why should we not form a secret society with but one object the furtherance of the British Empire and the bringing of the whole uncivilized world under British rule for the recovery of the United States for the making the Anglo-Saxon race but one Empire." The secret society was to place members in positions of influence, including colonial legislatures. There, the "Society should attempt to have its members prepared at all times to vote or speak and advocated the closer union of England and the colonies, to crush all disloyalty and every movement for the severance of our Empire." In order to found this secret society, Rhodes willed his small but growing fortune to his friend, the colonial administrator Sydney Shippard, as well as to the British Secretary of State for the Colonies (Flint

1974, 249–51). Rhodes modified his will many times over the course of the next two decades, but this notion of a secret society strikes most historians as the germ for the Rhodes Scholarships at Oxford University.

In 1880, Rhodes began his political career when he was elected to the Cape Colony's parliament from the district of Barkly West, a small town near Kimberley. His direct participation in parliamentary politics distinguished him from most Anglo-American tycoons of the late nineteenth century. Rhodes became a force in Cape politics, building a coalition of English and Dutch members that advocated for the unity of South Africa under the aegis of the British Empire. As early as 1883, Rhodes admitted to the Cape parliament that "I believe in a United States of South Africa, but as a portion of the British Empire" (Vindex 1900, xxvi). For the next decade, this position mollified Cape Afrikaners, who tended to want unification without the British Empire. Be that as it may, Rhodes used his alliances to push legislation that was favorable to the diamond industry, including draconian laws against diamond smuggling, known as "I.D.B" for "illicit diamond-buying," a key way in which African and European workers resisted and even profited from the mining regime.

Working together with a fellow diamond industrialist, Joseph B. Robinson, who was also elected to parliament in the same year, Rhodes promoted the seating of a special committee to investigate I.D.B. Witnesses to the committee recommended preventive measures intended to make it more difficult for workers to steal diamonds. Some recommended forced body searches; one mine owner even suggested that Africans work in the nude. Others suggested that I.D.B. be punished by flogging, and the creation of worker compounds was discussed as well. On the committee's recommendation, the Cape parliament passed the Diamond Trade Act of 1882. The act created a special detective department for I.D.B. and eliminated the presumption of innocence for perpetrators. Jury trials were also eliminated. With long-standing rights and procedures thus overturned, I.D.B. cases were to be tried by a special judge working with two magistrates. Those convicted could receive heavy punishments. Europeans were eligible for fifteen years of imprisonment, fines of £1,000, and banishment from the diamond fields. African workers were also subject to flogging (Worger 1987, 133–35). Flogging was a relic punishment—even the Royal Navy had abolished it in 1879. It was interesting to see this resurfacing in the South African context, where racially discriminatory legislation had been technically unconstitutional until the passage of a gun-control law, the Peace Preservation Act, in 1878 (Storey 2008). Just a few years earlier, then, others had opened the door to legal discrimination; Rhodes now began to push the door wide open.

Rhodes, the elected representative of diamond mine owners, thus reshaped colonial legislation. At the same time, Rhodes helped to unseat the prime minister, Gordon Sprigg, whose efforts to disarm Lesotho had plunged the Cape into a costly war. To ease his political efforts, Rhodes secretly bought a controlling interest in the colony's newspaper of record, the *Cape Argus* newspaper, which was owned by the old-school liberal politician, Saul Solomon. Rhodes's control of a key node of information technology symbolized and affected the wider shift in Cape liberalism toward more racial restrictions.

In the mid-1880s, Rhodes remained in the Cape parliament. He worked closely with the representatives of Afrikaner and British rural areas, playing an important role in the development of "native policy" and in the fostering of colonial development through the support of agriculture, railroads, and telegraphs. His political ventures (not coincidentally) tended to support his mining ventures. Better railroads would bring even cheaper African labor to the mines, while better communications assured closer management and easier responsiveness to global market conditions. With an eye toward controlling routes to the north and gaining access to gold and labor, Rhodes actively supported efforts by the Cape Colony to annex Botswana as well as the smaller Afrikaner territories of Stellaland and Goshen. While he incorporated these two into the Cape Colony, the Tswana chiefs succeeded in keeping themselves independent of the Cape by obtaining indirect rule by Great Britain itself. In spite of this setback for Rhodes, his political vision was clear—South Africa needed to be built by unifying the Cape Colony with its neighbors to the north, maintaining a British connection without British domination (Parsons 1998).

The Tswana chiefs had more success in resisting Rhodes than did the mine workers in Kimberley. As the mining companies consolidated, the mines went deeper and became more dangerous. Mine workers went on strike in 1884. On April 25, white and black workers walked off the job to protest a further tightening of searching policies, including a new rule against workers complaining during searches. One striker spoke out against the mine owners, saying that "in the early days of the Fields a capitalist would arrive from Europe with a capital of £5, and after being here a little while would commence buying the claimholder's diamonds from the claimholder's boys [i.e. illegally from African workers]; in a year or two the capitalist would get the claimholder's claims out of him, when the unfortunate claimholder would be turned into an overseer. This had actually taken place, and now the grateful capitalist wanted to strip the overseer of his clothes as a wind up to the whole affair" (Worger 1987, 182).

In his view, theft was the basis of owner power. Strikers charged the pumping works at the Kimberley Central mine, defended by armed police. The police fired into the crowd of strikers, killing six and wounding six. With order defended by the mine owners, the strike lost steam. Mine owners and politicians alike were against the workers. Rhodes's friend, John X. Merriman, one of the more liberal members of parliament, attributed the strike to the interest of workers in I.D.B. Rhodes himself noted in parliament that the struggle in Kimberley was not between labor and capital, or between white mine owners and white mine workers, but between white owners, on the one hand, and white and black workers, on the other hand: "white men (and IDB'ers at that) supported by natives in a struggle against whites" (Worger 1987, 182–6).

While trouble with labor simmered in the background, Rhodes and his DeBeers partners achieved a spectacular triumph at the diamond mines. In marketing, DeBeers was decisively shaped by one partner, Alfred Beit, the financier and commercial genius who envisioned and achieved a merger between the Kimberley diamond-mining company and the London diamond-buying syndicate. During the mid- and late-1880s, Beit, Rhodes, and DeBeers won the battle to control the production of all the Kimberley mines as well as the marketing of most of the world's diamonds. Pricing and production were now linked, bringing the prospect of stability to the business. And stability depended to a large extent on the recruitment, transport, training, and regulation of African workers, in addition to continued faith in the productivity of the mines.

Rhodes and his partners believed that diamonds would continue to be found deeper in the rock formations known as Kimberlite pipes. This correct geological surmise underlay Rhodes's persuasion of his backers, such as the Rothschilds, who were needed to finance ever more expensive schemes. Rhodes's vision of more diamonds under the ground also involved imagining a new method of production. Kimberlite pipes are shaped like cylinders. The cylinders contain the yellow ground, on the top, and the blue ground, on the bottom, and are surrounded by worthless rock. The best way to dig is straight down into the cylinder, but digging an open, cylindrical pit, poses some challenges. The rock around the cylinder starts to collapse and crumble over the pit. After several years of coping with rock slides it became clear to Rhodes and his engineers that the best way to access the diamonds deep in the Kimberlite pipes was to dig tunnels. For this reason, the final phase of claim amalgamation, starting in 1885, coincides with the commencement of underground mining (see fig. 2.2). Tunneling involved working with professional mine engineers, most notably the American progressive Gardner Williams, formerly consultant to the Rothschilds, who became the DeBeers

THE KIMBERLEY MINE   PRESENT DAY

F.N.H.

2.2. Kimberley Central Mine, ca. 1895. The open pit, known as the "Big Hole," was excavated by 1888. Note the strata of earth as well as the cylindrical shape of the Kimberlite pipe. At this point in time, diamond mining is taking place in tunnels that lay beneath the open pit. Photo courtesy of the McGregor Museum, Kimberley. MMKP 5334.

manager at Kimberley. As the tunnels were dug under the big, open-pit mines, the costs rose. Mine engineering, management, finance, and politics became increasingly intertwined in a high-stakes effort to extract diamonds as more business partners came to share the imperialist and racist imaginary.

By the late 1880s, it is apparent that, although Rhodes would continue to visit Kimberley's mines, he was developing a broader vision for South Africa. Together with Gardner Williams he began to engage in urban and rural planning. Their most notorious piece of urban planning involved the compounds for the African mine workers at Kimberley. Initially migrant workers from around southern Africa lived in informal settlements near the diggings. In spite of the passage of the Diamond Trade Act in 1882, I.D.B. was still thought to be so widespread that in 1885 and 1886 Rhodes and his fellow capitalists developed closed compounds, from which African migrant workers could not leave for the duration of their contract. Figure 2.3 shows that these compounds resembled closed barracks, with netting over them that prevented workers from tossing diamonds to confederates on the outside. While diminishing theft, the compounds also helped to stabilize the owners' access to a reliable workforce at a time when more capital was being invested in ever deeper mines. The deep mine shafts required more laborers, as well as more skilled and experienced laborers. Incarcerating them for six months at a time made production more efficient, taking pressure from investors off mine managers (Turrell 1987).

White business owners tended to see these compounds as symbols of social and technical progress. Images of the compounds were even distributed as picture postcards. Through postcards, tourism, and media reports, the word spread throughout middle-class, white South Africa that industrial and urban areas could be made safer and more productive by the recruitment and compounding of African workers. In his book, Gardner Williams boasted of the open spaces in the West End compound, where 3,000 African workers could enjoy fresh air in the courtyard as well as a swim in a pool. He did not brag about the sleeping quarters, which were crowded at 300 cubic feet per resident. On account of the close quarters, the rates of mortality were disturbing. In 1878, when workers lived in camps and open company compounds (in which workers could come and go more or less freely), the mortality rate was bad enough: 80 per 1,000, mainly due to infectious disease and exposure. In 1889, with all African laborers residing in closed compounds, the rate climbed even further to 100 per 1,000. When the mines reopened after the South African War of 1899–1902, the rebuilding of the compounds resulted in a lowering of mortality to 20 out of 1,000, which is still high compared with the figure considered acceptable

2.3. Compound for African workers at the DeBeers Mine, Kimberley, ca. 1896. This photo was taken from the surveillance tower. Note the wire mesh that covered the compound so that workers could not toss diamonds out to smugglers. Photo courtesy of the McGregor Museum, Kimberley. MMKP 829.

in England, 3 per 1,000 (Turrell 1987, 158–65). Even so, the practice of compounding became widespread, not only in diamond-mining areas such as Kimberley, but also in the much larger gold-mining industry that developed around Johannesburg. As mining came to dominate the South African economy, compounding became a normative form of residential segregation.

Rhodes and his fellow mine owners' vision for African workers extended into an intensive and unhealthy mode of surveillance. Not only were workers watched closely by guards. Workers were routinely subjected to strip searches and cavity searches, so that the owners could "see" any diamonds hidden inside their bodies as well as in their pockets and shoes. At the same time as DeBeers was housing African migrants in compounds, Rhodes also envisioned and created a suburb for white mine workers called Kenilworth. To outward appearances, it seemed idyllic. In 1888 and 1889, Rhodes worked personally with an architect, Sidney Stent, on designing the model village. Kenilworth lay three miles north of Kimberley. The avenues were lined with eucalyptus imported from Australia (see fig. 2.4). The houses

Surely this must be the Arcadia of the Diamond Fields L ᗡᖇ

2.4. "Surely this must be the Arcadia of the diamond fields."
Picture postcard of the main avenue leading to Kenilworth,
ca. 1900. Rhodes himself selected the eucalyptus trees. Courtesy
of the McGregor Museum, Kimberley. MMKP 7544.

were done in the classic, colonial style, with verandahs and gables. They were built with bricks made from the distinctive local clay with imported iron rooftops. Nearby Rhodes planted an orchard with a wide variety of fruit trees as well as a large grape arbor. A steam tram connected Kenilworth to town. Rhodes called Kenilworth his "hobby" (Roberts 1976, 271).

Rhodes was hardly creating a romantic or utopian community with attractive housing and streets lined with shady trees. In fact, by concentrating white mine workers in one location it was easier for DeBeers to conduct surveillance over them as well, to prevent I.D.B. Workers could not own houses in Kenilworth—they rented from DeBeers—which gave the company a further degree of leverage. To Rhodes, attractive cottages built in the Arts and Crafts style were instruments of control and surveillance. Rhodes even responded to public criticism of his motives during an 1889 parliamentary election speech, reported in a local newspaper, in which he "anticipated the fears that would be expressed to the effect that the project only meant an indirect way at eventually effecting the compounding of white men, and emphatically denied there would be any ground for such fears" (*Daily Independent*, March 20, 1890). He protested too much.

While building intensive capitalism, engineering, and social control at Kimberley to suit the needs of the diamond business, Rhodes also became involved in early gold mining on the Witwatersrand, near the town that would become the city of Johannesburg. In 1887 he helped to create the Gold Fields Company. Initially production was hampered by the costs and technical problems associated with deep-level mining. When expensive techniques had to be implemented in an unfamiliar political setting—the Transvaal, officially called the South African Republic—Rhodes began to push for a new, more efficient administration. The Boer republic not only excluded European immigrants from citizenship for ten years, it also monopolized dynamite to the disadvantage of the foreign miners. The Transvaal had tax, labor, and transportation policies that the foreigners found uncongenial.

Rhodes worked together with the other early Randlords to use their diamond-based finance—as well as the latest in technology and the sweat of African migrant laborers—to get the gold out of the ground. The environmental impact of early gold-mining efforts was tremendous. Not only were holes and tunnels dug, but tons of rocks were crushed by giant, coal-powered stamp mills. In 1888, on the Rand there were 688 stamp mills crushing 180,000 tons of ore into 200,000 ounces of gold. By 1894, 2,000 stamp mills crushed 2,750,000 tons of ore to produce 3,000,000 ounces of gold. At that time, mine owners started to use the Macarthur-Forrest process, in which the yield from the ore was enhanced by treatment with cyanide. Immigrants poured in from Europe and around the world. So did African people. In 1890, there were 14,000 African migrant workers in the vicinity; by 1899 there were 100,000, mostly living in closed compounds. The resulting social disruption and environmental pollution were legendary (Johnson 1987, 25–29; Van Onselen 1982).

Rhodes and his partner, Charles Rudd, formed the Gold Fields Corporation in 1886 and began to buy property around the Witwatersrand on the advice of mining engineers. The initial investments did not pan out. In 1889, Rhodes and Rudd admitted to their shareholders that their mines were not producing much gold. They sold the properties, investing again in diamond mining. Rhodes's gold-mining engineer, Percy Tarbutt, tried to convince Rhodes to buy different properties on the Rand, but Rhodes declined. It was only the intervention of Rhodes's partner in DeBeers, Alfred Beit, that got Gold Fields back into the gold-mining business. Gold Fields merged with Tarbutt's own company as well as several finance companies to found Consolidated Gold Fields. Rhodes, Rudd, and their new partners created a new kind of company, a mining holding company that bought majority interests in mines as well as exploration companies. The holding company provided finance, administration, and technical advice, through offices in both London and Johannesburg. Rhodes and his partners in Consolidated Gold Fields made a fortune through rationalizing mining in this way. Several other companies followed suit, so that by the 1900s, gold mining on the Rand became concentrated in the hands of a half dozen similar companies. Rhodes, by working with partners from Kimberley, had helped to create an approach to the financial and technical challenges of gold mining that still to this day shapes lives in South Africa's most important city and industry (Johnson 1987, 28–29).

In his various enterprises, Rhodes was never about just making money. His gold enterprise was created in large part to generate funds for the exploration and conquest of territories farther to the north. In other words, Rhodes's involvement at the Rand was a reflection of his growing interest in politics outside of the Cape. To extend his influence, he now had to maintain relations with the Afrikaner leadership of the Transvaal, who were rightly suspicious of his motives. Rhodes turned his attention northward to the most important northern ruler, Lobengula, the Ndebele chief who controlled most of what is today Zimbabwe. Lobengula's success derived partly through his leadership of the military and partly through his careful approach to Europeans. In 1888, when Lobengula signed a treaty of friendship with John Moffatt, the assistant British administrator of Botswana, Rhodes sent Rudd and several other emissaries to visit the chief. Lobengula kept them waiting, as was his custom, but was eventually persuaded by Rudd to give Rhodes and his associates access to the mineral wealth of his kingdom. Legend has it that Lobengula believed that the "Rudd Concession" would amount to a hole in the ground, while Rudd and Rhodes recognized that the agreement gave them virtual control of all the land of south-central Africa. In 1889 Rhodes incorporated the British South Africa Company (BSAC), with an eye to establishing control of the present-day countries of Zimbabwe, Zambia, and Malawi, and on October 29, 1889, he was granted a royal charter by the British government to pursue these ends. Rhodes's grand vision of an extensive British territory in Africa—that was practically his own personal estate—was on its way to becoming a reality (Galbraith 1974).

In 1890 Rhodes sent the BSAC's private army, the "Pioneers," to establish a fort on the site of present-day Harare. The settlement was called Salisbury, and soon the colony was known as Rhodesia. For three years, miners, traders, and speculators came to the country, producing conflict between settlers and Ndebele. In 1893, a war broke out between the BSAC and Lobengula that resulted in the death of the chief and of many of his followers. At around the same time, Rhodes and his engineers began to realize that a "second Rand" was not to be found in Zimbabwe. Instead, the country would be transformed into an agricultural colony of settlement, with Rhodes playing the lead settler and fostering the development of agriculture, railroads, and telegraphs. This would be a different society from Kimberley and Johannesburg, where destructive mining grew hand-in-hand with urban life. Zimbabwe also contrasted strongly with the long-established, cosmopolitan center of Cape Town, where Rhodes spent much of his time.

Bestriding southern Africa with one foot in Zimbabwe and another in Cape Town, in 1890 Rhodes became prime minister of the Cape Colony. As premier, Rhodes continued to balance the interests of Afrikaners and English, hoping that the colony could become more unified. During his term of office he also steered through two of the most notorious racial laws in South African history. The Franchise and Ballot Act of 1892 tightened the financial and educational requirements imposed on African voters. The Glen Grey Act of 1894 was an attempt to restructure life in a remote district of the Eastern Cape, with the hope, eventually, of enacting the same legislation throughout rural areas all over southern Africa. The bill increased taxes as a way to encourage individual land tenure and primogeniture. In the Cape legislature, Rhodes stated that his intent was to force African people without any more access to land to work for whites in mines and on farms. In the "great preserve" of the Eastern Cape, Africans "lived about in sloth and laziness, and never went out to work." The Cape government needed to "give them some gentle stimulants to go forth and find out something of the dignity of labour." Rhodes hoped that "in the future nine tenths of them will have to spend their lives in daily labour, in physical work, in manual labour" (Bundy 1987, 139–41). In the late nineteenth century, more white South Africans were willing to imagine a grim future for African people, in which they were stripped of rights and property. Rhodes's vision brought together antiliberalism with industrialization, fostering a new sociotechnical imaginary.

As Rhodes worked with business and political partners to transform African societies, the negative effects of his policies inspired much resentment and resistance. After the introduction of the Glen Grey Act, Rhodes toured the Eastern Cape to hear grievances and defend the legislation. Listening to a group of Mfengu in Nqamakwe in March 1895, Rhodes became defensive and began to berate the crowd. When he walked away, a heckler shouted, "Go, go, we won't pay it!" (Bundy 1987, 138).

The year 1895 saw Rhodes at the apex of his power in business and politics, yet it became a year of frustration. As influential as he was, his schemes to unite southern Africa in one, racially exclusive Afrikaner-British union were thwarted by the Boer republics in the Orange Free State and the Transvaal. The Transvaal, which controlled the gold fields of the Witwatersrand, denied rights of citizenship to Africans and foreigners. The government of the fundamentalist frontiersman Paul Kruger did not provide a friendly operating environment for English capitalists. During 1895, Rhodes pressed his contacts in Britain and southern Africa for change in the Transvaal. Probably with the collaboration of the British colonial secretary, Joseph Chamberlain, Rhodes and his partner, Dr. Leander Starr Jameson, organized a raid by the troopers of the BSAC on Johannesburg. There they hoped that the raid would spark a rebellion by the city's British residents. The raid failed. Rhodes did not participate in the raid personally, but Jameson and the other troopers were captured. Chamberlain successfully denied his involvement, leaving Rhodes hanging out to dry. Rhodes was forced to resign the Cape premiership in January 1896 as well as his directorship of the BSAC. Rhodes had begun 1895 very close to achieving his vision of South African political unity; he began 1896 with his vision nearly extinguished. Not only had he been relegated to the sidelines, a unified southern Africa seemed even less likely.

The failure of the Jameson Raid inspired a revolt against the BSAC in Rhodesia. The "rising" or *chimurenga* saw both Ndebele and Shona attacking white settlers and African collaborators. Rhodes seized the opportunity to be appointed a colonel over BSAC forces and led irregular cavalry in a brutal campaign of repression. Rhodes participated personally, killing rebels and torching homes. The campaign drew out his most violent side, revealing the extent to which he believed in a "pacified" countryside and in white racial dominance. Yet he also ended the war by approaching the Ndebele leadership personally and negotiating a settlement.

Rhodes's personal participation in the "dirty work" of empire set him apart from most nineteenth-century industrialists. While reestablishing BSAC rule—with increased British supervision—he also remained active in DeBeers and in farming. He took possession of two large clusters of estates, one in BSAC territory, in the Matopo Hills, and one in the suburbs of Cape Town. In the Matopos, he established himself as a "feudal lord"—in the words of historian Terence Ranger—settling British and Ndebele tenants on his land while engaging in scientific agricultural experiments (Ranger 1999, 76). In Rondebosch, he refurbished the old Dutch farmhouse and estate at Groote Schuur. The house was rebuilt by the colonial architect Herbert Baker, who followed Rhodes's instructions to decorate in two styles, Cape Dutch and the Arts and Crafts, both of which emphasized handcrafted simplicity. By contrast the estate was grand: the grounds were even populated by spectacular game animals, herded by Ndebele gamekeepers. His vision to unite the north with the south was accomplished in miniature on his personal estate. He now engaged more heavily than ever in farming, his personal way of showing that the new, racially exclusive imaginary for industrializing South Africa had one foot in the countryside, where labor could be put into more effective production, while the great visionary could play the landed squire.

Rhodes paid close attention to his estates in the Matopos and the Western Cape. At the Cape, he even started to buy old Afrikaner vineyards. They were bought cheap—the *phylloxera* pest was attacking vineyards locally as well as globally. Rhodes did not have a romantic or idyllic vision for vineyards. He hired an English engineer, Harry Pickford, who had worked in the California fruit industry, instructing him to make the farms profitable. Pickford introduced progressive agriculture to the Rhodes estates, rehabilitating vineyards by grafting vines onto American root stock, while planting orchards to supply the London market with fruit in the off-season, a concept made possible by the recent introduction of refrigerated shipping. Workers were recruited from nearby farms, and once again, on the surface, it seems that Rhodes offered them a better situation. Instead of paying workers partly in cash and partly in cheap wine, which was the local custom, Rhodes paid only in cash. He hired Herbert Baker to design and build the workers' houses (see fig. 2.5).

Rhodes intended for his farms in Zimbabwe to be productive as well. He introduced new plants and animals in the hope of being a good, improving landlord, very much in the tradition of the English landed gentry. He also made sure that the new colony, then called Rhodesia, had good access to the railroad. In fact, the construction of a railroad had been a component part of the agreement to give the BSAC the crown charter. The British government gave Rhodes crown lands for mining and railroads in Botswana, through which the Rhodesian railway would need to pass. In 1890, Rhodes negotiated a deal with the Cape Colony's prime minister, Gordon Sprigg, that gave the Cape the eventual ownership of the railway, as well as low freight rates, in exchange for financial support. Construction proceeded slowly,

Pay Day, Languedoc.

2.5. Farmworkers on Rhodes's estate, Languedoc, near Stellenbosch, Western Cape.
Workers are being paid in cash. Herbert Baker's cottages may be seen in the background.
Photo used by permission of the Rhodes Trust and the Bodleian Library, University of Oxford.
RHL MSS. Afr. S. 12/1.

much to Sprigg's consternation. Only in 1897 did the railway reach Bulawayo. In spite of the initial delays, Rhodes did push hard for the railroad in
1896 and 1897, at the same time as he was negotiating with Britain's High
Commissioner for South Africa, Lord Alfred Milner, for a settler government to replace BSAC rule. The Jameson Raid had weakened the base of
Rhodes's power in the BSAC yet had strengthened his reputation among the
white settlers. The railroad to Rhodesia underpinned the colony's nascent
economy, helping to establish eighty-two years of rule by the white minority.
Along the lines ran the associated technology of the telegraph, a key element
of "command and control" in business as well as in politics.

In characteristic fashion, Rhodes linked such technical developments
to visions of the landscape. In 1897, with the railroad reaching Bulawayo,
Rhodes was keen for it to continue northward, two hundred miles north to
Victoria Falls, where he hoped that the line could be built so close to the
spectacular chasm that riders would be sprayed by the mist. His ultimate
plan, famously, was to build the railroad as far north as Cairo, uniting the

farthest reaches of British rule in Africa, while funneling faraway labor to the mines (Rotberg 1988, 302–3, 369, 575–7, 591–2).

While developing his rails, telegraphs, and farms, Rhodes continued his involvement in the Cape's parliamentary politics, even after he was forced to step down as prime minister. From 1897 to 1899, he moved away from his former allies, the Cape Afrikaners and the British settlers who sought conciliatory unity with them. Now Rhodes embraced the "imperialist" party who supported Britain's aggressive moves to incorporate the Boer republics to the north. During the South African War of 1899–1902, Rhodes opened his grand Cape residence, Groote Schuur, to a steady stream of officers and their families, while decamping himself to a simple seaside cottage at Muizenberg. He waited for Baker to build him a similar cottage in the wine district, near Paarl. Yet at the very moment when Rhodes was coming close to realizing his dream of a united South Africa, he was also suffering from serious cardiovascular disease. He died in 1902 at the age of forty-nine. In Rhodes's famous will, the mountainside estate of Groote Schuur was left to the state. The mansion is the Cape Town residence of the president of South Africa; the grounds are now occupied by the University of Cape Town and the Kirstenbosch National Botanic Gardens.

What did Cecil Rhodes imagine, and what did he put in place? He envisioned nothing less than a new South Africa, with profitable industries based on progressive engineering, racial segregation, and corporate surveillance, all sanctioned by a unified state. His personal vision evolved as he worked with others and responded to the challenges of material conditions. That vision emerged from notions that were at once romantic and industrial, reflecting dominant strands of thought and action imported from his home country. He created the business, and political institutions took up his vision and worked with others to transform it into a full-blown sociotechnical imaginary. Rhodes fostered this development mainly by close collaboration with friends in the Cape parliament, as well as with business partners and technical experts ranging from engineers to architects. As his vision extended to affect the lives of many others, many different actors rose in protest—Tswana chiefs, Mfengu peasants, and Kimberley miners alike. In 1895, Rhodes engaged in a very high stakes game of social and political engineering—an attempted coup against the less-than-progressive Boer republic in the Transvaal. Yet the failure of the Jameson Raid—and even Rhodes's premature death in 1902—only slowed but did not stop the development of the capitalist, segregationist sociotechnical imaginary in southern Africa.

Rhodes's legacy for South Africa was to pioneer a way of conducting industrial production, business, and politics that lasted until the end of apartheid. The diamonds and gold were gotten out of the ground, processed, and sold, by means of increasingly technical methods. The mines were amalgamated. Africans were segregated and exploited. A new, unified, racist state was created. Throughout the entire process, Rhodes had a hand in all of it, from digging in the mines to fighting in the field, from deal making in London to governing the Cape Colony. He led South Africa by working with others in business and politics to set in motion and restrict African and European labor. Rhodes and his partners led South Africa toward a distinctive, racially segregated sociotechnical imaginary, in which technical expertise and migrant labor were harnessed to extract value from the soil. Migration, production, and destruction intensified, at the same time as segregation and surveillance became more widespread. Shareholders lined their pockets, while Rhodes bequeathed his own fortune to the cause of Anglo-Saxon unity.

References

Bundy, Colin. 1987. "Mr Rhodes and the Poisoned Goods: Popular Opposition to the Glen Grey Council System, 1894–1906." In *Hidden Struggles in Rural South Africa*, edited by William Beinart and Colin Bundy. London: James Currey.

Daily Independent (Kimberley). March 12, 1890. "Kenilworth, a Model Miners' Village."

Dubow, Saul. 2006. *A Commonwealth of Knowledge: Science, Sensibility, and White South Africa, 1820–2000*. Oxford: Oxford University Press.

Flint, John. 1974. *Cecil Rhodes*. Boston: Little, Brown.

Galbraith, John S. 1974. *Crown and Charter: The Early Years of the British South Africa Company*. Berkeley and Los Angeles: University of California Press.

Jasanoff, Sheila. 2004. "The Idiom of Co-Production." In *States of Knowledge: The Co-Production of Science and Social Order*, edited by Sheila Jasanoff. London: Routledge.

Jasanoff, Sheila, and Sang-Hyun Kim. 2009. "Containing the Atom: Sociotechnical Imaginaries and Nuclear Regulation in the U.S. and South Korea." *Minerva* 47(2): 119–46.

Johnson, Paul. 1987. *Gold Fields: A Centenary Portrait*. London: St. Martin's Press.

Parsons, Neil. 1998. *King Khama, Emperor Joe, and the Great White Queen: Victorian Britain through African Eyes*. Chicago: University of Chicago Press.

Ranger, Terence. 1999. *Voices from the Rocks: Nature, Culture, and History in the Matopos Hills of Zimbabwe*. Oxford: James Currey.

Rhodes House Library, Oxford University. RHL MSS Afr. S 1647, Hildersham Hall Papers (Rhodes Family), Box 1, 1868–1902, March 17, 1871, pp. 61–62.

Roberts, Brian. 1976. *Kimberley: Turbulent City*. Cape Town: David Philip.

Rotberg, Robert. 1988. *The Founder: Cecil Rhodes and the Pursuit of Power*. Oxford: Oxford University Press.

Scott, James. 1998. *Seeing Like a State: How Certain Schemes to Improve the Human Condition Have Failed*. New Haven: Yale University Press.

Storey, William Kelleher. 2008. *Guns, Race, and Power in Colonial South Africa*. Cambridge: Cambridge University Press.

Turrell, Robert Vicat. 1987. *Capital and Labour on the Kimberley Diamond Fields, 1871–1890.* Cambridge: Cambridge University Press.

Van Onselen, Charles. 1982. *Studies in the Social and Economic History of the Witwatersrand, 1886–1914.* 2 vols. Johannesburg: Ravan Press.

Vindex [Paul Verschoyle]. 1900. *Cecil Rhodes: His Political Life and Speeches.* London: Chapman and Hall.

Williams, Basil. 1926. *Cecil Rhodes.* London: Constable.

Williams, Gardner. 1905. *The Diamond Mines of South Africa.* 2 vols. New York: B. F. Buck.

Williams, Rosalind. 2008. *Notes on the Underground: An Essay on Technology, Society, and the Imagination,* 2d ed. Cambridge, MA: MIT Press.

Worger, William H. 1987. *South Africa's City of Diamonds: Mineworkers and Monopoly Capitalism in Kimberley, 1867–1895.* New Haven: Yale University Press.

THREE

Our Monsters, Ourselves: Reimagining the Problem of Knowledge in Cold War America

MICHAEL AARON DENNIS

Sociotechnical Imaginaries in the Early Cold War

Understanding the nature and character of US military science policy during the Cold War demands an answer to a seemingly simple question: Were researchers simply technicians working to solve problems and develop technologies of interest to their military patrons, or were the newly relevant scientists and engineers the moral equals of their new employers, freely choosing their research areas and forms of technological development to the mutual benefit of themselves, their patrons, and their nation? Different answers to this question yielded rather different policies relating to the integration of science and the Cold War state. Illuminating those differences displays the analytical purchase we gain by thinking with sociotechnical imaginaries (STI) as defined by Jasanoff (this volume).

In what follows, I delineate the coming into being of the Cold War military STI by identifying the monsters that invariably inhabit any imaginary. Monsters are well-known in science and technology studies (Jasanoff 2005b) and the history of science (Galison et al. 1992), but in this chapter I turn to them for a distinctive analytical purpose. If dirt is matter out of place, then monsters are often seen as the problematic and disturbing images that challenge and threaten the performance and reaffirmation of desired social orders.[1] An example from the history of American science might illuminate my point. Prior to World War II, funding for academic research fell to private philanthropies, corporations, and wealthy individuals (Kevles 1977a; Kohler 1991). The federal government played a nominal role in the support of academic research; the bulk of its support went to researchers in government agencies such as the Department of Agriculture, the National Bureau of Standards, the Census, the Geological Survey, and the Naval Research

Laboratory. During the Great Depression, academic researchers joined the long lines of citizens seeking federal monies. Despite the best efforts of the National Academy of Sciences and its Rockefeller-funded Science Advisory Board, it proved nearly impossible to support academic researchers with federal dollars (Auerbach 1965; Kargon and Hodes 1985; Genuth 1987). Yet, even when the federal government and private universities could agree on research worth funding, they could not always agree on contractual terms to support the work. It was no small historical irony that in 1933, Vannevar Bush, then the Massachusetts Institute of Technology (MIT) Dean of Engineering, found himself unable to sign a contract with President Franklin D. Roosevelt's Tennessee Valley Authority (TVA) for the support of research on long-distance power transmission, because MIT and TVA could not come to terms on the ownership of the project's intellectual property (Owens 1990).

At their moment of greatest need, American academics such as Bush refused government money because they feared the control that *might* come with a contract. Federal patronage was one of the monsters troubling the interwar imaginary of American scientists; academic researchers coveted government funding, but they feared the possibility of state control, while paradoxically embracing the support of philanthropists and corporations.[2] Vannevar Bush, as head of the wartime research and development programs, would play a fundamental role in taming this monster during the wartime emergency, while articulating a particular means of avoiding it in the era of government patronage to come (Zachary 1997).

The monster lurking in the early Cold War American STI, as conceived by scientists, was Lysenkoism, representing all that had gone wrong in the Soviet experiment and serving as a powerful reminder of paths not to take in forging the postwar relationship between science and the state. The historical reality of Lysenkoism in the Soviet Union did not really matter to Western understandings of the concept (Dejong-Lambert and Krementsov 2012; Krementsov 1997, 54–92, 158–90; Wolfe 2010; Krementsov 1996). Lysenkoism as imagined represented the state's intrusion into the practices and life worlds of scientists in two distinct, yet related, senses. First, the notion implied that certain technical beliefs were in synchrony with associated political ones. In the Soviet case, this was notably the argument that Mendelian genetics was incompatible with Soviet Marxism. Political disagreement via technical beliefs had the potential for drastic consequences, including exile and death in the famous case of geneticist Nikolai Vavilov.[3] Second, Lysenkoism represented a breach in a fundamental tenet of the American STI—that the content of objective technical knowledge remains separate from the contexts in which it is discovered and developed (Jasanoff 2005a;

Ezrahi 1990). Lysenkoism by contrast implied that the autonomy of science was a political choice, not a timeless given.

For Americans and many Britons, Lysenkoism was a telling example of boundary breaching in which politics contaminated the production of knowledge with catastrophic outcomes. In Britain, the Society for Freedom in Science had its origins, in part, as a protest against J.D. Bernal's views on planning scientific research, as well as against Nazi and Soviet transgressions with respect to issues as diverse as Aryan physics and Lysenko's attack on Mendel.[4] Michael Polanyi, among the society's most famous members, articulated his idea of tacit knowledge as part of a larger critique of a state that might interfere in the production of knowledge (Nye 2007; Dennis 1997). By definition, tacit knowledge, the unspoken knowledge of science acquired through practice, was knowledge the state could not possess, much less claim to direct. More relevant for our purpose was the way in which American scientific statesmen casually invoked the Soviet case when dealing with domestic politics. Take a 1950 *Foreign Affairs* essay by Harvard University president and Manhattan Project leader James B. Conant, entitled "Science and Politics in the Twentieth Century." Ostensibly about conflict over technical opinions with respect to weapons development, in this case, thermonuclear weapons, Conant's essay quickly moved to contrast the US and the Soviet Union, even while acknowledging that the evaluation "of technical opinions must be as perplexing in Moscow as in Washington." For Conant, Lysenkoism and the Soviet experiment in general demonstrated that "the wholehearted acceptance of science by politicians can lead to the curtailment of the work of scientists" (Conant 1950, 190, 192). In turn, American politicians, aware of the power of science, needed to remember that it was science, not technology, that had delivered the war's wonder weapons—radar, the bomb, and the proximity fuse. More importantly, if science had become a "horse worth backing," it was vital that politicians bet on "another type of gamble, adventures in pure science" to ensure that the United States would have even more knowledge in the future to produce powerful weapons (Conant 1950, 198).

Managing, rather than resolving, technical debate seemed Conant's primary concern; he proposed a tribunal of experts, a setting in which those members of the government who would overturn a technical decision might clearly see the weight of the evidence they would discard. He ended his inconclusive essay with an evocative, haunting image: "An historian 50 years from now writing of the mid-twentieth century will certainly record that science and politics were by 1950 no longer to be regarded as two totally unrelated activities. He might well add that scientists and politicians were in this period to be found sometimes in amicable cooperation, sometimes in violent disagreement; only one thing seemed certain, the type of society in which each could go his own way with only a polite bow to the other had disappeared as irrevocably as the American buffalo from the plains" (Conant 1950, 202).

Hunted to near extinction, such would be the fate of an American science that tried to remain separate from politics, but it was also the fate of a science utterly subsumed by politics as in the Soviet Union. Avoiding an American Lysenkoist moment was the goal of postwar science leaders; preserving and policing the boundaries of science and politics was a means to achieve that end. Conant (1950, 189) had argued that science and politics had become "intermeshed" and "at times the grinding of the gears produces strange and disturbing noises" (Conant 1950, 189). Originally, those sounds and noises were words and arguments, exchanged among men, and they were almost always men, about weapons, yields, blast radii, and all the measures that a democratic polity would have to enact to assure its future in a potentially never-ending confrontation with its totalitarian other. A reality in which American science worked to achieve national security goals was as close to an unacknowledged Lysenkoist moment as any of the wartime elite might allow.

Still, there was an interesting slippage in Conant's essay. The title used the words *science* and *politics*, but the memorable conclusion addressed the relations of *scientists* and *politicians*. Just what was the nature of that relationship? Were scientists, military officers, and politicians normative equals? That is, were they all professionals, masters of their respective domains and entitled to a form of mutual respect in articulating the goals of research, if not the wariness that Conant evoked? If they were not equals, then were their relations akin to those of management and labor? Vannevar Bush's whole emphasis during the war and in his failed postwar planning was based on the idea that scientists were peers of their military partners. This assumption flowed naturally from the organizational fact that the Office of Scientific Research and Development (OSRD) had been a separate civilian organization linking academic researchers with the armed services. There was no OSRD equivalent in the postwar world; instead, the armed services became the dominant patrons of academic science. Wartime military project management manifested itself most clearly in the Manhattan Project. Within that project, researchers learned that they were welcome to develop a powerful new weapon but that they were not welcome to play a substantive role in the weapon's actual military use. Manhattan Project scientists, such as Leo Szilard and others at the Chicago Metallurgical Laboratory, learned that they were employees with limited means to contest managerial decisions (Schweber 2000; Price 1995).

Here was another dimension of a potential Lysenkoist moment in America. The successes of wartime research provided citizens with two different and radically divergent understandings of the relationship of scientists to politicians and the military. In Vannevar Bush's version, scientists as independent actors drew upon the existing stockpile of basic research to develop the new and powerful weapons that made the Allied victory possible. Another equally plausible version, however, was that when researchers were given enough public money, it was their job to develop new and more powerful weapons. The very ideal of a socially responsible science, championed by West Virginia Senator Harley S. Kilgore, rested upon such an understanding (Kevles 1977b; Wang 1995). In the latter case, researchers were nothing more than hired hands, technicians, whose training supported by government funding would yield national security when needed.

Think of the difference between these two imaginations in the context of fears of an atomic secret that scientists might deliver to the Soviets. Believers in Bush's version of the relationship understood that the secret, to the extent that there was a secret, lay in the tacit knowledge embodied in the bomb (Kaiser 2005). Anyone seeking to build an atomic bomb would have to redo all the experiential learning that had taken place in the Manhattan Project. There was no formula to hide in a safe or any place so secret that erasing it from a map might offer protection. Secrecy was a process, one that the Smyth Report (1945), *Atomic Energy for Military Purposes*, addressed as know-how. Believers in Kilgore's version of the relationship, however, saw the secret as something that might be hidden in a safe because scientists were simply technicians following a recipe. To see scientists as technicians was not simply to misunderstand the intellectual richness and thickness of knowledge production, engineering development, and industrial production, it was a strategy in which scientists were "on tap," but never "on top," and bore no responsibility for the uses of their research (Mukerji 1989). Let us return to Conant's "grinding of the gears" in the machinery of state. Those noises, not easily diminished through the use of cash, that most penetrating lubricant, were the thrashing out of the proper role and place of the scientist in the early Cold War state's administrative structure. And the task of reimagining the problem of knowledge creation in early Cold War America yielded the following paradox—only by becoming wards of the state, especially of the armed services, could American scientists preserve the freedom to pursue their science while keeping the Lysenkoist monster at bay.

Sam Schweber began his eloquent comparative biography of Hans Bethe and Oppenheimer with a discussion of the problem of the scientist as simply the state's hired hand (Schweber 2000). He made a powerful case that this was a central issue for science in the early Cold War but that it lost its salience in succeeding years. In what follows, I take Schweber's insight and examine a text that sought to allay the fears of the monsters lurking in the dominant postwar American STI—Don K. Price's 1954 work, *Government and Science: Their Dynamic Relation in American Democracy* (hereafter, *G&S*). Price's text, unlike Schweber's subjects, offers a fruitful site of inquiry because it made the point of who was employing whom its central focus. Accepting this modest, but important, point demands that we see science as simply another form of useful knowledge and the scientist as another kind of worker (Shapin 2008).

Reading *Government and Science*, Again

Overshadowed by his later work, especially *The Scientific Estate* (Price 1965a, 1965b), Price's first book was an important and well-received articulation of the themes that would dominate his writings, especially the development of an administrative cadre within the American state capable of dealing with the problems wrought by scientific and technological change. Originally presented at the 1953 James Stokes Lectures on Politics at New York University, *G&S* served as an apologetic for the new technical and political order emerging from World War II, what the volume's dust jacket described as "how to achieve the indispensable benefits of government-supported science without also incurring the intolerable calamity of a government controlled science."[5] The copywriter's pithy text captured the book's content and tone, while also making clear for our purposes that this was an attempt to address the early Cold War monster of Lysenkoism. Focusing on this particular volume affords us a chance to see how the past invariably and sometimes invisibly shapes our present.

Price was among the founders of American science policy as an academic discipline—what Sheila Jasanoff memorably called the Charles River School—as well as the founding dean of Harvard's John F. Kennedy School of Government.[6] As a young policy analyst in the Bureau of Budget, Price had championed an administrative form for the National Science Foundation that differed dramatically from the politically insulated structure proposed by the National Research Foundation's prime proponent, Vannevar Bush. Still, Price saw himself as Bush's ally in securing academic science's place in the federal firmament; in turn, Bush eagerly commented upon Price's chapters and seldom questioned their merit or analysis.[7] There was good reason for their agreement, even their occasional agreement to disagree. Price was an effective apologist for the world fashioned by Bush's wartime and postwar organizations. Bush had organized science for war with the assumption that the postwar world would very much resemble the world left behind in 1940; when it became clear that the future would not resemble the past, he attempted to shape the future, unsuccessfully.[8] Most importantly for our purposes, Bush's never-realized National Research Foundation included a division devoted to military research and development (Bush 1945). The failure to enact that vision provided the armed services with the opportunity to become the dominant patrons of postwar American science and technology. Price graciously avoided mentioning this failure, but strikingly, Price understood some things that Bush and other members of the wartime elite were blind to, not the least of which was that secrecy and the creation of a classified world might prove far more valuable than transparency in crafting a public image of science. Moreover, Price saw the decentralized organization of American science as a powerful antidote to socialism and a template for government intervention in the larger economy (Price 1954, 200).

Price contrasted the "mood of the scientific community" with that of the general public delight at war's end and the concomitant belief in a "postwar utopia of new gadgets" powered by wartime research. If the public saw a future illuminated by electricity too cheap to meter, scientists saw a future in which their dependence upon the federal government for research support might come at an unbearable cost—loss of "the freedom of science" (Price 1954, 1). Nor was a retreat to the prewar world of private patrons a choice, since the world's fate rested upon which nation's research program could produce the next "revolutionary advance in military tactics, following those already made by radar, jet propulsion, and nuclear fission." Equally important were the industrial benefits, the new firms and commercial technologies that would emerge from government support of science. Price sought to persuade his audience that American science need not find itself caught between a potential Communist triumph extinguishing academic freedom and an American government's deep pockets hopelessly corrupting the practice of research. America's disaggregated system of government would serve as protection for both parties.

While acknowledging the novelty of the postwar relationship of scientists and the federal government, Price did not agree that this was "an unhappy shotgun marriage"; instead, it was the upshot of a set of historical processes beginning with the American Revolution. Still, the book's central mission was to articulate and deconstruct the fundamental tension of the postwar era: "American scientists are now required to work in a complicated network of secret and confidential data, and to communicate on many subjects only with those who have been officially investigated and cleared. Then, too, science has been accustomed in the past to rely for its support (and incidentally for its independence) on a great variety of local and private institutions. Yet it is now obvious to everyone that the structure of scientific research in American universities and industry has come to depend heavily on federal grants and contracts" (Price 1954, 2). Or more precisely, how could the state that science was required to defend with its new knowledge and weapons simultaneously serve as guarantor of the freedom of science?

Science itself was part of the answer, for science was "the most explosive force in modern society," shattering the authority of sovereigns since the eighteenth century while playing a vital role in the nation's republican revolution and providing the motive power for the dynamics in the book's subtitle (Price 1954, 3). While not invoking the scientific method, Price did believe that scientists, unlike politicians, lawyers, or clergy, possessed a common attitude, one based in action rather than in faith or a respect for precedent. Nor was this attitude limited to scientists; over time it spread to "the allied professions and the 'mechanic arts,'" making it impossible to limit science's impact upon the polity to the influence of scientists alone. "The modern factual and objective way of thinking" that Price attributed to science had found its way into the foundation of the United States, where the census, an enumeration of the nation's citizens, was the very basis of sovereignty (Price 1954, 4–5). History offered Price one means of arguing that science and the American polity were far from strangers before World War II; they were joined in practices of "factual and objective" thought. He took great pleasure in observing how much social science the government sponsored, especially through its support of the Census Bureau, but he accepted the proposition that support of academic science had not been a government priority before the war.[9] History also provided Price with the mass against which science's explosive power might react—the decentralized American state.

Arguing that sovereignty was not unitary, Price proposed the idea of "federalism by contract," the key concept in his vision of the right relation between science and the state, since it was literally through contracts that science and the American state might together constitute a mutually beneficial future. Contracts between military departments and academic researchers or private corporations thus were not monsters to be feared but generative instruments capable of producing not only new technical knowledge and weapons but new organizations that might manage science better. Price was not simply legitimizing a system in which public monies supported private institutions, although he did note that there "had been no organized objection from scientists and educators to the general development of federal aid," wryly adding that "nobody shoots Santa Claus" (Price 1954, 67). Contracts, in Price's understanding, did not produce dependence but rather a set of relationships effectively erasing traditional, and no longer tenable, distinctions between public and private. Over time, universities and corporations developed a "stake in federal programs," but they did not do so for money alone; rather it was a "combination of patriotic sentiment plus the interest and excitement of taking part in the greatest and most challenging enterprise of the age—an organized effort that makes any private program, even the dealings of the greatest captains of industry, look trifling by comparison" (Price 1954, 79–80). Seldom has the project of arming and defending the Cold War American state been so glowingly described, but it was a terribly serious business. It reveals something we easily forget—for Price the freedom of science was not necessarily the most important thing at stake. The very existence of the United States rested upon the success of science in developing new weapons to assure an American victory in case the Cold War turned hot.

As a generative social technology, contracts had for Price a five-fold taxonomy. First, they might be used to literally contract for goods or services, such as the improvement of an existing weapon or the provision of a specific service. A single technology might encompass a system of innovations requiring extensive management, and in such instances new organizations would arise to manage these multiple-tiered systems. Second, the military produced "master contracts" with universities and corporations such that each new project did not require a new contract but only recognition as part of "the overarching contract." Third, the military might sponsor studies of particular strategic or tactical problems, such as MIT's famous summer studies—Project Hartwell on antisubmarine warfare and Project Charles on continental defense. Here the military outsourced to civilians what had once been the domain of senior strategists. Fourth, universities might establish special laboratories or divisions to manage specific tasks for the government—witness Argonne's reactor work managed by the University of Chicago for the Atomic Energy Commission (AEC) or MIT's Lincoln Laboratory, which dominated the air force's continental defense research effort. Finally, the private research corporations established to work on government problems, such as Associated Universities, Inc., for the AEC or the RAND Corporation, which served as the air force's first think tank long before that term became part of the vernacular.

What Price called "federalism by contract" can today be seen as a networked state in which connections were multiple and multiplying. Price was aware of the novelty of this political form for it not only erased and effaced distinctions between public and private but bound together elites in diverse institutions in a set of common endeavors. And elites they were: the Department of Defense acknowledged that it placed 63 percent of its research and development contracts in ten institutions, while the AEC put 88 percent in the same small set (Price 1954, 89). Contracts were the means through which the government acquired access to scarce intellectual and technical resources it could not provide for itself; it was recognition "of the pressure of the Pentagon on American universities to organize their best brains to support the development of weapons" (Price 1954, 37).

The networked nation imagined through federalism by contract remained explicable within the standard language of political science and administrative regularity as long as one attended only to the direct connections or the hubs. Hence, Price could offer as an example of change the creation and maintenance of physical standards. Even the establishment of weights and measures, a constitutionally mandated government function, had become an enterprise demanding significant new instrumentation, such as atomic clocks, that would provide a framework for new, industrially important standards. Such measures would not emerge in the isolation of government labs but in a constellation of facilities, some in government hands and others in universities and industry. While it was possible to worry about conflicts of interest among suppliers of the apparatus used to produce new standards, Price noted that this was not the main problem. Rather, the issue was whether the government possessed the administrative competence to avoid "dangerous political pitfalls," while shaping a complex yet coherent technical program. Until the nation was successful in "educating men who combine some appreciation of scientific problems with an understanding of the problems of policy administration in a government setting," the new federalism would have to serve as the foundation of the state's technical work (Price 1954, 92–93).

Price's call for a new layer of administration made sense given his own lifelong belief in the need for a powerful administrative mechanism for matters related to science and technology. Yet the administrative problems he addressed in *G&S* were raised more urgently by the loyalty and security hearings, especially those related to the atomic bomb, and matters of advice given by that most important but little studied institution of the early Cold War, the Department of Defense's Research and Development Board (RDB). Both were sites where the Lysenkoist monster reared its head. The hearings were a site where the state sought to assess the reliability and loyalty of scientific and technical personnel on the basis of their political beliefs. The Department of Defense RDB was a site where the state not only adjudicated the validity of knowledge claims but standardized the processes and outcomes of knowledge production. Tellingly, Price passed over these realities in silence.

Loyalty and Security Revisited

In a chapter on "Security and Publicity Risks," Price discussed what he found important about the loyalty and security hearings. If his readers expected to encounter revelations about the various hearings underway, they were bound for disappointment and a very dull evening. Price did not mention physicists or other scientists under attack, nor even Joseph McCarthy by name, but he did make much of Congress's inability to criticize its own committees. One suspects that his contemporary audience left this chapter wondering just what mattered, but after summarizing Price's argument, I suggest that his proffered solution, a more robust administrative structure to protect science, might better be read as an attempt to turn a bug into a feature, that is, to turn the massive disciplining machinery of the loyalty and security hearings into an effective demonstration of why scientists were better off as employees than as peers of their military and political patrons or even as comanagers involved in a common research program.

American scientists, Price explained, were "still struggling to reconcile their eighteenth-century devotion to science as a system of objective and dispassionate search for knowledge and as a means for furthering the welfare of mankind in general, with the twentieth century necessity of using science as a means of strengthening the military power of the United States" (Price 1954, 96). War's end and the belief that American security rested upon an atomic monopoly produced a situation in which researchers required constant surveillance by the Federal Bureau of Investigation (FBI) lest they, inadvertently or through misplaced idealism, reveal the atomic secrets (Price 1954, 98). Price saw the entire loyalty and security controversy as a striking example of the pitfalls of boundary crossing. First, the problem lay in a failure to recognize the difference between the content of scientific research and the actual use of that knowledge. As Price put it, "the political trouble comes in the phase where scientific findings are being applied to practical problems" (Price 1954, 103). A world in which a firm boundary between basic and applied research afforded researchers political cover was inconceivable since science could not "exist on the basis of a treaty of strict nonaggression with the rest of society" (Price 1954, 106). Nor could America hope to separate basic from applied research since scientists and engineers were encouraged to move into positions of executive and administrative responsibility where they might actually translate knowledge from theory to practice. Unless the United States could defend the application of science to public affairs, we would prove unable to "maintain support for basic science and defend its right to freedom of inquiry."

One could not argue for the freedom of science on the basis of researchers only engaging in basic science, disconnected from worldly affairs, since, as Price observed, we had also given up the idea of sorting society into discrete "estates." Still, arguing for scientific freedom demanded recognition that freedom was a "fundamental value in political society" and of the historical evidence that "only free science can play a dynamic role in furthering human welfare." Price found proof of this claim in a simple observation: "The scientists of Germany, who thought that science was so separate from politics that it could prosper no matter what political philosophy dominated the government, discovered their mistake under Hitler" (Price 1954, 107). Nor was the solution "some science or body of scientists" who would control the application of science to public affairs. Here the briefly noted historical example was "the Lysenko case" and the "whole history of Communist experiences that it typifies" (Price 1954, 107). With the competition quickly dispatched, all that remained was to devise "a political system by which the freedom of research can be defended and its results applied to practical problems under the guidance of responsible democratic processes" (Price 1954, 108).

Differentiating between "unchecked central authority" and "responsible" central administration," Price informed his readers that political attacks on the integrity of science in the United States resulted more from a lack of central administrative power than from too much central authority. "Irresponsible special interests" rather than those who recognized the need for "objective and unbiased research" were the source of such attacks. The solution was simple: "an intervening layer of administration between science and politics, to protect science and to make their relationship more smooth. The lack of this layer has been as responsible as any personal malevolence for the political attacks that have been made on the integrity of science" (Price 1954, 108–9).

Reverting to the language of mechanical parts that Conant used in his article, we might wonder if Price's proposed administrative layer was a lubricant or a muffler. His solution was distance—a separation of science from politics allowing reasonable administrators to discern who was or was not a security risk and what knowledge should remain secret. "Our freedom,"

Price declared, "would be furthered in general, and particularly that the objectivity and integrity of government research would be best protected, by a more authoritative and responsible executive supported by a strong and stable career service" (Price 1954, 122). The American military afforded Price the only model for this administrative vision; but he was not arguing for a military state, only a state in which the executive branch was as well run as the military with its emphasis on discipline, responsibility, and well-recognized lines of authority. In Price's America, individuals working in the national interest would almost always reach the same conclusions about the loyalty of individuals or the reasons for declaring some knowledge secret.

That Price invoked the British "mystique of monarchy" as the force under which authority might unite and the University Grants Committee as the exemplary institution creating an "absolute" separation of university research from politics gives one a sense of Price's anglophilia as well as of his sense of the mechanisms through which governments fashion consensus. America had no such mystique or mechanism for making absolute distinctions. Nor could it. The very idea of a networked state that Price articulated through "federalism by contract" was at odds with a strong central administration, let alone a strong central authority. Instead, Price anemically concluded that by making the US government more competent and responsible one would assure the freedom and integrity of science in the American future.

Price's approach to the issue of the loyalty-security hearings reads as amazingly bloodless, devoid of genuine emotion save for an odd refusal "to accept the dilemma between national security and individual freedom" (Price 1954, 110). As early as 1948, mathematician Oswald Veblen declared that "we are now living to a large extent under a police state" (Wang 1999, 190). None of that outrage makes its way to Price's belief that an additional layer of government staffed by thoughtful administrators could effectively insulate science from politics. Perhaps he imagined that those administrators would be products of the Kennedy School of Government where Price later served as dean, but the ideological and political costs of Price's solution were staggering. Except for a passing mention of the "humiliations" entailed by FBI investigations, Price never came out and addressed the reality of the loyalty and security hearings. We are tempted to view them in the context of political liberty and rabid anticommunism, viable and necessary interpretive contexts, but they were also a massive, concerted effort at disciplining a workforce unused to political discipline (Thorpe 2002). Whatever the intended meaning of the loyalty process, the effects were clear—the state was in charge, and scientists were cogs in its well-oiled machine.

Price's analysis contrasted sharply with that of Vannevar Bush. Writing in the wake of the Gray Board's public destruction of J. Robert Oppenheimer, Vannevar Bush made clear why such a separation between knowledge makers and the uses of their knowledge was unacceptable: "The position that the scientist occupies today is certainly not that of a technician who makes gadgets and leaves their entire economic or military significance to others. The scientist does not enjoy being called a technician. The scientist is not beyond his field in considering modern military strategy and tactics. Every such question has its military and also its scientific aspects, and only by a fusion of these can it be adequately analyzed" (Bush 1954, 62). To reduce scientists to technicians remained for Bush part of the American Lysenkoist moment, since it effectively bounded scientists' ability to participate in the workings of the polity. For him, this was an unacceptable situation—he had spent the entire postwar period fighting this very interpretation of both the war and the postwar era. Science might be the servant of man as one historian of science explained, but surely science wasn't simply the military's hired hand (Cohen 1948). Price's solution—a new layer of administration— was merely a polite veneer for disciplining scientists to make them aware that their technical knowledge was of value to the Cold War state; but their political views were either their own business or else seriously inconvenient and potentially worthy of prosecution or dismissal. Bush's angry intervention was the belated recognition that if the state could strip Oppenheimer of his clearance then scientists were mere employees whose only recourse was not to work for their all-powerful employer. Of course, Bush was fighting a losing battle; Price's networked state was the only game in town.

The tension between scientists as equal to their new patrons and the scientist as employee was best captured in the Manhattan Project when the Chicago Metallurgical Laboratory researchers sought to play a role in deciding how to use the revolutionary new weapon they had delivered. As Matthew Price (1995) shows, struggles between labor (scientists) and management (military and industrial elites) took place during the course of the war, and the famous Jeffries and Franck reports were only the final products of a long struggle over the control of the new weapons technology. Similarly, during the debate over the Acheson-Lilienthal and subsequent Baruch Plans for the international control of nuclear weapons, politicians dismissed the concerns of researchers such as Oppenheimer (Herken 1988, 151–71). Physicist Philip Morse made the point most clearly in his December 1950 essay, "Must We Always Be Gadgeteers?" Morse argued that physicists solved problems as scientists, but they could not "*directly*" contribute to making decisions about the use of the technologies. Morse wanted physicists as scientists to make a contribution to running the "operations of civilization." (Morse

1950, italics in original) He wanted the physicists to become policy makers rather than advisers, with knowledge rather than an electorate serving as the basis of their political power. Hence, operations—whether the movement of goods from factory to market, the bombing of cities, or the design of communication systems—became objects of study, and clarifying the factual basis upon which decisions should be made became the physicists' way to ensure the wisdom of decision makers. Morse's Operations Research (OR) offered scientists a bigger role in using the laboratory's technical knowledge; it should come as no surprise that Don Price's audience, too, was soon to learn about OR's administrative potential.[10]

The Research and Development Board

Price embraced OR as part of his call for administrative reform, but he nested it within his discussion of the recently dismantled RDB. Only in this section did Price actually address current events, especially the debate over the hydrogen bomb and the leaks that had brought the topic from the classified world into public view. Originally established by Vannevar Bush as a way to manage and monitor military research and development, the RDB quickly became a bureaucratic nightmare. Among its many problems was lack of statutory authority to change the research and development budgets of the various services as well as an unwieldy bureaucratic structure. Initially, Bush likened the board to a district attorney keeping an eye on the services. It was a metaphor drawn from criminal law, and it accurately captured the struggles at war's end as the armed services realized that they would control their own research portfolios rather than share that authority with the proposed National Research Foundation. Little studied and seldom examined as an integral aspect of early Cold War American science, the board was declared a failure and eventually eliminated with Bush's best wishes in 1953 in a Department of Defense reorganization that created undersecretaries for research and development.[11]

Price's discussion of the board and OR took place within a larger discussion of scientific advice for the government. Advice remains an important dimension of the relationship between researchers and the federal government, but Price recognized that a focus upon advice also allowed for a return to one of his most basic points—that scientific knowledge was a necessary, but far from sufficient, basis on which to make political decisions. For Price, any decision that scientific advice might settle was distant from serious politics. Take the H-bomb. Building it was far from a nakedly technical decision, but instead one involving a sense of how such a weapon might fit within a portfolio of technological and strategic possibilities. Even OR could not provide a simple answer to the issue, since OR required some statement of the assumptions of the world in which the operations would take place. In turn, researchers using OR techniques might develop the optimal approach to the design of a continental defense program, but they could not make a decision on the wisdom of building such a technology. Once again we are back to the problem of the role of the scientist in government; for Price, that role was profoundly circumscribed.

This entire framework rested upon an artificial separation of decisions based on science from decisions grounded in values, a view that even Price accepted might not prove "real" (Price 1954, 168). What makes his admission so remarkable is that he did it while discussing the RDB as an advisory institution. Certainly that was one way to understand the RDB, but it did not do justice to the sheer enormity of the RDB's imagined mission. Here was an American institution, part of the National Military Establishment and subsequently the new Department of Defense, which was to provide the military with a "five-year plan" for American science and technology. Bush was adamant about the need for such a plan and seemingly oblivious to how it echoed the language of the nation's great new enemy. Not only was the board expected to create a census and atlas of American research and development at a granular level, giving every program its own project card and IBM punch card. The board also worked to standardize the vocabulary of new fields, with bound volumes carrying titles like *Glossary of Guided Missile Terms* (1949), while simultaneously adjudicating abstract technical debates over technologies that were only beginning to be designed. With these functions, the RDB actively constituted the very objects researchers would design and develop under government contracts. National security served as a cover, obscuring the very operation of what might appear as the Lysenkoist monster of government control.

A well-known debate over the feasibility of inertial guidance for long-range aircraft and ballistic missiles found itself under adjudication by a special panel of the Guided Missile Committee, known as the Sub-Panel on the 84 Minute Pendulum Problem (MacKenzie 1990; Dennis 1994). This group of experienced researchers debated whether or not it would prove feasible to develop technologies that took advantage of the fact that a pendulum with a period of 84 minutes was one whose length was the same as the radius of the earth. If it were possible to build technologies incorporating such an insight, one might be able to subtract the acceleration of gravity from the acceleration of a moving vehicle, or so proponents claimed. Although this might fall under the heading of advice, it was also something more—the panel's verdict would become authoritative. The facts in this particular case are outside the scope of this paper. My point is simply that Price saw the RDB through the lenses of advice and bureaucratic labor rather than through its actual practices and goals. To use Price's own vocabulary, the RDB did not separate science from values since science strictly adhered to the military's values of being useful in the coming conflict with the Soviet enemy.

The RDB Price described reinforced the very view that Bush had wanted to eliminate—that scientists were after all mere technicians, working at the behest of military officers and politicians. What gave their advice value in Price's imagination was that it was delivered in confidence and seldom made public. In fact, for Price the anglophile, secrecy functioned as an administrative tonic, protecting those who offered advice and those who professed to take it (Price 1954, 181). What was given in secret would never become public and neither embarrass nor vindicate either party. Politics ruled, as it should in a democracy, but it did so at a cost to transparency and in support of the paradoxical vision that government patronage would provide the surest path to preserving the freedom of science. More succinctly, the monsters that loomed so large and threateningly for Bush were in Price's imaginary just that: imaginary. Within the state's cash-laden embrace, there was as he saw it only the promise of scientific and technological development and compelling intellectual challenges; government patronage was not a treacherous pathway to a monstrous Lysenkoist moment but the key to ensuring the freedom and value of American science.

Endgame

What gave Price's message its power was its urgency—that there might not be much time to harness science in the service of the United States:

> The restless energy of the scientist and the engineer has broken the constraints of red tape and supplied a dynamic drive to the development of government programs, as well as to the productiveness of private industry. But the problems the United States faces today cannot all be solved by rebellious independence and restless energy. The role of world leadership is an uncomfortable one; it requires a steadiness of purpose, an economy in the use of our energies, and a breadth of philosophy that have never been characteristic of the American temper. We may well pray that we shall be given time to develop them. (Price 1954, 203)

If a final conflict with the Soviet Union were to occur soon, one might argue that the monsters of the early Cold War STI were the least of America's problems, since even Price's networked state might not have survived global thermonuclear war. Nonetheless, the monsters of that STI have haunted American science and its students. Witness all the ink spilt and trees killed in the contest between internalists and externalists in the histories of science and technology (Shapin 1992) or the science wars of the 1990s. Those discipline-shaping events stemmed in large part from a fundamental inability to reconcile the image of scientists as employees laboring on behalf of state and industry with their disciplinary identities as seekers after the truth. Internalists and science warriors wanted to believe in a world in which scientists were untainted by worldly affairs, while failing to admit that worldly affairs—empire, warfare, commerce—were the only things that kept science functioning.[12] To protect that world, Vannevar Bush had argued for the creation of a new category, basic research, pursued and funded without thought of its utility (Bush 1945; Dennis 1997). But as we know from the example of the humanities in US higher education, funding for pure knowledge does not pay very well, certainly not well enough to support the successor institutions of Cold War American science and technology.

Let us return to the wartime furnace that did so much to forge our world and a brief epistolary exchange between Vannevar Bush and his good friend Karl Compton, chair of the OSRD's Office of Field Services as well as the president of MIT. Bush had served as one of Compton's most trusted lieutenants during the latter's radical refashioning of MIT to meet the challenges of the Great Depression. He had only left Cambridge for the chairmanships of the Carnegie Institution of Washington and the National Advisory Committee for Aeronautics when it became clear that his friend would not soon retire from overseeing MIT (Zachary 1997). However, the letters they exchanged in early February 1945 were only tangentially related to their wartime efforts; letters marked "Personal" articulated a distinctive view of the future of science and its management. With war still raging in Europe and the Pacific, the two men were involved in activities ranging from the deliberations of the various committees that would form the appendices to Bush's famous report, Science—The Endless Frontier, to planning the OSRD's endgame in the Pacific to Compton's own role as head of the short-lived Research Board for National Security. Despite all this and more, including the big and still unfinished business outside Santa Fe, New Mexico, the two exchanged a remarkable pair of letters in which they carefully reviewed the wartime activities of some of the OSRD's many members and outlined their possible careers. Bush's letter of February 5, 1945, "commenting on some of the outstanding personnel who have demonstrated their abilities in connection with OSRD activity" received Compton's whole-hearted endorsement as well as further elaborations of their individual prospects.

The Bush-Compton letters expressed a shared view of a future in which junior colleagues received their due and more. While the rewards were mainly academic positions, as well as important jobs in American industry, there was also the expectation that the military's role in American science would require some talented members of the war generation to remain in harness. For example, Compton suggested that Edward L. Bowles, Bush's first graduate student and an MIT professor of electrical engineering, join Compton at the Research Board for National Security. Embedded in these plans was the unexamined assumption that the future would largely be continuous with the past. From our privileged position in the future that Bush and Compton did not imagine, we have little sense of how war and preparations for war would affect all that was to come. Much of their vision would remain in the darkness of closed file cabinets and Hollinger boxes, as evidence of the vast industrial effort that historian Michael Sherry called, "preparing for the next war."[13]

What are we to make of these letters? Are they messages in a bottle from an imaginary that the authors could not successfully engineer? Or should we read them as warnings, evidence of the ease with which even the best-laid plans are overwhelmed by the complexities in which individual visions are resources for both stasis and change? If persuasive, this essay demonstrates the power of the concept of STI in rendering historical choices and practices visible against the assumptions of both the past and present.

In a generous and thoughtful *éloge* for J. Robert Oppenheimer, Price (1967) observed that "[t]he new powers that science had conceived and engineering had delivered had destroyed the innocence and the sense of freedom of the scientist. Henceforth, the scientist could never profess a lack of responsibility for the fate of society; yet whenever he responded to the call to political action, he would have to deal with problems that far transcended his specialized scientific competence." Cast in a different idiom, the problem was how to assume responsibility as employee, citizen, and scientist. We see this problem expressed today in the new worlds of genetics and pharmacology as researchers struggle to preserve their identities with the pharmaceutical firms underwriting so much of contemporary biomedical research (Jasanoff 2005b). Understanding the early Cold War STI with respect to science and technology means that we have to understand that the issue of time, so haunting in Price, might have meaning for us as well. Only the enemy of scientific freedom is no longer an armed enemy nation state angling for our destruction, but we ourselves.

Notes

1. We can identify at least three different strands of monsters in science studies and the history of science. Most famously there is the strand associated with the work of Donna Haraway (1991), especially her essay "A Cyborg Manifesto: Science, Technology and Socialist-Feminism in the Late Twentieth Century." A second strand is found in the work of historians of science Katharine Park and Lorraine J. Daston (1981) and Lorraine J. Daston and Katharine Park (1998). Finally, there is the work of Mary Douglas and those historians and sociologists who endeavored to make grid/group theory a framework for understanding the production of knowledge. In particular, see Douglas (1982). A powerful example of using Douglas' ideas appears in Bloor (1978). More recently, Marina Warner (2012) eloquently observed that "one might say in the era when the Humanities are under such stress, thinking with monsters shows how an understanding of Nature, and of medicine, law and custom is impossible without cultural expression."

2. Of course, this did not always work out well. For an excellent example of what could go wrong with corporate patronage, in this case from the ubiquitous Sperry Gyroscope Company, see Galison et al. (1992).

3. As Krementsov (1997, 78–80) argues, we still do not know the exact reasons for Vavilov's arrest in 1940. The timing lends credence to the argument that he was arrested because of his contacts with British researchers given that Britain was then an enemy of the Soviet Union.

4. On the British case, see McGucken (1978, 1984). In Britain, Bernal's famous work on the social function of science also served to inspire Polanyi's works. What is striking is that the British founders of this society were opposed to any form of planning, whether in science or economics, even when neither was actually a viable political option in the United Kingdom. On the profound conservatism of the group, see Edgerton (2006, 221–2). Despite what appear to be similar conservative politics, Polanyi and Popper did not get along; see the interesting materials in Jacobs and Mullins (2011)

5. So says the dust cover of my copy, from the third impression, dated May 1958.

6. Jasanoff uses this phrase in a cover blurb for Guston and Keniston (1994).

7. See the correspondence in the Bush Papers in the Library of Congress, box 94, folder 2147 (2).

8. One might usefully compare Bush's attitudes to those who feared a return of the Great Depression following cessation of hostilities. Pent-up consumer demand as well as the existence of an intact manufacturing base meant that the United States would not return to the world of insufficient aggregate demand, massive unemployment, and economic despair. On Bush's plans, see Dennis (2004).

9. Clearly one reason that Price relished all the mentions of government support of social science lay in Bush's desire to bar the social sciences from his National Research Foundation and later, the National Science Foundation. For Price, Bush's argument was very much an example of arguing over something that had long since been decided and funded. On the social sciences in the National Science Foundation, see Kleinman and Solovey (1995) and Solovey (2013).

10. Of course, the most famous discussion of the powerlessness felt by American (and Soviet) scientists in the formation of actual government policy was Leo Szilard's (1992) short story of 1961, "The Voice of the Dolphins," in which researchers perform a complicated feat of interspecies ventriloquism and have dolphins design the policies that allow for global nuclear disarmament.

11. On the district attorney metaphor, see Dennis (2004); other studies discussing the RDB are Friedberg (2000), Hogan (1998), and Needell (2000). Perhaps the best discussion of the RDB, including its five-year plans, is an unpublished speech by Bush from March 1948. See March 4, 1948, "Address by Vannevar Bush," Vannevar Bush Papers, box 131, folder 3081, Library of Congress.

12. Surely this is one reason for the title of Steven Shapin's (2010) collected essays: *Never Pure.*

13. See Sherry (1977, 120–158) for plans relating to science. One of the biggest plans, for universal military training, came to naught in the postwar era, but it offers us an interesting thought experiment in how such a movement would have affected postwar history.

References

Auerbach, Lewis E. 1965. "Scientists in the New Deal: A Pre-War Episode in the Relations between Science and Government in the United States." *Minerva* 3(4): 457–82.

Bloor, David. 1978. "Polyhedra and the Abominations of Leviticus." *British Journal for the History of Science* 11: 245–72.

Bush, Vannevar. 1945. *Science—The Endless Frontier.* Washington, DC: GPO.

———. June 13, 1954. "If We Alienate Our Scientists." *New York Times Magazine*, June 13, pp. 9, 60, 62, 64, 65, 67, 68, 71.

Cohen, I. Bernard. 1948. *Science, Servant of Man: A Layman's Primer for the Age of Science.* Boston: Little, Brown and Company.

Committee on Guided Missiles. 1949. *Glossary of Guided Missile Terms.* Washington, DC.

Conant, James B. 1950. "Science and Politics in the Twentieth Century." *Foreign Affairs* 28: 189–202.

Daston, Lorraine J., and Katharine Park. 1998. *Wonders and the Order of Nature, 1150–1750.* New York: Zone Books.

Dejong-Lambert, William, and Krementsov, Nikolai L. 2012. "On Labels and Issues: The Lysenko Controversy and the Cold War." *Journal of the History of Biology* 45(3): 373–88.

Dennis, Michael A. 1994. "Our First Line of Defense": Two University Laboratories in the Postwar American State. *Isis* 85(3): 427–55.

———. 1997. "Historiography of Science: An American Perspective." Pp. 1–26 in *Science in the Twentieth Century*, edited by John Krige and Dominique Pestre. Amsterdam: Harwood Academic Publishers.

———. 2004. "Reconstructing Sociotechnical Order: Vannevar Bush and US Science Policy." Pp. 225–53 in *States of Knowledge: The Co-Production of Science and Social Order*, edited by S. Jasanoff. New York: Routledge.

Douglas, Mary, ed. 1982. *Essays in the Sociology of Perception.* London: Routledge & Kegan Paul.

Edgerton, David. 2006. *Warfare State: Britain, 1920–1970.* New York: Cambridge University Press.

Ezrahi, Yaron. 1990. *The Descent of Icarus: Science and the Transformation of Contemporary Democracy* Cambridge, MA: Harvard University Press.

Friedberg, Aaron. L. 2000. *In the Shadow of the Garrison State: America's Anti-Statism and Its Cold War Grand Strategy.* Princeton: Princeton University Press.

Galison, Peter, Bruce Hevly, and Rebecca Lowen. 1992. "Controlling the Monster: Stanford and the Growth of Physics Research, 1935–1962." Pp. 46–77 in *Big Science: The Growth of Large Scale Research,* edited by Peter Galison and Bruce Hevly. Stanford: Stanford University Press.

Genuth, Joel. 1987. "Groping Towards Science Policy in the 1930s." *Minerva* 25: 238–68.

Guston, David. H., and Kenneth Keniston, eds. 1994. *The Fragile Contract: University Science and the Federal Government.* Cambridge, MA: MIT Press.

Haraway, Donna J. 1991. "A Cyborg Manifesto: Science, Technology and Socialist-Feminism in the Late Twentieth Century." Pp. 149–82 in *Simians, Cyborgs, and Women: The Reinvention of Nature,* edited by Donna J. Haraway. New York: Routledge.

Herken, Gregg. 1988. *The Winning Weapon: The Atomic Bomb in the Cold War, 1945–1950.* Princeton: Princeton University Press.

Hogan, Michael J. 1998. *A Cross of Iron: Harry S. Truman and the Origins of the National Security State.* New York: Cambridge University Press.

Jacobs, Struan, and Mullins, Phil. 2011. "Relations between Karl Popper and Michael Polanyi." *Studies in History and Philosophy of Science* 42: 426–35.

Jasanoff, Sheila. 2005a. *Designs on Nature: Science and Democracy in Europe and the United States.* Princeton: Princeton University Press.

———. 2005b. "In the Democracies of DNA: Ontological Uncertainty and Political Order in Three States." *New Genetics and Society* 24(2): 139–55.

Kaiser, David. 2005. "The Atomic Secret in Red Hands? American Suspicions of Theoretical Physicists during the Early Cold War." *Representations* 90: 28–60.

Kargon, Robert H., and Elizabeth Hodes. 1985. "Karl Compton, Isaiah Bowman, and the Politics of Science in the Great Depression." *Isis* 76: 301–18.

Kevles, Daniel J. 1977a. *The Physicists: The History of a Scientific Community in Modern America.* New York: Knopf.

Kevles, D.J. 1977b. "The National Science Foundation and the Debate over Postwar Research Policy, 1942–1945." *Isis* 68: 5–26.

Kleinman, Daniel L., and Mark Solovey. 1995. "Hot Science/Cold War: The National Science Foundation after World War II." *Radical History Review* 63: 110–39.

Kohler, Robert. E. 1991. *Partners in Science: Foundations and Natural Scientists, 1900–1945.* Chicago: University of Chicago Press.

Krementsov, Nikolai. 1996. "A 'Second Front' in Soviet Genetics: The International Dimension of the Lysenko Controversy, 1944–1947." *Journal of the History of Biology* 29(2): 229–50.

———. 1997. *Stalinist Science.* Princeton: Princeton University Press.

MacKenzie, Donald. 1990. *Inventing Accuracy: A Historical Sociology of Nuclear Missile Guidance* Cambridge, MA: MIT Press.

McGucken, William. 1978. "On Freedom and Planning in Science: The Society for Freedom in Science, 1940–1946." *Minerva* 16(1): 42–72.

———. 1984. *Scientists, Society, and State: The Social Relations of Science Movement in Great Britain 1931–1947.* Columbus: Ohio State University Press.

Morse, Phillip M. 1950. "Must We Always Be Gadgeteers?" *Physics Today* (December): 4–5.

Mukerji, Chandra. 1989. *A Fragile Power: Scientists and the State.* Princeton, NJ: Princeton University Press.

Needell, Alan A. 2000. *Science, Cold War and the American State: Lloyd V. Berkner and the Balance of Professional Ideals*. London: Harwood Academic Publishers.

Nye, Mary Jo. 2007. "Historical Sources of Science-as-Social-Practice: Michael Polanyi's Berlin." *Historical Studies in the Physical and Biological Sciences* 37(2): 409–34.

Owens, Larry. 1990. "MIT and the Federal 'Angel:' Academic R&D and Federal-Private Cooperation before WWII." *Isis* 81: 188–213.

Park, Katharine, and Lorraine J. Daston. 1981. "Unnatural Conceptions: The Study of Monsters in Sixteenth- and Seventeenth-Century France and England." *Past and Present* 92: 20–54.

Price, Don K. 1954. *Government and Science: Their Dynamic Relation in American Democracy*. New York: New York University Press (*G&S*).

———. 1965a. *The Scientific Estate*. Cambridge: Harvard University Press.

———. 1965b. "Escape to the Endless Frontier." *Science* 148(3671): 743–49.

———. 1967. J. Robert Oppenheimer. *Science* 155(3766): 1067.

Price, Matthew. 1995. "Roots of Dissent: The Chicago Met Lab and the Origins of the Franck Report." *Isis* 86(2): 222–44.

Schweber, Silvan S. 2000. *In the Shadow of the Bomb: Bethe, Oppenheimer, and the Moral Responsibility of the Scientist*. Princeton: Princeton University Press.

Shapin, Steven. 1992. "Discipline and Bounding: The History and Sociology of Science as Seen through the Externalism-Internalism Debate." *History of Science* 30: 333–69.

———. 2008. *The Scientific Life: A Moral History of a Late Modern Vocation*. Chicago: University of Chicago Press.

———. 2010. *Never Pure: Historical Studies of Science as If It Was Produced by People with Bodies, Situated in Time, Space, Culture, and Society, and Struggling for Credibility and Authority*. Baltimore: Johns Hopkins University Press.

Sherry, Michael S. 1977. *Preparing for the Next War: American Plans for Postwar Defense, 1941–1945*. New Haven: Yale University Press.

Smyth, Henry D. W. 1945. *Atomic Energy for Military Purposes*. Washington, DC: GPO.

Solovey, Mark. 2013. *Shaky Foundations: The Politics-Patronage-Social Science Nexus in Cold War America*. New Brunswick, NJ: Rutgers University Press.

Szilard, Leo. 1992. *The Voice of the Dolphins and Other Stories*. Expanded edition. Stanford: Stanford University Press.

Thorpe, Charles. 2002. "Disciplining Experts: Scientific Authority and Liberal Democracy in the Oppenheimer Case." *Social Studies of Science* 32(4): 525–62.

Wang, Jessica. 1995. "Liberals, the Progressive Left, and the Political Economy of Postwar American Science: The National Science Foundation Debate Revisited." *Historical Studies in the Physical and Biological Sciences* 26(1): 139–66.

———. 1999. *American Science in an Age of Anxiety: Scientists, Anticommunism, and the Cold War*. Chapel Hill and London: University of North Carolina Press.

Warner, Marina. 2012. "Review of Wes Williams, *Monsters and Their Meaning in Early Modern Europe*." *Times Literary Supplement*, February 10, 2012, pp. 7–8.

Wolfe, Audra J. 2010. "What Does It Mean to Go Public? The American Response to Lysenkoism, Reconsidered." *Historical Studies in the Natural Sciences* 40(1): 48–78.

Zachary, Gregg P. 1997. "Endless Frontier: Vannevar Bush, Engineer of the American Century." New York: Free Press.

FOUR

Imagining a Modern Rwanda: Sociotechnological Imaginaries, Information Technology, and the Postgenocide State

WARIGIA BOWMAN

Introduction

The reopening of the Nyamata telecenter on August 17, 2007, was a grand affair. The Deputy Director of the Rwandan Information and Telecommunications Authority, Mr. Patrick Nyirishema, cut the red ribbon to much applause. In attendance were representatives from the national electric, gas, and water companies, development representatives, local government officials, and various businessmen. An audience of local citizens filled every folding chair in the room. Many of the women were attired in Rwanda's simple yet elegant national dress: a silk sleeveless top covered by a floor-length piece of patterned chiffon draped over one shoulder and belted at the waist.

In the manner of formal African celebrations, the eminent personages were seated at a high table on the side of the room. Speeches were delivered in Kinyarwanda and simultaneously translated into English. A representative of the Rwandan Development Gateway noted that the nation had a goal to build two hundred telecenters and told the assembled crowd that "telecenters will show that ICTs can benefit people of all walks of life, not just the very educated." Ten Rwandese schoolchildren sang a song in Kinyarwanda. Accompanied by cardboard props representing technology, the children then performed a humorous skit that they had written themselves about the ways in which their lives would be transformed by the ability to use computers.

Although Nyamata is close to Kigali, poor roads and limited access to public transportation make the town more isolated than its distance from the capital would seem to indicate. Nevertheless, the Nyamata telecenter has grown from its beginnings in 2004 when it had six desktop computers, six chairs, six tables, and two employees. The telecenter provides valuable services in high demand in the community. For a small fee, community members may utilize a veritable cornucopia of services. Two full-time staff members provide secretarial services, lessons about what the Internet is and how to browse the Internet, small business support services, receipt and delivery of federal express packages, translation of documents, and even charging of cellular phones.

Despite the opportunities and resources it provides, the Nyamata telecenter faces serious challenges. Of four telecenters established by the Academy for Educational Development in 2004 with support from USAID, three (in Gitarama, Nyanza, and Nyagatare) collapsed because they were unable to cover their expenses. Only Nyamata (founded in October 2004) remains active, making it the oldest telecenter in Rwanda. The technical components of the telecenter, including the computers, the scanner, and the fax, are not fully utilized, in part because few people in Nyamata read English. Rwandan literacy rates are low even in Kinyarwanda.

In his speech, Nyirishema Nyamata telecenter founder Paul Barera referred to infrastructural challenges that occasionally stop the telecenter from operating. The Internet network at the center often goes down owing to power shortages. Outages lasting up to two days mean that telecenters cannot always recharge the batteries for their generator on a daily basis. As a result, electricity supply remains uneven, cutting deeply into profitability, as without electricity computers are simply inert metal boxes. Internet connectivity is expensive, more than most community members can pay. Despite the best efforts of Barera to expand the telecenter's revenue stream, the services provided are costly compared with the purchasing power of the local people. The problem is not a lack of vision, energy, or commitment. Rather, the questions are whether these noble efforts are financially sustainable and whether the capacity of indigenous Rwandans is sufficient to maintain this technology over the long term.

The Rwandan state took responsibility for the development of information and communications technology (ICT) infrastructure with the intent of rebuilding a shattered economy, turning Rwanda into a second world country, and modernizing the nation. Most of the early work of the Rwandan Patriotic Front (RPF) government, headed by the charismatic president and former RPF general Paul Kagame, was devoted to rebuilding and reconstruction. The government's work also included explicit exercises in envisioning a new nation. As part of its reconstruction effort, the government created a national consultative process in 1998–99 aimed at answering questions like, How Rwandans envision their future. What kind of society do Rwandans want to live in? What transformations are needed to emerge from a deeply unsatisfactory social and economic situation?

These are fascinating questions. Rwanda's future depends on eliciting meaningful responses from the people, but, as this chapter suggests, that effort faces serious difficulties because of institutional deficiencies that an ideology of technology-driven modernization may be insufficient to correct. ICT development is a central element in the government's imagination of a modern Rwanda. As such, ICT policy is a key site for investigating the role of sociotechnical imaginaries in rebuilding that shattered nation.

In the early twenty-first century, President Paul Kagame and the Rwandan state were in the midst of simultaneously building a variety of imaginaries, both political and sociotechnical. First, Kagame wished to knit Rwandans into a nation not divided by ethnicity or caste but unified in imagining a shared national identity. Second, he wished to rebuild a nation shattered by civil war and depressed by poverty into a modern state with strong economic prospects. Third, he wished to show that his government could provide for all Rwandans in terms of security, prosperity, and economic advancement. Kagame used ICT in particular to build this sociotechnical imaginary. He publicly performed a vision of Rwanda as a modern and prosperous country, projecting a desirable future for a country haunted by a bloody past. He believed, and indeed evangelized, that Rwanda could attain a positive social order of peace and unity, modernity and prosperity, through vigorous advances in science and technology.

Past and Future

In 1994, humanity lost its way in Rwanda, a small, hilly, densely populated, landlocked country in central-east Africa (Uvin 2001). A civil war[1] between the RPF, based in Uganda, and the Habyarimana government of Rwanda turned into genocide against Tutsis and moderate Hutus. Violence directed largely at civilians on the basis of ethnicity took "appalling, barbarous forms" (Allen 1999, 369). The Rwandan génocidaires put their hands on the levers of state power, with deadly results (Mamdani 2001; Semujenga 2003; Dallaire 2003; Gourevitch 1998). The international community, including the United Nations (UN) and the United States, stood by and watched passively (Dallaire 2003; Melvern 2004; Allen 1999).

Engulfed in violence and warfare, the Rwandan state and political system collapsed completely (Allen 1999). More than a million people were killed, two million fled the country as refugees, and approximately three million were internally displaced out of a total population of about eight million. Further, the country suffered a disastrous drop in professional capacity, particularly in science, technology, and public administration, losing a generation of teachers, entrepreneurs, civil servants, and doctors (UNDP 2004). The gross domestic product was halved in a single year. The nation's productive physical infrastructure was completely destroyed (Republic of Rwanda 2000).

Despite Rwanda's devastating history and its status as one of Africa's poorest countries, the nation's self-proclaimed "ICT Champion," Paul Kagame, imagined that one day Rwanda would be the information technology hub of the region. Following Kagame's ascension as president in 2000, his government implemented a sweeping agenda to reshape the Rwandan economy and society. The government believed that information and knowledge, powered by ICT, would be the primary source for job creation, wealth generation and redistribution, and rapid economic development. Just as in the South Korean national imaginary nuclear power was viewed through the lens of "atoms for development" (Jasanoff and Kim 2009, 121), so in the Rwandan national imaginary telecommunications and the Internet are viewed as ICT for development.

Kagame's government envisioned a Rwanda in which citizens coexist peacefully and collaboratively to build an economically productive future. For Rwanda to move beyond the history of bloodshed and the failures of the postindependence period, the state would endeavor to create a unified nation out of two groups who are extremely close culturally but yet have experienced bitter and long-standing social and political divisions. The government reimagined a Rwandan nation in which ethnicity was not the key factor in inclusion or exclusion and in which real power sharing occurred between the Tutsi and the Hutu. Kagame and the RPF government attempted to provide the Rwandan people with a common narrative of who they were, where they had come from, and where they were headed (Jasanoff, introduction to this volume). This new narrative emphasized linguistic and cultural similarities, attempting to elide the differences so sharply delineated and reified by the Belgians.

The ICT policies embraced by Kagame attempted to implement three goals that together constituted Rwanda's sociotechnical imaginary of modernization and development. The first was the construction of Rwanda as an "African Singapore," thoroughly modern, wealthy, and powered by ICT. The second was the formation of an inclusive and nonracial state. The final—

and most externally directed—element of this imaginary was that of changing the political culture of the fractured nation, perhaps even moving toward a more participatory democracy. Technology, more specifically ICT, represented a "crucial constitutive element" in the emerging Rwandan national imaginary (Jasanoff and Kim 2009, 133). Indeed, the RPF government could claim great success in distributing ICT infrastructure throughout the country. The government prioritized state reconstruction (Ottaway 1999), rebuilding heavily damaged telecommunications infrastructure, such as damaged phone lines, and simultaneously upgrading and replacing them with infrastructure that can carry data.

At the same time, Rwanda resisted adapting to a vision of its political future largely conceived by international donors. Donor money and advice flooded Rwanda in the late 1990s, aiming to create a Western-style democracy from the top down, as was also attempted in Iraq and Afghanistan. RPF leaders feared that this externally generated vision of governance in the Western mold would conflict with the nation's peace and security. A discourse of democracy was adopted in Rwanda, but elections were not heavily contested nor competitive at the national level. On its face, a Western observer might cry foul; yet a persuasive argument could be made that simple reliance on majority rule would result in permanent disenfranchisement—or worse—for the minority Tutsi, the main victims of the 1994 genocide. Like the rapidly developing countries known as the "Asian tigers," the ruling government in this small East African country determined that peace and economic progress were more desirable than Western-style democracy, at least in the short term.[2]

Since the Rwandan state has not embraced participatory democracy as understood in the sense of the European Union or the United States, the Rwandan people have had little input into whether ICT is a good way to move forward for development; they also had little input into how it should be configured or where it should be distributed. They were not empowered to put forth alternative imaginaries. Instead, the government of Rwanda, firmly steered by the RPF, laid out a clear plan about how the building of ICT would proceed, a plan with little room for discussion and debate.

So, will a top-down imaginary take hold in Rwanda? What is needed for it to become the shared goal of many rather than the vision of one leader or an elite few? Kagame's vision of Rwanda as a modern, technologically advanced, middle-income country was clearly popular with both the RPF political machinery and donors. How could it be made popular with the common Rwandan? The former Soviet Union may provide a model for this kind of spread of a top-down ideology to the common people. The Soviet

Union—particularly under Stalin—had a similar sociotechnical imaginary that combined manufacturing, technology, and production. Like Rwanda, which had just gone through a genocide, the Soviet Union had gone through the devastating Second World War, in which millions died brutal deaths on the Western front. Yet Russians today longingly remember the stability, the order, and the pride of the old Soviet Union. Stalin tapped into two key themes that tugged at the heartstrings of his fellow countrypeople: pride and reconstruction. If Kagame can build a similar semblance of national pride and identity (which at this point is very much still a possibility) and overcome the vicious ethnic divisions of the past, there is a chance that his sociotechnical vision may take deeper root. However, this is no easy task. Kagame and the RPF government have had only decades to imagine a new nation, build a national identity, reconstruct the nation's economic capacity, and push forward the image of a scientifically based economy. The Rwandan leadership has been simultaneously involved in creating "an imagined political community" (Anderson 1983) and a "collectively held, institutionally stabilized, and publicly performed vision of a desirable future" based on ICTs (Jasanoff, Introduction). To make these two intertwined imaginaries take root together is not impossible, but the task may be Herculean.

Embedded Hierarchy: Centralization as an Artifact of History

ICT policy in Rwanda reflects and reinforces the nation's tradition of hierarchical rule as well as the new economic and political visions for its future. Centralization in Rwanda reaches back centuries. In contrast to its neighbors, Rwanda existed as a fairly coherent whole from precolonial times (Fisiy 1998, 20). The Kingdom of Rwanda was a monarchy with an administrative structure that emanated from the court as early as the mid-sixteenth-century (Melvern 2004, 5; Fisiy 1998, 20). A strong military system powered by an expansionist state began to develop by the eighteenth century. During the nineteenth century, the Rwandan state embarked on empire building and reached a height of centralization, becoming a powerful state respected by neighboring rulers (Semujenga 2003, 15). The state system of Rwanda before the genocide has been characterized as "hierarchical, omnipresent and forceful" (Van Leeuwen 2001, 639). Scholars note that Rwanda had an "entrenched culture of obedience," which facilitated the swiftness of the genocide (Paluck and Green 2009).

Before Rwanda came under Belgian rule after the First World War, there were deep cultural commonalities across the nation that made many ethnic distinctions meaningless (Mamdani 2001; Fisiy 1998). All Rwandese speak one language, Kinyrwanda, share one style of religious celebration, and perform the same set of traditions and rituals.[3] Perniciously, however, the Belgians constructed artificial "racial" distinctions between the Hutu and Tutsi. Under colonialism, the Hutu were brutally discriminated against in all walks of life, and social divisions were reified and strictly enforced (Mamdani 2001, 92–105; Semujenga 2003, 16; Melvern 2004, 5). These actions privileged the Tutsi socially and economically (Fisiy 1998, 20).

Rwanda attained independence from Belgium in 1962, but the colonial ordering of Hutu and Tutsi into a visible socioeconomic hierarchy poisoned the period following independence. As colonialism came to an end, the Hutu majority called for a change in Rwanda's power structure that would accord them more rights (Semujenga 2003, 16). In 1959 the Hutu Revolution called for majority rule and overthrew the Tutsi chiefs with the support of the Belgian government. In response to the rise of political majority in the 1959 "peasant revolution," a first wave of killings against Tutsis took place, resulting in the deaths of 20,000 Tutsi. Those Tutsi who survived were expelled from political life and became stateless refugees. Many fled to Uganda (de Lame 2004).

Kagame's family belonged to this massive wave of refugees. Kagame was born into an aristocratic Tutsi family with ties to King Rudahigwa of Rwanda (Grant 2010), but the family lost all wealth and status as refugees. As a child, Kagame remembered houses burning as a Hutu death squad ran toward his family's car. His family fled from Rwanda, spent time in the Democratic Republic of Congo (Zaire) and Burundi, and finally landed in Uganda in 1960. Kagame was raised in a Nshungerezi refugee camp in southern Uganda (Thompson 2004). As a refugee, he had to queue for food and study under a tree (Grant 2010). As Kagame reached manhood in Uganda, a 1973 military coup installed Major General Juvenal Habyarimana, an ethnic Hutu, into power in Rwanda.[4] The coup led to a period of relative calm between the Hutu and the Tutsi; however, under Habyarimana, the Rwandan economy declined, corruption escalated, and health and education services collapsed.

In the 1980s, Kagame fought in the bush in the Ugandan civil war alongside the leftist National Resistance Army organized by Yoweri Museveni, and, as a military intelligence officer, he helped Museveni ascend to the presidency. Museveni arranged training for Kagame, first in Cuba in 1986 and later in 1989 at the US Army Command and Staff College. Seeing the success of Museveni's armed struggle, in the late 1980s, Kagame and his childhood friend from the Ugandan refugee camp, Fred Rwigyema, began building an army of Rwandan exiles within the Ugandan army, which would become the RPF. Despite Museveni's support for Kagame, Rwandan refugees were not granted Ugandan citizenship. The Ugandan president encouraged the departure of Rwandan refugees back to Rwanda, because they had become a political liability in a country with its own long history of ethnic civil war.

In October 1990, the RPF crossed the border into Rwanda, where they were routed, but a "low-intensity civil war" began (Reyntjens 2006). Kagame assumed command and rebuilt the RPF in the Virunga Mountains. On July 17, 1994, the RPF, led by Major General Kagame, defeated the remnants of Rwandan government troops and declared the end of the civil war. Four days later, under the Arusha Accords, a multiethnic Government of National Unity was formed.

From the start, one of the most pressing problems facing the RPF government was how to enact the imaginary of a Rwandan nation distinct and separate from its bloody past and comprising a newly unified future. Under Kagame, leadership shifted to the Tutsi minority, which at the time comprised approximately 15 percent of the population; by contrast, the Hutu majority comprised 84 percent, and only 1 percent was Twa (Central Intelligence Agency 2007). As Mamdani noted early on, "Rwanda's key dilemma is how to build a democracy that can incorporate a guilty majority alongside an aggrieved and fearful minority in a single political community" (Mamdani 2004, 266).

Kagame's experiences as a refugee, as a stateless man alienated from his homeland, as a military commander raised up and then rejected, and as a rebel fighting for his people's place in their homeland forged his iron personality. A passionate yet pragmatic man, his life was dedicated to righting an injustice, yet above all to surviving. He was a military man through and through, trained and mentored under Museveni, himself a nation builder and an autocrat. Kagame's background and associations groomed him to be authoritarian. He also inherited institutions that had been crafted before, during, and after colonialism to support a style of governance that favors centralized, hierarchical rule. At the same time, the Rwandan state also inherited a nation whose civil society and administrative leadership had been destroyed by war. The most effective societal bulwarks against authoritarianism were missing from the start of Kagame's presidency.

A Phoenix from the Ashes? The Reconstruction Government

Critics observe that Rwanda's postgenocide government remains militaristic and authoritarian (Uvin 2001, 184). The Tutsi-dominated RPF agreed to share power with the moderate Hutu parties, the Democratic Republican Movement, the small Social Democratic Party, and the Liberal Party

(Lemarchand 2007). Over time, however, power became concentrated in the hands of the RPF. In 1995, Prime Minister Faustin Twagiramungu resigned and became an exile, as did President Bizimungu and other leading Hutu members of the coalition. Later that year, the moderate Hutu party, the Mouvement Democratique Republicain was banned on grounds of "divisionism." Respected Africanist and scholar Rene Lemarchand states that "Rwanda had become for all intents and purposes, a one-party state" (Lemarchand 2007, 7).

On April 22, 2000, Kagame became the fourth president of independent Rwanda. He was reelected in 2010 in an election that was considered "transparent and efficiently run" by some (CNN Wire Staff 2010). That he received 93 percent of the vote, however, called into question the fairness of the election. In his governance style, Kagame emerged as a player of established Rwandan political scripts as well as a ruler in the well-worn style of the African Big Man. Rwanda's policy apparatus provides neither a tradition nor an established set of practices for participation by average Rwandans or, more strikingly, for the opposition to have a voice. Regardless of whether one is an admirer or a critic of the RPF government and its leader, Rwanda's hesitant return to governability in the 2000s displayed few systemic checks and balances on the ruling party (Allen 1999, 179).

As the leader of a minority government, Kagame was under pressure to demonstrate that he intended to treat all areas of the country and all citizens "fairly," partly in accordance with its mission of reconciliation and partly to ensure its own survival. According to former US Ambassador Herman Cohen, "In Brussels and Paris, Hutu intellectuals continue to plot revenge" against the RPF. Kagame himself observed, "Without successful reconciliation, political stability and security, private investors will not develop confidence in the country." Despite this decorous statement, more is at stake than just the level of private investment in Rwanda. The very existence of the Rwandan nation is at risk because of external threats on the Congo border and internal challenges to Kagame's rule. Inside the country, growing numbers of political opponents view armed resistance as a real option. Heirs to the Hutu Power movement led an insurrection at the end of 1997 in the northwestern provinces. Enemies mass at Rwanda's borders. Burundi and the Congo house large contingents of exiled génocidaires who would be happy to see Kagame overthrown.

On taking office, Kagame developed a strong national agenda (Ottaway 1999). His government aggressively pursued policies for eliminating the ethnic distinctions that led to the genocide (International Crisis Group 2002). It brought relative order, security, and stability to a country that had endured almost forty years of ethnic bloodshed. Eight political parties joined in the government of national unity. Kagame consistently respected the central tenets of the Arusha Accord and treated the accord as a fundamental law (International Crisis Group 2002). His supporters note that his government has been highly effective in rebuilding the country and responsive to the needs of the majority. The electoral system pays careful attention to the representation of various vulnerable social groups in both the parliament and ministries. Members of parliament are mainly elected indirectly by organizations of youth, women, and the disabled. On a positive note, as of 2004, Rwanda had the highest number of female parliamentarians in the world. Further, Hutu held at least fifteen of the twenty-nine positions in the 2004 government and thirteen of the eighteen ministerial portfolios.

Participatory democracy, however, has been slow to take root in the new Rwanda. The International Crisis Group states that political parties are "only tolerated if they agree not to question the definition of political life drawn up by the RPF." The party has banned local political meetings and grassroots meetings. It has banned opposition political parties and imprisoned political opponents. Indeed, Mrs. Victoria Ingebire, a moderate Hutu whose brother was killed in the genocide, returned to challenge Kagame in the 2010 presidential election but was placed under house arrest and accused of "genocide ideology." Human rights observers have expressed serious concerns over government efforts to muzzle the press, which has resulted in the flight of some editors, including those of the formerly pro-RPF *Imboni*, into exile. One critical newspaper editor died mysteriously in an automobile accident and was found with his head nearly severed. The vice president of the Democratic Greens died of machete wounds days before the election.

Dreaming of a Rebuilt Nation

Nearly two decades after the genocide, Rwanda has crushing social and economic problems. In a country that ranks among the world's least developed, approximately 60 percent of Rwandans earn less than a dollar a day (Baldauf 2007; Barigye 2008). Life expectancy at birth is only forty-five years. Rwanda's population is set to double by 2030. The economy is predominantly agricultural, and 91.1 percent of the population is actively involved in agriculture, with only 1.7 percent working in the industrial sector of the economy. The Human Development Index ranks Rwanda 161 out of 177 countries.

Against this background, the sociotechnical imaginary of ICT for development represents a crucial ingredient for Rwanda to transform itself from an agricultural third world country into a technologically driven second world country. In 2008, Kagame announced his aim to "use the power of science and technology to transform" Rwandan society (Kagame 2008). He sought rhetorical inspiration from the United States, promising to leverage science and education to permit "a more rapid socioeconomic transformation" and help the country make better development choices (Kagame 2008). He argued that he had a "developmental vision" and that he aimed for Rwanda's public sector to play a leading role in this matter. Enormous support from the outside world for this vision translated into significant resources from donors. Rwanda's economy began growing at 7.5 percent a year. It had a competent government with a transformative vision (Kinzer 2008). At the heart of Kagame's vision for the country's economic rebirth was a new technological tool: ICT.

The RPF government believes, in a manner reminiscent of Walt Rostow, the father of modernization, that ICTs offer Rwanda the opportunity to "leap-frog the key stages of industrialization and transform her subsistence agriculture dominated economy into a service-sector driven, high value-added information and knowledge economy that can compete on the global market."[5] Kagame asserts that "Rwanda is at risk of being . . . marginalized if she fails to embrace these technologies to transform her economy and society."[6] He believes the potential of ICT can help achieve the "vision of a modern economy for Rwanda."[7] Albert Butare, Minister of State for Energy and Communications, seconds Kagame, calling ICT "an indispensable tool for . . . modernization" (Bowman 2007b).

According to Dr. Pius Ndyambaje—the president's ICT advisor in 2004—Rwanda's first ICT policy is borrowed directly from Malaysia's Vision 2020[8] and Singapore's vision of transforming the country into an "Intelligent Island,"[9] using ICT as the main engine for promoting accelerated development and growth (Dzidonou 2002). It is hard to determine whether this vision is in competition with other competing choices because the government of Rwanda since 1994 has not been transparent. Indeed, Rwanda's reconstruction has occurred through central planning, massive national consultations, and elaborate white paper guidance documents that combine Soviet-style planning with American-style motivational mission statements. Whereas in the American and to some extent the Korean cases the educational and scientific capacity to attain the desired sociotechnical imaginary was not at issue (Jasanoff and Kim 2009), in Rwanda the national vision of ICT as a socioeconomic engine did not evolve over time. Instead it emerged fully formed from the forehead of a Zeus-like United Nations Economic Commission for Africa (UNECA), in detachment from questions of the population's actual capacity.

Getting ICT on the Agenda

The American discourse of the digital divide seeped into Africa gradually via multiple authors and international organizations. Former UN Secretary General Kofi Annan firmly believed that, by overcoming the divide, ICT could help Africa's developing nations to "modernize" while allowing them to pursue social welfare goals. Annan stated "Information and Communication technologies can help us turn the [potential for investment growth in developing countries] into concrete opportunities that will help the poor work their way out of poverty while, at the same time benefiting the world community as a whole."[10] The African Information Society Initiative (AISI) was launched concurrently with the rise of ICT as a social goal in 1996, only two years after the genocide ended and Rwanda's Government of National Unity was formed. According to the UNECA, "Africa needed a common vision for its quest not only to bridge the digital divide between Africa and the rest of the world but more importantly to create effective digital opportunities to be developed by Africans and their partners, and to speed the continent's entry into the information and knowledge global economy" (UNECA 1996).

UNECA—a major donor body—encouraged the Organisation of African Unity (OAU) Heads of State Summit to adopt the AISI in 1996. The initiative supported the efforts of twenty-eight African countries to develop "national information and communication infrastructure" policies. In March 2001, under Annan, the UN established an Information and Communication Technologies Task Force in part to support Africa's drive for self-development. This project presumed that ICT can be used to contribute to the elimination of poverty, human development, the elimination of gender disparities, and the combating of disease. Indeed, Goal 8, Target 18, of the UN Millennium Development Goals urges the international community to distribute the benefits of ICT more equitably.[11]

African and global activists participated in internationally sponsored ICT forums such as the 2001 World Summit on the Information Society, organized by the UN's International Telecommunications Union and endorsed by the UN General Assembly. The summit drew together civil society members, private sector organizations, governments, UN organizations, and other donors. Social activists observing the emergence of the new set of technologies recognized ICT as a potential economic and political resource and, importantly, one that acts as a vehicle to discuss long-held social justice objectives such as combating poverty, empowering women, and improving education and health care in a modern context. These activists followed the lead of US actors and set about creating a discourse that emphasized the need to distribute that resource equitably.

This commitment garnered results. By 2012, the International Telecommunication Union had named Rwanda among the top six developing countries in the world in terms of the strength of its ICT market (Buteera 2012). Kagame again was in the lead. As cochair of the UN Broadband Commission for Digital Development, he called on world leaders to place access to broadband on the policy agenda of the Millennium Development Goals. His vision coupled ICT's technological potential with imagined solutions to the region's most urgent problems: "As we look to the future we realize that we need to do more and faster, the world is waiting, and our people are counting on us, whether it is central databases of crop yields and market information for farmers, integrated school curricula for pupils, and entrepreneurial opportunities for youth" (Kagame 2008).

"Flying Geese" versus "Utopian Computing"

As Kagame's words make clear, Rwandan advocates and policy makers imagine ICT as an instrument of social and political betterment, not just as a technological tool. A closer investigation suggests that two somewhat different imaginaries are at work in the Rwandan context: that ICTs will improve productivity and lead to economic growth and that ICTs will improve social and development outcomes. I refer to the first as "flying geese" and the second as "utopian computing."

Economists and technocrats predominantly focus on ICT's contribution to manufacturing productivity. These advocates believe ICT can improve African participation in the global marketplace (Dzidonou 2002) as well as economic participation in the domestic marketplace for small businesspeople. This vision of ICT focuses on nationwide infrastructure and connectivity—particularly in remote and rugged areas—meaning electricity, copper wires, satellite towers, very small aperture terminals, and fiber-optic cables. The vision of ICT as a rural economic facilitator also requires improvements in the educational infrastructure. Students will have to become computer literate—no small task on a continent where most classrooms lack glass windows and where sums are done on blackboards with chalk.

A more utopian vision, held by civil society activists, sees this technology as a tool that can help develop social justice and perhaps democracy. Many practitioners and scholars have argued that ICT may contribute to improved social, economic, and development outcomes in poor areas of the West, in developing nations in general, and in Africa in particular (Eggelston 2002). According to these visionaries, access to ICT can empower women and give the poor increased economic opportunities while improving the quality of education and the delivery of medical services in rural areas. This most utopian version will require African governments to move from paper files to electronic files, invest in a well-developed infrastructure of electricity, fiber-optic cables, satellite towers, and copper wires in both cities and rural areas and utilize healthy doses of political will at the national, regional, and multinational levels.

The rapid economic growth of the Asia Pacific region between the 1960s and 1990s, termed by some economists as an economic miracle, provided the main source of inspiration behind Rwanda's ICT planning. Following this flying-geese model, Rwanda seemed to be aiming to move up in technological development by following in the pattern of the countries ahead of it in the development process (Radelet and Sachs 1997, 52). On the basis of both his rhetoric and his planning priorities, it is clear that Kagame's focus was on ICT as "a symbol of the power of science and technology that [a nation] should actively seek to acquire in order to develop into a strong, modern nation," just as the Korean state once viewed nuclear energy (Jasanoff and Kim 2009, 131). Yet the RPF vision of ICT as a socioeconomic driver lacks contestation, lacks debate, and lacks discussion with the grass roots. ICT policy in Rwanda has been developed and implemented by donors and the state, not crafted by citizens.

Citizen Participation in ICT Policy Making

Remarkably, Rwanda developed the first ICT policy in the East African region in 2001, but modernization was not a grassroots creation. It was the result of direction from the topmost levels of government. As of 2007, Rwanda had already issued two ICT policies organized in five-year increments. This process produced a detailed document with clear implementation indicators and detailed time frames widely discussed and admired by other governments throughout the East African region.[12] One adviser to the Rwandan government notes that "This country is very hierarchical, and whatever the government decides to do, it will do, and society will follow in a [. . .] disciplined way" (Baldauf 2007).

The representative from Duhamic—the main umbrella organization that oversees the nongovernmental organization (NGO) sector in Rwanda and one of the only strong NGOs visible in Rwanda—Mr. Innocent Benineza, stated his belief that civil society "had not been consulted enough in the National Information and Communications Infrastructure plan (NICI) process" (Bowman 2007a); however, he qualified that statement by saying that "many NGOs do not understand the importance of ICT" (Benineza 2008). Benineza believes that the government does listen to civil society but also asserted that the government's lack of engagement with civil society results from the weak and inactive nature of Rwandan civil society. Benineza asserts that the government needs to make more concerted efforts to involve civil society in the policy process.

One government supporter, a Rwandan with impressive foreign academic credentials who lives abroad, cautioned me to focus on the effectiveness of the government's implementation, not on its autocratic nature. After all, there are many African governments that are autocratic yet not effective. For example, he remarked, "What would a farmer say about ICT policy anyway?" He observed that there are certain prerequisites for any person to participate in a consultation and that significantly more education and capacity building are needed in Rwanda to allow citizens to exercise democratic rights, such as participation. Drawing on democratic theory, an observer could contend that perhaps the citizens are not as well qualified as they might be, but the correct response nevertheless—particularly in a country like Rwanda, which is rebuilding society from the rubble—is to furnish citizens the opportunity to understand the ends and means of their interests, not to exclude them from decision making altogether (Dahl 1989). In fact, this type of an approach informed the *gacaca*[13] process.

Although a small number of "stakeholders" were recruited during the ICT policy development process, even the most generous account by government officials suggests that participation by the average Rwandan, or even elite representatives of major social sectors, was low. By contrast, the Rwandan government, multilateral donors, and the multinational private sector enjoyed high levels of participation and influence in the development of the Rwandan ICT policy.[14] The second round of policy development increased citizen participation to the level of "placation" (Arnstein 1969), in the sense that a few handpicked representatives of civil society and the Rwandan private sector participated on the National Task Force. In the words of one official at the Rwandan Information Technology Authority, "There were stakeholders, but [the process] was not engaging the stakeholders. There were stakeholders in the writing. One consultant led the NICI I process. He engaged, but not as much as one would wish."[15]

As shown above, the ICT policy-making process supports arguments that the RPF government's idea of citizen participation is one of guided and controlled consultation. Rwandan citizens do not have an effective way of voicing concerns about decisions the government is taking, with respect to ICT.

Some argue that Rwanda is a society based on censorship and that those who speak out against the government are punished, imprisoned, or worse (Reyntjens 2006). Some observers believe that the government of Rwanda is guilty of cooptation or repression of independent forces in civil society. Indeed, according to the International Crisis Group, civil society in Rwanda "exist[s] between repression and coercion" (International Crisis Group 2002). Accordingly, there is no social organization that can provide a voice of dissent or even mild critique and, for that matter, criticism regarding the direction, pace, and design of Rwanda's ICT policy and implementation or any other policy issue.

Equity, Education, and Human Resource Capacity in Rwanda

The centerpiece of the Rwandan government's effort to distribute ICT to the Rwandan population is education. Education is one of the eight pillars in the nation's ICT policy. Education-related ICT projects are wide-ranging, and span training teachers, rolling out computers, installing Internet connectivity, pursuing monitoring and evaluation, writing content in Kinyarwanda, and digitizing math, biology, chemistry, and physics. Rwanda has made primary school education free and has extended this free education to the first three years of secondary school. As of 2008, the Rwandan government was spending 1.6 percent of its gross domestic product on the promotion of science.

The government places especially strong emphasis on science and technology in education. In the words of the director of planning for the Rwandan Ministry of Education, "We want to use ICTs for education. We want a skilled workforce."[16] One of the government's key goals is to deploy the technological resources needed to implement educational reform and ICT initiatives.[17] A specific objective is to "transform Rwanda into an IT literate nation" and improve the educational system over a period of ten years. As part of the process of attaining digital literacy, the government initiated a comprehensive program to deploy and "exploit" computers in schools.

This effort included the placement of computers in schools, work to bring the Internet to schoolchildren, attempts to put a computer science curriculum in place, and a program to train thousands of teachers in basic computer literacy in cooperation with Microsoft. In 2006, the Rwandan government began placing computers in schools, both public and private. The aim was, by 2007, to identify every single school in the country and give each an identical number of computers, regardless of its size and location. Primary schools received one laptop each, while secondary schools were slated to receive ten laptops each. In addition, each of Rwanda's thirty districts was to receive precisely one telecenter.

This distributive strategy essentially ignored a school's needs in allocating hardware such as laptops. Giving each school exactly the same number of computers regardless of "need" maximized blame avoidance, but at the expense of educational utility or the efficient use of resources. Computers were distributed with little regard for general literacy, computer literacy, staff conditions, electrification, or student needs. To some extent, this blueprint for distributing technological artifacts evenly could have the perverse effect of actually reinforcing existing inequalities. Areas of the country that already had more schools would have more computers than areas that were educationally more deprived. At the same time, schools with extremely high student populations would receive too few laptops for their size. The adoption of such a numerically egalitarian notion of equity underscores the state's power but also the perverse constraints created by the imaginary of equity in postgenocide Rwanda.

The government was aware of the concern that sending out the physical artifacts of ICTs before attempting to accomplish other core social objectives, such as reducing poverty,[18] or ensuring universal literacy (Baldauf 2007). But Education Ministry officials have a ready response: "Europe did not wait until everyone had a car before building airplanes." Echoing these sentiments, Rwanda's Minister of State, Energy, and Communications Albert Butare recounted in a 2007 interview an exchange with Rwanda's development partners: "'You are too ambitious. Do you really need computers and the Internet or [do you need] sufficient drinking water, good shelter and food?' We said, [they are] not exclusive. We need all of them" (Kimani 2008).

Given Rwanda's experience with ICTs during the genocide, particularly with radio, it is no accident that Kagame wished to keep his hands firmly on the levers of ICTs. The deposed Hutu government used both official and unofficial radio sources to incite the genocide (Metzl 1997). The Rwandan media, encouraged by the *Akazu*—a powerful circle around the widow of President Habyarimana—attempted in 1994 to convince Hutu that they would soon be victims of a genocide mounted by the Tutsi (Chalk 1999). The Arusha Accords, signed in 1993,[19] barred the government-owned Radio Rwanda from inciting hatred, so the *Akazu* created their own private radio station, the RTLM, blending African music, talk radio, and coded attacks on Tutsi and their allies, deliberately targeting youth gangs like the *Intera-hamwe* (Chalk 1999). RTLM was founded in part in response to reforms that allowed moderates to take positions inside the Ministry of Information, which controlled Radio Rwanda (Metzl 1997). One may speculate that Kagame wished to keep control of the means of communication, including radio, television, and information technology, firmly within his own grasp so he could control the message. It was, however, precisely such centralized government control of the radio that facilitated the genocide. The RPF's top-down approach gives Tutsis more access to ICTs without addressing the structural dangers of centralized, top-down communications systems.

Conclusion: Constructing Alternative Imaginaries

As the first government to be elected to power after the genocide, the ruling RPF regime adopted a transformative vision of the role of information and communication technologies in rebuilding Rwandan society. The goal of nationwide connectivity proved successful in terms of securing the access of the average Rwandan to ICTs. However, the Rwandan planners envisioned and built a top-down, centrally controlled, state-run Internet. The Rwandan state succeeded in distributing hardware but was less successful in building capacity or local ownership and buy-in of the technology.

Kagame and the government of Rwanda have repeatedly announced their ambition to be an African Singapore, but what would this mean in practice? Singapore is wealthy and technologically advanced, but it is also autocratic and, with regard to human rights, repressive. The Singaporean government relies on the Internal Security Act "to hold, without charge or judicial review, those suspected of subversion, espionage, and terrorism." Freedom of expression is sharply limited in Singapore. Government authorities curtail rights to freedom of expression, association, and assembly. They deny legitimacy to associations of ten or more, if they deem the groups "prejudicial to public peace, welfare or good order." The government requires police permits for five or more people planning a public event, and it uses contempt of court, criminal and civil defamation, and sedition charges to rein in critics (Human Rights Watch 2013). Accordingly, imitating Singapore economically comes with an implied imitation of its less than democratic process.

Kagame imagined ICTs as a gift from the state to the people, not as a collaborative or, even more profoundly, a grassroots effort. This approach was implemented at the cost of determining what a town or a school actually needed or what was appropriate for people in particular localities. Increased participation might have slowed deployment, especially if it meant including people with little prior understanding of or use for new technologies. Nonetheless, the effect would have been to construct a more contentious but possibly more productive set of state-society relationships from those that the RPF found tolerable. Further, the effect would have been to educate the public and get them involved in their own development decisions. Given that Kagame selected a technology that evolves most quickly, and most innovatively from the users themselves, it seems a crucial oversight to keep Rwandan users in the dark about the power of this potential engine of socioeconomic transformation.

Yet Rwanda remains in a very delicate situation, with a sharply divided population and enemies massed at the border. Multiparty democracy, implemented at a breakneck speed, for its own sake, could easily lead to new outbreaks of violence. Singapore offers Rwanda the idea of stability, which gives the economy time to develop and the population time to become more unified and to gain the educational and technical capacity they need to compete. Transition to multiparty democracy and a free press shook neighboring Kenya to the core in 2007, nearly plunging the country into a bloody civil war. Indeed, to avoid ethnically based violence, the Kenyan press voluntarily chose to limit itself to messages of peace for nearly one week before the 2013 election. What alternative could Rwanda follow? A managed transition toward democracy on the model of countries like Ghana, South Africa, and Kenya might be an effective way to move Rwanda toward a more democratic future, while avoiding the backward slide of neighboring Uganda.

Consistent with such a plan, an alternative sociotechnological imaginary could have focused more on infrastructure and hardware sustainability. Given that fewer than 10 percent of Rwandans had access to electricity in 2008, the government of Rwanda's rapid effort to distribute resources such as laptops seems at best loosely connected to a broader effort to create interconnected systems that could enhance the ability of the Rwandan people to communicate (Majtenyi 2008). This failure to put necessary infrastructure in place before distributing the artifacts of proposed development is particularly problematic in the case of ICTs, which quickly become obsolete and require sophisticated human resources for optimal use.

An imaginary rooted in concerns for sustainability might also have placed greater emphasis on grassroots participation. Local people need to be listened to, not just "consulted" or "educated." The citizens of Nyamata could be talked to and incorporated into a decision-making process about how to make their telecenter more sustainable and what infrastructure as well as educational requirements they need to keep that telecenter running and to allow schools and hospitals to most effectively utilize new technologies. Criticism and input could actually strengthen the development of a durable, technologically sophisticated framework, as well as a durable system of governance that further develops with use. The creativity on display by the children of Nyamata on opening day could be harnessed to help design Rwandese software, write local content in Kinyarwanda, and acquire know-how to repair and maintain the computers.

Under Kagame, dramatic changes were wrought in postgenocide Rwanda. The country elected more women political leaders than any other country in Africa and possibly the world. Progress has been made in attaining the political imaginary of Rwanda as a single nation uniting once fiercely divided social groups. Nonetheless, rumblings of discontent indicate that the process of national unification will take time to complete. Kagame and the RPF accused and jailed political opponents for "genocide ideology" and "ethnic divisionism." Some Hutu leaders feel that Kagame—who appeared to be president for life, not unlike his neighbor Yoweri Museveni—has reinstalled the Tutsi as a de facto politically, if not socially, privileged group.

ICTs can, of course, be governed in a top-down fashion, as is the case in China and Iran, or from a bottom-up fashion, as is largely—although not exclusively—the case in the United States. Because ICTs emerged in the West as an academic experiment, although one with government funding, ICT governance remained an unregulated and even anarchic domain for many years. Indeed, there are at least two types of ICT governance: the governance of the technology inside the technological community, for example, the distribution of domain names like Rwanda.go.rw; and the more overt regulation of gateways, access points, and information providers. In the Rwandan case, a largely authoritarian government created an interesting paradox. On the one hand, the government promoted broad access and national immersion in this innovative technology. On the other hand, access was provided, and controlled, by the state and not the private sector. And what the state giveth, the state can also take away.

In many ways, Rwanda's ICT policy presumed more capacity and more unity than actually existed. For the government's ambitious efforts to be successful in the long run, dramatic strides must be made in developing the capacity of the Rwandan population to understand, manage, maintain, and make choices about both ICT and their own governance. This is, in its essence, an endeavor to nurture the capacity of the grass roots. Technological systems are social as well as material, and people need to have a deep enough understanding of and intellectual investment in technology to fully utilize it, maintain it, and prevent it from decaying and becoming obsolete. Similarly, for the sociotechnical project of a sustainable state to succeed, Kagame and the government of Rwanda must gradually step back from centralized control and make space for creativity, independence, and innovation, lest their efforts to build a new, modern, economically successful Rwanda decay owing to peoples' lack of participation in and maintenance of the institutions of governance.

Notes

1. The Rwandan Civil War began in 1990.
2. My analysis is based on several visits to Rwanda over a four-year period spanning 2004–2008 during which thirty-five ethnographic interviews were conducted with high-ranking Rwandan policy makers, private sector participants, and nonprofit activists. In addition, I have consulted numerous primary and secondary sources.
3. In precolonial times, the designations of Hutu, Tutsi, and Twa were historical social roles representing forest, pasture, and field, with social mobility among these groups.
4. The Habyarimana regime, which drew its strength from the Hutu of the north, broadened its discriminatory lens, discriminating against the Tutsi as well as against Hutu from the south of the country.
5. National Information Communications Infrastructure Plan (NICI) I, preamble, paragraph 5; NICI II, Foreword.
6. NICI I, preamble, paragraph 8.
7. NICI II, Foreword.
8. Information on Malaysia's plans for information and communications technology is available at http://www.american.edu/initeb/ym6974a/nationalictpolicies.htm (accessed December 12, 2008).
9. Mahatir Mohamad, "Malaysia on Track for 2020 Vision," speech given on January 10, 1999.
10. "ICT: A Priority for Africa's Development," remarks by Kofi Annan to the Opening of the third meeting of the United Nations Information and Communication Technologies Task Force.
11. "Information and Communication Technologies: A Priority for Africa's Development," A Statement by H.E. Kofi A. Annan, Secretary General of the United Nations, contained in ICT Task Force Series 2 at xv.
12. The membership of the NICI-2005 Plan Steering Committee can be found in Appendix 2 of the NICI I Plan. The final plan was produced by Clement Dzidonou, a Ghanaian, in consultation with a steering committee consisting primarily of Rwandan government officials. Appendix 2: The NICI-2005 Plan Steering Committee, "An Integrated ICT-Led Socio-Economic Development Policy and Plan for Rwanda, 2001–2005: The NICI-2005 Plan.
13. The *gacaca* court is part of a system of community justice inspired by tradition and established in 2001 in Rwanda, in the wake of the 1994 Rwandan genocide.
14. As noted above, donors were crucial in the planning and development phases of the NICI process. The most important donor was UNECA. Donors have also played a key role in implementing the NICI process. Most other donors began their involvement in implementation of the NICI process after 2001. The donors who have been most involved after implementation in Rwanda are the United Kingdom Department for International Development, which has worked with the Rwandans on education

and capacity building in the realm of ICT; the Swedish International Development Cooperation Agency, which supported the establishment of Rwandan Information Technology Authority and has provided strong support for the National University of Rwanda; and the United Nations Development Program, which provided significant financial support to the Rwandan Ministry of Infrastructure for the formulation of the NICI II plan and for the development of a feasibility study for the construction of telecenters in Rwanda. Indeed, the United Nations Development Program worked with RITA, the Rwandan Ministry of Infrastructure, and the UNECA to supervise the completion of NICI II and to design Rwanda's telecenters, relying heavily on high-priced international consultants. The World Bank did not begin showing serious interest in Rwanda's ICT plans until 2005 but became very engaged at that point in the E-Rwanda program.

15. Anonymous RITA official, August 16, 2007.
16. Bowman 2007c.
17. NICI I, Strategy E. According to NICI I, the government of Rwanda also aims to develop human resources in ICTs; develop ICT applications for education; computerize the civil service, particularly within the Ministry of Education; develop the necessary standards for deployment of ICTs in schools; and create conditions that allow ICT to be fully utilized in education.
18. Approximately 60 percent of the Rwandan population lives below the poverty line.
19. Somewhat confusingly, the Arusha accords began on August 4, 1993, before the genocide was completed, between the then government of Rwanda and the then rebel RPF, to end the three-year-old Rwandan Civil War. They were an international effort to bring peace to Rwanda that was not completely successful. The genocide began on April 7, 1994, against Tutsi and Hutu moderates, after Hutu president Habaryimana was killed in a plane crash on April 6, 1994.

References

Allen, Chris. 1999. "Endemic Violence, and State Collapse in Africa." *Review of African Political Economy* 26(81): 367–84.

Anderson, Benedict. 1983. Imagined Communities: Reflections on the Origin and Spread of Nationalism. London: Verso.

Arnstein, Sherry R. 1969. "A Ladder of Citizen Participation." *American Institute of Planner's Journal* 35(4): 216–24.

Baldauf, Scott. 2007. "Rwanda Aims to Become Africa's High-tech Hub." *Christian Science Monitor*, October 17.

Barigye, Tony. 2008. Rwanda Focused on Her Population. *New Times*, July 11.

Bowman, Warigia, 2007a. Interview with Nongovernmental Activist, Innocent Benineza, July 15.

———. 2007b. Interview with Rwandan Minister of Education, Albert Butare, July 17.

———. 2007c. Interview with Eugene Karangwa, Rwandan Head of ICT, Mineduc and Yisa Claver, Rwandan Director of Planning, Mineduc. August 13.

Buteera, Fred. 2012. "Global Report Names Rwanda Most Dynamic in ICT." *Digital News*, October 15.

Central Intelligence Agency. 2007. "Country Profile for Rwanda." *World Factbook*. Washington, DC: Central Intelligence Agency, September 6.

Chalk, Frank. 1999. *Radio Propaganda and Genocide*. Montreal: MIGS.

CNN Wire Staff. 2010. "Kagame wins re-election in Rwanda, official results show." CNN World Online. August 11. Available at http://www.cnn.com/2010/WORLD/africa/08/11/rwanda.elections/.

Dahl, Robert A. 1989. "A Theory of the Democratic Process." In *Democracy and Its Critics*. New Haven: Yale University Press.

Dallaire, Romeo. 2003. *Shake Hands with the Devil*. New York: Carol and Graff.

de Lame, Daniel. 2004. "Mighty Secrets, Public Commensality, and the Crisis of Transparency: Rwanda through the Looking Glass." *Canadian Journal of African Studies* 38(2): 279–317.

Dzidonou, Clement. 2002. *A Blueprint for Developing National ICT Policy in Africa*. Nairobi: African Technology Policy Studies Network.

Eggelston. Karen. "Information and Communication Technologies, Markets, and Economic Development." In *The Global Information Technology Report: Readiness for the Networked World*. New York: Oxford University Press, 2002.

Fisiy, Cyprian. 1998. "Of Journeys and Boarder Crossings: Return of Refugees, Identity, and Reconstruction in Rwanda." *African Studies Association* 41(1): 17–28.

Grant, Richard. 2010. "Paul Kagame: Rwanda's Redeemer, or Ruthless Dictator?" *Telegraph*, July 22.

Gourevitch, Philip. 1998. *We Wish to Inform You That Tomorrow We Will Be Killed with Our Families: Stories from Rwanda*. New York: Farrar Strauss and Giroux.

Human Rights Watch. 2013. "World Report 2013:Singapore." http://www.hrw.org/world-report/2013/country-chapters/singapore. Accessed April 30.

International Crisis Group. 2002. "Rwanda at the End of the Transition: A Necessary Political Liberalisation." *Africa Report No. 53*. Nairobi/Brussels.

Jasanoff, Sheila, and Sang-Hyun Kim. 2009. "Containing the Atom: Sociotechnical Imaginaries and Nuclear Regulation in the US and South Korea." Minerva 47(2): 119–46.

Kagame, Paul. 2008. "Challenges and Prospects of Advancing Science and Tec: The Case of Rwanda." *Science Magazine* 322(5901): 545–51.

Kimani, Mary. 2008. "Information Technology Helps Rwandan Clinics Reach Out." *New Times*, June 1.

Kinzer, Stephen. 2007. "After So Many Deaths, Too Many Births." *New York Times*, February 11.

———. 2008. *A Thousand Hills: Rwanda's Rebirth and the Man Who Dreamed It* Hoboken: John Wiley and Sons.

Lemarchand, Rene. 2007. "Consociationalism and Power Sharing in Africa: Rwanda, Burundi, and the Democratic Republic of Congo." *African Affairs* 106(422): 1–20.

Majtenyi, Cathy. 2008. "Rwanda Strives to Become High Technology Hub for Africa." *Voice of America*, June 3.

Mamdani, Mahmood. 2001. *When Victims Become Killers: Colonialism, Nativism, and the Genocide in Rwanda*. Princeton: Princeton University Press.

Melvern, Linda. 2004. *Conspiracy to the Murder: The Rwandan Genocide*. London: Verso.

Metzl, Jamie. 1997. "Rwandan Genocide and the International Law of Radio Jamming." *American Journal of International Law* 91(4): 628–51.

Ottaway, Marina. 1999. *Africa's New Leaders: Democracy or State Reconstruction?* Washington DC: Carnegie Endowment for Peace.

Paluck, Elizabeth, and Donald Green. 2009. "Deference, Dissent, and Dispute Resolution: An Experimental Intervention Using Mass Media to Change Norms and Behavior in Rwanda." *American Political Science Review* 103(4): 622–44.

Radelet, Steven, and Jeffrey Sachs. 1997. "Asia's Reemergence." *Foreign Affairs* 75: 44–59.

Republic of Rwanda. 2000. *Rwanda Vision*. Kigali: Ministry of Finance and Economic Planning, July.

Reyntjens, Filip. 2006. "Post-1994 Politics in Rwanda: Problematising 'Liberation' and 'Democratisation.'" *Third World Quarterly* 27(6): 1103–17.

Rostow, W. Walt. 1960. *The Stages of Economic Growth: A Non-Communist Manifesto*. Cambridge, UK: Cambridge University Press.

Semujenga, Josias. 2003. *Origins of Rwandan Genocide*. New York: Humanity Books.

Thompson, Mark. 2004. "Pacific Asia after 'Asian Values': Authoritarianism, Democracy, and Good Governance." *Third World Quarterly* 25(6): 1079–95.

UNDP, Kigali Rwanda. 2004. "UNDP Country Cooperation: 2004–2008, Project Profiles, Government of Rwanda." United Nations Development Program.

Uvin, Peter. 2001."Difficult Choices in the New Post-Conflict Agenda: The International Community in Rwanda after the Genocide." *Third World Quarterly* 22(2): 177–89.

Van Leeuwen, M. 2001. "Rwanda's Imidugudu Programme and Earlier Experiences with Villagisation and Resettlement in East Africa." *Journal of Modern African Studies* 29(4): 623–44.

FIVE

Keeping Technologies Out: Sociotechnical Imaginaries and the Formation of Austria's Technopolitical Identity

ULRIKE FELT

"Sire, now I have told you about all the cities I know."

"There is still one of which you never speak."

Marco Polo bowed his head.

"Venice," the Khan said.

Marco smiled. "What else do you believe I have been talking to you about?"

The emperor did not turn a hair. "And yet I have never heard you mention that name."

And Polo said: "Every time I describe a city I am saying something about Venice."

—Italo Calvino, *Invisible Cities*, 1972/1974, 86

Understanding how technologies matter in the formation of national identities and how identities in turn frame the governance of innovation has been a long-standing, yet notoriously difficult, issue for scholars in science and technology studies. Benedict Anderson's (2006 [1983], 163) reflections on how certain technologies as "institutions of power" contribute to the formation, stabilization, and development of nationalist imaginations have been an important starting point. He successfully shaped our understanding of "the nation" as an "imagined community" whose coherence is created through specific economic, cultural, and political practices. He was much less concerned, however, with the continual exercises needed to maintain shared imaginations. This latter aspect, that ideas about national identity "must be actively cultivated in order to persist" (Hecht 1998, 12), as well as the central role technological choice can play in the formation of national identities, has been convincingly elaborated in Gabrielle Hecht's analysis of

French nuclear politics. Not only does she point at the deep entanglement of "technopolitical regimes"—that is, "linked sets of people, engineering and industrial practices, technological artefacts, political programs, and institutional ideologies"—and national identity, but she also stresses that imaginations of nationhood need to be "articulated and rehearsed" on a regular basis (1998, 16).

While these concepts have done important work in understanding the building of national identities and how technologies can play a key role in this, and indeed have inspired some of the thinking in this chapter, they do not capture in a satisfactory manner the messy, long-term processes through which national technopolitical identities are created and maintained through the (non-)uptake of certain technological developments. The concept of "sociotechnical imaginaries," as developed in this volume, seems better suited to this purpose. It sensitizes us to how profoundly technologies are entangled with national technopolitical cultures (Felt et al. 2010)—in this case that of Austria—and how the (non-)development of specific technoscientific projects, on the one hand, and imagined preferred ways of living, value structures, and social order, on the other, are mutually constitutive. The concept also invites us to consider the multiple ways in which sociotechnological idea(1)s and experiences of different actor constellations matter when making choices about which societal futures are to be attained.

This chapter investigates what I call an "imaginary of the absent." I focus attention on a case in which a national identity, a specific kind of "Austrianness," became tied to an imaginary of technological choice, namely, keeping a set of technologies out of the national territory and becoming distinctive as a nation precisely by refusing to embrace them. This analysis offers insights into the work needed to construct this kind of sociotechnical imaginary, to nourish and keep it alive as well as to naturalize it. It also sheds light on the multiple ways and moments in which citizens actively participate in the coproduction of technological and societal developments (Jasanoff 2004). Furthermore, I show the broader impact that an imaginary of absence may have on the ways in which new technologies (can) get imagined, practiced, assessed, and governed—in short, how it manages to become a powerful force ordering both technology and society. Finally, the chapter challenges standard perceptions of technological resistance as technophobia, which threatens the innovation-friendly climate deemed crucial to the development of contemporary Europe.

The starting point of this paper is a series of debates on nanotechnologies (Felt 2010)[1] organized to investigate the ways in which Austrian citizens use broader cultural resources in performing the work of sense making and anticipation in this domain.[2] The chapter then focuses on the role of specific discursive and material practices around nuclear and agrobiotechnologies, both "banned" from Austrian territory in the process of constructing the nation's identity. Revisiting these manifest, acknowledged, and explicit technological absences (Law 2004) allows us to understand how they managed to occupy space in the national (discursive) arena, to develop forms of materiality, and become woven into mythical narratives that participate in the formation of a specific kind of national identity—namely, a technopolitical one. The chapter then elaborates on the processes at work in the formation and stabilization of a sociotechnical imaginary and develops a model of how an imaginary is gradually built across different technological sectors in a long-term process of national identity formation. In conclusion, I argue that efforts at keeping certain technologies out can be seen as an important form of innovation governance (Felt et al. 2007).

Citizens' Imagination of Technoscientific Futures

In early 2010, the Austrian government launched an "Austrian Nanotechnology Action Plan" (ANAP 2010), driven by the desire to become a player in this emerging field and to harvest its promised economic benefits. In this policy document we encounter the effort to simultaneously create a bright future imaginary of this technological field and to frame "the public as the problem" (Felt et al. 2007)—a technophobic Austrian public, which might potentially show "'innovation resistance' in the light of former neo-technologies" (ANAP 2010, 20), that is, nuclear energy and agricultural biotechnologies.

As this vision of the public is a rhetorical figure often shared by policy makers and scientists alike, it seems relevant to ask *how* Austrian citizens actually assess nanotechnologies, especially in the light of prior experiences with new technologies. This was traced in four 4-hour public discussion workshops in which nanotechnology-related futures were explored.[3] Each workshop focused on a different field of nanoapplication—medicine, food, information and communication technologies (ICTs), and everyday consumer products—thus covering a broad range of potential applications.[4]

While the debates in the four workshops took different directions, shared reference points were clearly evident. In all groups, although to varying degrees,[5] Austria's strong opposition to nuclear power plants since the 1970s (e.g., Nowotny 1979; Torgersen 2002), as well as to genetically modified food/crops about two decades later (e.g., Seifert 2003), served as a shared frame of reference. A tacit agreement seemed to exist at the table about the importance of these cases for Austrian (technopolitical) culture. Very little explanation was required when referring to these cases, and short allusions were sufficient to elicit collective reflection. Speakers objecting to other group members' strong positions of resistance to the nuclear power plants or genetically modified organisms (GMOs) were aware that they were arguing against the mainstream. For example, a woman pleading for a more "positive attitude towards innovation" stated—not without some cynicism—that she "is absolutely pro nuclear energy and thus [. . .] absolutely evil" (NF, 1247–48). Another participant who tried to argue in favor of nanoencapsulated food additives met quite strong emotional opposition from other participants—a plea for a technological fix in the food sector being pictured as being against "Austrian nature."

But how did Austria's prior sociotechnical experiences affect the ways participants addressed the issue of nanotechnology and society? Four observations are crucial.

Power Struggles

Power relations were a recurrent issue, with resistance to nuclear energy being perceived as *the* key moment when Austrian citizens made a highly visible statement against the political establishment. One workshop participant appealed to this element of collective memory when asking for a more precautionary approach to the nano case: "One should at least set a sign, shouldn't one? It is similar to our (sic!) non-active nuclear power plant. Maybe the others laugh, but it is a clear statement. It says: 'it also works without' or 'citizens can be against' [the technological mainstream]" (NEP, 2515–17). She invited the others to remember that while the Austrian way of handling nuclear energy was regarded as irrational back then, the situation was different now. Against the pressures exerted by the political establishment, she underlined, it sometimes is necessary for citizens to set a precedent and to consider technological issues from a long-term perspective. Similarly, another participant reflected that "back in the 1960s [. . .], nuclear power plants were the new solution for all problems of mankind etc. Let's look at the situation today: we have no clue where to put the fuel rods" (NM, 278–80). Both participants thus recalled past nuclear futures—an activity Brown and Michael (2003) aptly call "retrospecting prospects"—in order to demonstrate the limited anticipatory capacity of policy makers and to argue for the legitimacy of questioning such developments. They did not see the future as "an empty space waiting to be filled with our desire, to be shaped, traded or formed according to rational plans and blueprints" (Adam and Groves 2007, 11). Instead, they acknowledged the complexity and uncertainty linked to any activity of "futuring," pointed at the importance of past experiences, and called for prudence in doing so (Callon et al. 2009).

As in earlier research on genetic testing (Felt et al. 2008), references to economic lobbies also highlighted underlying power issues. Lobbies were viewed as invisible drivers of technological development, "sitting in the same boat as legislators" (NF, 2211–12). One discussant explicitly compared gene technology and nanotechnology: "But the basic question is: [. . .] *Cui bono*, who profits? [. . .] Gene technology [. . .] only profits specific companies, we know it. [. . .] with nanotechnology it is similar. And if I think about all these [applications] that were mentioned, then there is not a single one [. . .] where I really think it is useful for mankind" (NF, 480–88). For this discussant, the issue was not merely one of addressing potential risks but rather asking whether or not such innovations make sense at all for wider society, thus implicitly shifting the frame from risk to innovation governance (Felt et al. 2007).

Expertise

Expertise, experts' habitus, and "institutional body language" (Wynne 1992) were important matters of concern in the debates. One participant stressed that his critique was "not directed towards nanotechnology as such" but rather against experts, who argue "We do not know a lot, but we will make the best out of it, trust us." He continued, "I am so old that I still remember Zwentendorf [the Austrian nuclear plant that triggered the anti-nuclear movement] and it is still clear in my memory today; this debate where the physicists said: 'what do you non-scientists pretend to know; we are nuclear physicists, we know.' [. . .] And if there is no counter-expert, then [their position] remains" (NF, 196–206).

Countering the classical concept of expertise, allusions were made to the intelligence of the crowd as having better anticipatory capability than that of the political establishment.[6] Citizens were thus implicitly represented as knowledgeable agents who "managed to stop it [nuclear power in Austria] in time" (NM, 2170). A similar argument came up in the GMO case when a participant posed the rhetorical question about who succeeded in "imposing" the label "without gene technology" in the food sector, immediately answering, "the masses" (NICT, 2369–70). Ultimately, these observations tie to questions of who can frame the agenda and legitimately participate in technoscientific debates, reflecting participants' imagination that they can contribute positively to the process of valuation and deliberation.

Fragile Relations and Continuous Struggles

The relations between technological innovation and deeply rooted cultural values were a third line of citizens' concern. Here arguments related to agro-biotechnology, that is, banning GMOs from Austrian fields, appeared as part of a repertoire of sociotechnical resistance. While the nuclear debate only gradually became an issue of an "alternative energy culture" in the Austrian context,[7] in the GMO case, the problem was framed in a more straightforward manner: agrobiotechnology was seen as threatening Austria's identity as a nation with a specific relation to nature. Participants in the discussion workshops pointed to the country's "pioneering role [in the organic food sector]" (NF, 844) and that "80% of the population would definitely not want to have [gene technology] in their food" (NF, 1052). Statements like "Nature hasn't produced it, and thus it should not enter our bodies" (NF, 2970) belonged to the discursive repertoire relating GMOs to nanofood.

The image of Austria's relationship to nature emerged as central to making good technopolitical choices. Critics stressed that an unreflective fascination for "the new" emerges through the unquestioned importance attributed to science and technology as a motor of progress. "When I buy an organic product," a participant explained, "it is not so much the fear that I get poisoned [. . .]" by "artificial food products" that drives the choice. Rather, the concern is about the "kind of treatment of land as a resource [that] is behind it" (NF, 1717–19). Here, contrary to the nuclear case, it is not so much an immediate health risk to individuals that is at stake but rather Austria's health as a land, thus threatening the very foundation of national identity.

While being convinced that "keeping certain technologies out" would be in principle the right choice, participants also reflected on whether this would be possible in the long run. In several instances, they noted that this proved more complex in the GMO case than for nuclear energy. While Austria could ban the planting of GMOs, international trade agreements would force the country to allow GM food into supermarkets. Labeling was seen as a way to confine such products with the help of "enlightened citizens" — they would not buy them, so participants assumed. Yet, it was also clear that GMOs could enter Austrian territory through animal feeding and thus land on people's plates through the food production chain. Keeping this technology out was therefore framed more as a continuous struggle and less as a clear-cut choice between competing products.

The GMO case also served as a successful precedent for labeling nano: "There should simply be a label: 'nanotechnology free'" (NEP, 1591–2). This idea received support—even though it was also debated critically in some instances—as Austrian consumers were perceived as valuing natural food. They would say, "Oops, this one is nano-free, while the other has nano inside. Maybe I don't take the latter" (NEP, 1597–8). For the ideal "enlightened consumer" the label would signal, "If there is nothing inside, then I'm on the safe side" (NEP, 1929–30). Thus, the potential containment of nano built strongly on the GMO experience and on people's (imagined) capacity to stand up for their values and make informed consumer choices.

Non-Austrianness

Finally, it is worth reflecting that people did not conceptualize nuclear energy and agrobiotechnology as "local entities" but rather as products of globalization intruding into Austria from outside. Alternatives were then described as rooted in local values and specific ways of handling the problem at stake. The Austrian nano case thus parallels what Hecht (1998) observed for France and nuclear power: a technology gets embraced either if it is perceived as national, and hence appeals to a general sense of tradition, pride, and history, or if it can be made national, for example, in our case by making a convincing demonstration of how it fits with Austrianness. This explains why technologies that can be constructed as "local" are embraced more easily (for similar observations in the case of biomedical technologies, see Felt et al. 2008). By contrast, those linked to the imaginations of powerful (often not clearly recognizable) actors—such as lobbies—are generally conceptualized as coming from outside or as not embracing the values that are the foundation of Austrian identity.

Summing up the nano debates, it seems that the GMO and the nuclear experiences have merged into a broader sociotechnical imaginary. Participants used this imaginary to claim a role for themselves in assessing new technologies based on previous technological decisions and—when necessary—to refer to those precedents in order to argue for keeping specific applications out of Austria or at least for keeping them contained. Second, the two earlier technological experiences were used in a rather differentiated manner. Nuclear energy and agricultural biotechnologies were regarded neither as straightforward analogies nor as incomparable.[8] Instead, elements of these complex technopolitical histories were used to make more focused, and necessarily partial, analogies in assessing nanorelated innovations. Thus, while acknowledging that "nano is much broader" (NM, 392) and differently structured, participants embraced prior technopolitical experiences as a testing ground for more general reflections on possibilities and also potential problems.

Third, the analysis highlights how sociotechnical imaginaries create a feeling of solidarity—a "we" experience—with a shared history and common frames of reference that need no further explanation. Even if imaginaries are questioned and counterarguments put forward, they remain astonishingly robust. In that sense, we could interpret what can be traced in this case as an "invention of tradition" (Hobsbawm and Ranger 1983, 2) in how to assess new technologies, that is, a "set of practices [. . .] which seek[s] to inculcate certain values and norms of behaviour" when developing a position towards a new technology and which "automatically implies continuity with the past." Being able to refer to such traditions makes "parts of social life [. . .] unchanging and invariant," an attractive possibility in a rapidly changing sociotechnical world.

Memory Practices and Technopolitical Identity

In order to understand the resources Austrian citizens draw on when making sense of new technologies, we need a reappraisal of the political and technopolitical work done to build Austria after 1955. I will start by sketching briefly the constitutional moment of establishing "a free Austria," which would form a building block for later technopolitical developments. Subsequently, the antinuclear movement became a foundational experience for a specific kind of positioning toward technology and in turn nourished the environmental debates that followed, above all the public controversy over GMOs in the 1990s. This account thus traces the making of a sociotechnical imaginary across different technological fields and episodes in history simultaneously pointing toward the symbolic organization and impact of collective (technology-related) memory practices (Anderson 2006 [1983]).

A Constitutional Moment and a Political Imaginary

On May 15, 1955, the Austrian State Treaty was signed at the Belvedere Palace in Vienna, bringing an end to the postwar Allied occupation of Austria. The Minister of Foreign Affairs presented the signed treaty from Belvedere's balcony to the waiting masses and supposedly uttered the sentence, "Austria is free!" The scene has become deeply etched in the nation's collective memory. It is inextricably linked to the formation of an Austrian national identity (Wodak et al. 2009). Interestingly, however, the sentence was never proclaimed publicly from the balcony but was spoken inside the Belvedere, during the ceremony of the signing of the treaty. Yet, through innumerable media rehearsals—in that sense media had written what Garde-Hansen (2011, 3) calls "the first draft of history"—and official reenactments on anniversaries of the date, as well as through frequent appearances in narratives of nationhood, the so-called balcony scene turned into one of the most robust political myths of postwar Austria. Both the scene and the spoken sentence became "a role-play of identity propaganda" (Wodak et al. 2009; de Cillia and Wodak 2005) and could therefore easily be transposed into contexts as different as artistic performances, advertising, or political demonstrations (Distelberger 2009). The year 1955, more than 1945,[9] became inscribed in collective memory as the moment of the "(ultimate) restoration of Austrian sovereignty" (Wodak et al. 2009, 162) and as the beginning of a national consciousness, one that positively embraced the role of being a small country. The imaginary of "Austria being free," as I will argue, could thus also be successfully deployed as part of a resistance narrative—in this case against specific technologies—depicting the country as making a difference in Europe, despite its smallness.

Nuclear Power—a Premiere in Technopolitical Identity Work

On the bank of the Danube, some thirty-five kilometers northwest of the Austrian capital Vienna, stand the ruins of a nuclear power plant, Zwentendorf. By consensus across all political parties, it was supposed to become the first of several Austrian nuclear plants, a symbol of Austria's technological development. Critical voices were dispersed and hardly audible at the beginning of the construction work in the early 1970s, but this changed with the formation of an umbrella organization, the Initiative of Austrian Nuclear Opponents, in 1975. Even then it was not possible to speak of a clear public position against Austria's nuclear plans (Nowotny 1980).

In 1976, two years before the plant's scheduled opening, the government began an information campaign. Following the logic of the deficit model (Wynne 1991), the planners assumed that lack of knowledge was the only reason for skepticism toward the project. The presumption was that giving voice to different positions and letting experts publicly debate the issues would lead to consensus and calm public concern. Not surprisingly, however, the campaign broadened public debate, and 1977 saw the first big anti-Zwentendorf demonstrations. By that time, the power plant was ready to go online; yet too many doubts had been raised—from plant safety to waste disposal—to simply push ahead.

Failing to reach agreement with the opposition parties and wanting to keep the issue out of the forthcoming elections, the strongly pronuclear left-wing government decided to hold a public referendum. The expectation, as commonly expressed, was that the majority would "vote for progress." Stark dichotomies dominated the battles before the vote, creating two scenarios for citizens to choose between: technoscientific rationality, economic independence, high living standards, and no brownouts versus irrationality, economic fragility, loss of jobs, and lack of electricity. The left's future-oriented discourse sought to make any antinuclear position look regressive. Opponents by contrast framed the problem in terms of risk to humans and the environment, proposed potential alternatives, and "reminded" Austrian citizens of "their" foundational values, among them their relation to nature (and God). Not only did experts and counterexperts fight public battles for the first time in Austrian history, but it was unclear who actually held relevant expertise. Nowotny (1980, 17) stresses the tension between a narrow definition of expertise, legitimized by the expert's position in academic institutional hierarchies, and opponents' claim to extend "the narrowly-conceived, traditional boundaries of science in order to include social and political concerns as equally legitimate." An intense phase of debate and protest began that was highly unusual for Austrian politics, which generally avoids public dissent.[10]

On November 5, 1978, with an extremely thin majority of 50.5 percent, voters said no to Zwentendorf, bringing the Austrian nuclear power program to a halt. The well-known writer and left-wing intellectual Günther Nenning commented on this event in a weekly magazine: "The Austrian people have won over its rulers; the committed over the bureaucrats; the penniless over the money-bags; the common sense over the I-know-it-all experts—a real pleasure that such a thing can happen in a democracy" (1978).[11] This storyline stressing citizens' assertion of power would play an important role in stabilizing the antinuclear position and become a cornerstone of a broader sociotechnical imaginary that still nourishes contemporary debates. Yet for such an imaginary to emerge, quite substantial work was still needed.

In December of the same year, the Austrian parliament unanimously enacted a law prohibiting the use of nuclear fission for energy generation in Austria,[12] and the issue seemed settled. Yet already the oil crisis of 1979 triggered a reconsideration. Both sides (re)addressed the issue over and over again in TV debates, newspaper articles, and two further referenda in

1980.[13] The hope for a pronuclear future was still alive and found expression, for example, in the words of a conservative Austrian political leader in 1985: "If one could buy public shares (*Volksaktien*) of Zwentendorf, I would buy them."[14]

It took more than two decades of struggle and the Chernobyl accident in April 1986 to stabilize the antinuclear position as an integral part of Austrian political culture. The debate shifted from potential risks to accounting for Chernobyl's actual consequences for people and the environment across Europe. The accident also highlighted that national territories imagined as free from nuclear dangers could quite easily be threatened by nuclear plants in neighboring countries. Thus, the 1990s saw substantial "transborder work" by Austria protesting against German, Czech, and Slovak nuclear plants and managing to rally politicians and citizens across all political camps. Politicians' public performances and visual images such as one showing Austria as a space surrounded by nuclear plants yet itself free of them underlined the identification of the nation as a nuclear-free zone.

The rather thin majority at the public referendum in 1978 was retrospectively constructed as *the* turning point in Austria's energy policy. The antinuclear narrative became a reference point for the fact that a direct democratic vote had put an end to a risky technological project and for the experience of power being exerted "from below" in different ways than imagined and practiced so far. A myth began to stabilize that Austrians had known better, that they had displayed anticipatory capacity and made the right choice not to support a problematic technology. Ultimately, this myth prepared the ground for the formation of new social movements in Austria (Gottweis 1997), and arguably for the constitution of a new technopolitical imaginary fundamentally different from what had been possible before.

With Austria's entry into the European Union in 1995, however, the issue of nuclear energy became relevant once more. This might explain why in late 1997 yet another referendum was held under the heading "Nuclear-free Austria." The aim was to trigger parliamentary action so as to upgrade nuclear legislation from a simple law to a constitutional provision, thereby further solidifying Austria's position on nuclear power. In 1999, after intense debate, the text of the referendum was transformed—with virtually the same wording—into the (federal) Constitutional Act for a Nuclear-Free Austria.[15] Thus, "Nuclear-Free Austria" moved from being a public campaign slogan into the denomination of a law, further stabilizing this imaginary. By the late 1990s, nuclear policy had become one of the rare issues uniting political parties across the spectrum. To take but one example, in the 2009 vote in the European parliament on the "Second Strategic Energy Review Report," which mentioned nuclear power as an energy option, Austria was the only country *all* of whose parties voted against the report.[16]

A story that began with ambivalence and was reopened several times stabilized into a naturalized claim: Austria is nuclear free! This dogma has been rehearsed so many times that it now seems unshakable. In March 2011, during the National Assembly debate on "Current Perspectives on Austrian and European Energy Politics after Fukushima," the Austrian chancellor affirmed "that probably nobody in the Austrian parliament would give a speech in favor of nuclear technology." And he continued insisting that it is "an obligation for Austria" to continue "stand[ing] up against the nuclear lobbies, [. . .]—in particular because we had a clear vote of the population in Austria against nuclear energy in 1978 and we have made ourselves Europe's spokesperson for *not* enforcing nuclear energy as a future technology."[17] In the chancellor's memory, it was the 1978 referendum, and not subsequent stabilization work, that stood for Austria's position on nuclear power. Despite the chancellor's confidence in the strength of the Austrian antinuclear stance, given the shape of energy politics in neighboring countries, Austria's naturalization of its antinuclear position cannot be taken for granted. It still needs performance (see Jasanoff's introduction and Hurlbut in this volume). A recent episode nicely illustrates this point. In order to ensure Austria's credibility with respect to nuclear politics on the European level, the Austrian chancellor proclaimed at the third anti–nuclear energy summit—a meeting of the government, energy industry, and environmental nongovernmental organizations—in April 2012 that it was essential for Austria to stop importing nuclear power from neighboring countries. "We have created a coalition of reason against nuclear energy," he stressed, on the one hand, "to be able to continue in the future the road of a nuclear-free Austria" and, on the other hand, to "be credible on the European level." That way, Austria would be the "spearhead against all those who pretend that nuclear energy is economically advantageous and a form of renewable energy."[18]

Agrobiotech: Sociotechnical Choice and National Identity Revisited

"'Austria is free of gene technology,' a modern Leopold Figl might announce today from the Belvedere," writes the daily newspaper *der Standard* on April 17, 2010. The article alludes to the fact that a political reality has been created through steady, media-supported performances: "We Austrians" are keeping the nation free of agrobiotech. Building on the myth of the balcony scene and the success story of nuclear refusal, being against GMOs started to become part of Austrian identity, part of an imaginary of technopolitical choice. Daily TV spots, such as one with a friendly farmer and his talking piglet pondering the "naturalness" of Austrian food, have come to embody Austria's new food culture. Dense visual discourses of untouched nature, happy animals, and healthy people nourish and perpetuate a vision of what makes the country unique. It seems "natural" to keep "green biotechnology out" in order to allow local organic food practices to develop and expand. This is "world-making by kind-making" (Hacking 1992) on a much larger scale: the new kinds created through genetic manipulation are seen as not fitting into the world imagined by many Austrian actors.

But how did this aspect of the sociotechnical imaginary emerge and stabilize? Public debates on gene technology began with the Austrian Gene Technology Law, which was enacted in 1994 and entered into force a year later.[19] This law regulated the release of genetically modified organisms for research and development as well as genetic testing and gene therapies for humans. Austria was among the first countries to adopt such explicit legislation. The process leading to the law was innovative, importing the German model of the *Enquete Kommission* ('Inquiry Commission') and showing awareness of the sensitivity and complexity of this issue in the Austrian context—even though at that time there was no broad public debate. The aim of the commission, under the heading "Technology Assessment Using the Example of Gene Technology," was to open up the debate beyond the positions of the political parties and only then bring the issue to parliament. That way, the commission hoped to avoid the controversies taking place in other countries, especially in Germany[20]—a plan that did not work out (Grabner 1999, 184; Seifert 2003).

It was only in the aftermath of this law's passage that an alliance critical of gene technology began to form. It grew rapidly, with NGOs getting active support from the most widely read Austrian tabloid, the *Neue Kronenzeitung*. In a nutshell, gene technology was represented as mainly profiting big industrial players and threatening local culture (Torgersen 2002). This framing made it possible to draw on the repertoire developed in the nuclear debate: citizens against the mighty gene lobby and small Austria against mighty economic actors. The weekly magazine *Profil* (February 4, 2006) portrayed the situation with an ironic undertone: "Everything would have been wonderful. Austria—an island of bliss, free of gene technology, a green analogue to the little nuclear-free country in the middle of the circle of unteachables."

The debate reached its height in 1997, when a referendum was held on three claims: "No food from genelabs in Austria; no release of genetically modified organisms in Austria; no patent on life."[21] In parliamentary debate in April of that year, explicit reference was made to Austria's antinuclear position: "As in the nuclear domain (preventing Zwentendorf), Austria should also adopt a pioneering role on the EU-level in the case of the 'uncontrollable' gene technology, in order to prevent negative consequences right from the start."[22] Those supporting a more liberal policy saw this connection of nuclear to gene technology as highly problematic, colorfully arguing that "in the propaganda-final for the gene technology referendum a dangerous horror-virus had been released: nuclear technology = genetechnology" (*Profil*, April 5, 1997).

The referendum, signed by more than 1.2 million citizens, became the second most successful in Austrian history. Heated discussion in parliament focused on the issue of interventions from the powerful "gene technology lobby," the lack of responsibility in case of a "biotechnology-Chernobyl," and insufficient protection of Austria being constructed as a "delicatessen store" with heightened concern for food quality.[23]

Campaigning throughout this phase of the GMO debate across all political camps was heavily influenced by the antinuclear imaginary—addressing the power relations embedded in technological issues (the lobby argument), the intelligence of the crowd, and the fragility of expert deliberations. Yet beyond linking plant biotechnology to nuclear power, the campaigns also addressed broader food issues. This allowed for a productive imaginary in which Austria, more than ever, would become a leader in Europe, setting an example for organic food production and respect for the natural environment. While nuclear energy policy was secured through the passage of a constitutional amendment, the GMO case proved much more complex and thus needed more enculturation and performance in order to reach stability. It quickly became clear that, in this case, international trade agreements would make it impossible to ban genetically modified food products from the Austrian market; instead, one would have to rely on consumer choice, even though it was hard to follow the multiple food chains and control where genetically modified ingredients would be used. The only solution was to make visible those food products that were free of gene technology through labeling. Such labels, addressing the rational citizen sharing the sociotechnical imaginary of the absent and thus making the right choice, would at least ensure containment and express Austrian sovereignty with regard to food.

This figure of "the label as a solution," together with the technopolitical identity of knowledge-*able* consumers, that is, consumers who can search for, handle, integrate, and rationally act upon information relevant to their choice (Jasanoff 2011), was also picked up in the discussion workshops on nano. And we reencounter it in more recent policy debates about Austria's nuclear future.[24] Drawing a parallel to the GMO case, a label was proposed for electricity, which would allow consumers to distinguish ecocurrent, that is, nuclear-free current, from current coming wholly or partly from foreign nuclear power plants—with the assumption that Austrian consumers would definitely prefer to buy the ecocurrent.

Making Sociotechnical Imaginaries: Rehearsal and Stabilization

The imaginary of Austria being free of a specific kind of technology ties into a complex of preferred ways of living and social order organized around the nondevelopment or the nonimplementation of specific technoscientific projects seen as foreign. Thus, it is important to trace the process of developing this imaginary across time and across distinct technological sites. The process of creating, nourishing, and stabilizing such a sociotechnical imaginary runs through several stages, which are represented in figure 5.1. In a first stage, we observe the assembling of the nuclear energy case. In a situation that seemed clearly pronuclear in the beginning, a controversy had to be carved out and staged to make space for opposing views. This controversy, temporarily closed through the 1978 referendum and the subsequent law, allowed for new (energy) options to be imagined and debated in public. Yet, it was also clear that the antinuclear orientation remained fragile. Thus, in order to produce a stable outcome, multiple rehearsals had to follow, anchoring the choice in different public arenas, in different actor constellations, and at different moments in time. Events such as the oil crisis, pressure from pronuclear lobbies, new referenda, elections, Austria joining the European Union, and many more demanded a continuous reperformance of keeping the nuclear out of the national territory. This, in turn, allowed complex, sophisticated, and multifaceted versions of resistance to develop, enculturing the antinuclear position and making it increasingly robust. Through specific kinds of memory practices such as iconic pictures, slogans, or stories, performed on anniversaries of important decisions and showing Austria to be free from direct nuclear threats, a sociotechnical imaginary gradually took shape that supported a specific kind of new Austrianness, nourishing resistance to a nuclear future but also opening up toward alternative visions.

A frequent form of rehearsal happens through the production of maps of Europe in which Austria is shown as an empty, technologically untouched space, while other countries either have nuclear power plants or use agricultural biotechnology. Thus, Austria is represented as surrounded by neighbors not sharing its vision but also as capable of resisting. These maps make

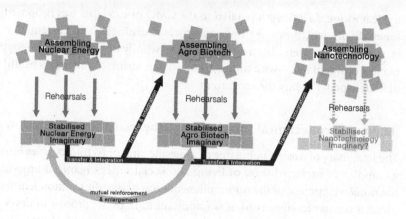

5.1. Sociotechnical imaginaries—rehearsals and stabilizations.

tacit reference to a broader political narrative, to the balcony scene and the famous slogan "Austria is free!" Media also translate this situation into other, more contemporary registers of popular culture. For example, referring to the famous Asterix and Obelix comic strips, Austria has been cast as "the famous Gallic village in Caesar's time." It is today "this clear white spot in Central Europe," showing the same unbreakable resistance as the Gallic village in Roman times (*Profil*, March 31, 2005).

Successful rehearsals gradually lead to stabilization, through the creation of a standardized history (Anderson 2006 [1983]) where few other interpretations are given space. The passage of a constitutional provision for a nuclear-free Austria in 1999 can then be seen as a key step in the stabilization of this sociotechnical imaginary. The lasting effect of this stabilization process can be seen in an assertion by the president of the Austrian National Council in spring 2011 on the occasion of the Fukushima disaster that "all the fractions are united by a clear 'no' to nuclear energy, which Austria decided in a public vote more than 30 years ago."[25]

Finally, once the emerging sociotechnical imaginary began to stabilize and gradually became a part of collective Austrian identity, elements could also be transferred and integrated into other technological debates, where they became important resources to feed people's imagination of potential relations between technological projects and preferred ways of living and social order. Even if the agrobiotechnology case followed its own specific processes of assembling and rehearsing, it could draw quite successfully on the nuclear experience. It had entered the "repertoire of the possible" that Austria could develop a different future based on alternative solutions, despite or maybe even thanks to its special size and place in Europe. The stabilization of GMO resistance, in turn, fed back into the nuclear story. Even though the two cases are structurally different and both problems and possible solutions diverge, the cases sustain each other in a common socio-technical imaginary—one that supports a culturally specific desirable future.

Finally, when participants in the discussion workshops were asked to imagine how they would position themselves toward nanotechnology, nuclear power and GMOs served as elements of their shared reference frame. Their imagination revealed an understanding of Austria being capable of keeping certain technologies out, of citizens being able to make meaningful (indeed correct) sociotechnical decisions, and of experts not necessarily having the final say, while at the same time realizing that dealing with new technologies is anything but easy.

Conclusion

I began this chapter with Italo Calvino's reflection on how deeply any narrative about the new is framed by the preexisting backdrop—symbolic, cultural, or material. Talking about nano in the Austrian context meant positioning it with regard to a preexisting sociotechnical imaginary of the absent, which had been gradually shaped through the nuclear and the agrobiotech experiences. Citizens performed an archaeology of socio-technical engagement, recollecting past encounters with technological developments, their struggles, and the gradual development of positions—whether their own or attributed to others.

Above all, as I have shown, this imaginary was the outcome of a gradual, long-term, bottom-up formation, always in need of rehearsal and (re)stabilization. These experiences worked as filters through which new elements were sieved and new sociotechnical developments were refracted. Keeping specific technologies out created the imagination of a well-delimited Austria, different in its sociotechnical practices from "the (equally imagined) others." Thus, a national technopolitical identity was created, a new self-understanding of Austria as a small nation that can choose a different sociotechnical trajectory from its more powerful neighbors. This carried the message that citizens' perceptions can make a difference when it comes to technological choice, and with it emerged a particular kind of "imagined community": of Austrians being "naturally" opposed to nuclear energy, refusing certain interventions into nature and food, and capable of making "the correct" value-oriented choices. Yet, as de Cillia and Wodak (2009, 28) remind us, such "we-discourses can never be understood as static, but as dynamic, contradictory, fragmented and historically malleable: multiple identities could thus be created discursively, situated and context dependent, as politics always asks for ever new coalitions" (my translation). National technopolitical identities are therefore neither stable nor monolithic. Rather, they have to be seen as developing out of a complex blend of intertwined histories, as being in a continuous process of reenactment and in need of nourishment and caring.

This analysis further suggests that citizens by no means misunderstand nanotechnologies by linking them to nuclear energy or agrobiotech—a fear frequently expressed by policy makers. Instead, they embrace a broader and simultaneously more fine-grained vision of what is at stake when assessing new technologies while trying to make sense of them against the backdrop of previous collective experiences. Citizens clearly differentiate between areas of application in the nano case and the corresponding issues at stake, between technological projects that fit with broader collective values and those that seem disruptive. They thus not only fulfill their role as knowledgeable agents making choices, but also as value-able citizens, that is, able to develop a set of valuing practices and integrate them into their deliberations. Participants actually embraced the task of innovation governance, which created space for pondering benefits to society and the public good, without narrowing the frame of reference to purely economic benefits or direct harms.

How then should we read the rhetorical figure of "Austria being free" in this context? Throughout the chapter, I aimed to show that this is a case in which the idea of national political freedom—expressed through protest, a popular vote, and many other means—gets coproduced (Jasanoff 2004) with freedom from technologies constructed as threats to sovereignty, territorial integrity, and foundational values. The political and the technological cannot be clearly demarcated in any reasonable manner, technology always being an agent of political production (Jasanoff 2011). "Free" thus emerges as a hybrid term, embracing both the material absence of contested technologies and the imagination of political freedom: Austria is thus "free of" and "free to" at the same time. This take on the relation between technology and politics, however, also makes clear that Langdon Winner's famous question "Do artifacts have politics?" (Winner 1986) by no means adequately captures the intricacy and fluidity of both artifacts and politics. We witness much more the emergence of a process of coproduction of the social and the technological across different technological fields and how deeply history matters in the gradual formation of a sociotechnical imaginary.

Finally, and consistently with the foregoing observations, resisting a technological innovation also means rejecting a mode of politics. Imposed from outside rather than developed from within, driven by lobbies rather than by the ideal of the public good, imposed from above rather than developed from below, artificial rather than natural—these were but the most obvious charges and countercharges that emerged during the discussions. The alleged rejection thus should much rather be read as a choice of one future over another, chosen by one set of political actors—citizens and popular media—over another—politicians, lobbyists, and powerful technoscientific actors. The refusal of some technological options allowed for the emergence of an alternative sociotechnical imaginary: one with Austria figuring as an alternative innovation space, with "clean and renewable energy" production predominantly through hydroelectric, wind and solar power at the core, as well as an organic food culture caring for its environment. Zwentendorf as a ruin on the banks of the Danube River has become an icon of technological choice as much as the labels proudly proclaiming that certain Austrian foods are organic and free of gene technology.

This chapter thus invites a different reading of technological resistance: While public questioning of technoscientific innovation is often interpreted as a sign of technophobia and as putting in danger the innovation-friendly climate that is so sought after by governments, it could also be seen as opening up alternative understandings of the public good. In the contemporary world, virtually no choices can be seen as antitechnology pure and simple because all choices are made against the backdrop of an already technologized past and with the prospect of a technologized future. In other words, public choices are not for or against technology but for or against particularly imagined forms of life—and these sociotechnical imaginaries are not given in advance but are constructed through the collective work of designing futures that seem to a nation's citizens worth attaining.

Notes

1. A paper is always a collective endeavor, even if written by a single author. Parts of the paper have been presented at several occasions from 2009 onward. It has benefited from many comments and questions posed at these occasions. My thanks also go to the team Simone Schuman, Claudia Schwarz, and Michael Strassnig, who have been working with me on the project *Making Futures Present: On Nano and Society in the Austrian Context* funded by the FWF (P20819) (Felt 2010), as well as to Maximilian Fochler, who has been working with me on questions of technopolitical cultures in the biomedical realm. Gernot Rieder is acknowledged for his careful reading and commenting on an early version of the text. Finally, my gratitude goes to Sheila Jasanoff for her generous and insightful comments, which tremendously helped sharpen the argument.

2. This could be seen as part of "civic epistemologies" that people share in a given national/cultural/political context (Jasanoff 2005).

3. For details on the methodological aspects of the discussion workshops see Felt et al. (2014).

4. Passages from these workshops are cited in the following analysis. The references indicate the discussion workshop, that is, nanomedicine (NM), nanofood (NF), nano in everyday products (NEP), and nano in relation to information and communication technologies (NICT) along with the lines in the transcript.

5. The ICT debate referred to a quite different repertoire of arguments than the other three. It did only marginally address governance-related issues, which reflects the rather uncritical stance toward the ICT sector in Austrian public debate. Consequently, we find little reference to the nuclear energy and genetically modified food cases. Dorothy Nelkin pointed at a similar discrepancy as early as 1995, stressing in a comparison between ICT and biotechnologies that "there is nearly total absence of organized public concern about a set of technologies with profound and highly problematic social and political implications" (p. 381).

6. This is much in line with some of the current enthusiasm for crowd sourcing rather than regulation. Yet it does not address adequately the complex normative questions about the consensus so achieved.

7. For a comparison with the German case on nuclear energy and biotechnology debates. see Radkau (1995).

8. For a discussion on the lessons to be learned from GMO for nanotechnology see, for example, David and Thompson (2008).

9. In fact, the end of the war was not perceived as a victorious moment by those adhering to the national socialist ideals. This is the reason why the signing of the State Treaty in 1955, together with the declaration of neutrality the same year, was a much more suitable date for the creation of an original founding myth. That way, the more troublesome parts of history were largely denied or silenced, a situation that only started to gradually change in the 1990s (Wodak et al. 2009).

10. For an overview on the media debate see Hubauer (2006).

11. For a broader reflection on emergence of popular movements around the antinuclear debate see Natter (1987).

12. BGBl No. 676/1978: *Bundesgesetz über das Verbot der Nutzung der Kernspaltung für die Energieversorgung in Österreich* ('Federal law for the ban of the use of nuclear fission for the energy supply in Austria').

13. In contrast to the public vote of 1978, these referenda merely oblige parliament to debate the issue if more than 100,000 signatures are collected. The pronuclear voices were actually much more numerous than the antinuclear ones. This, however, did not manage to create any impact.

14. In 1985, a number of leading politicians expressed their pronuclear position (Austrian Chamber of Commerce, 1985).

15. http://www.ris.bka.gv.at/Dokumente/BgblPdf/1999_149_1/1999_149_1.pdf.

16. Report of the Committee on Industry, Research and Energy, A6–0013/2009, http://www.europarl.europa.eu/sides/getDoc.do?pubRef=-//EP//NONSGML+REPORT+A6 –2009–0013+0+DOC+PDF+V0//EN; for a distribution of votes see http://sohalt .wordpress.com/2009/02/04/europa-fur-kernenergie-aufschlusslung-des-abstimmung sergebnis/; last accessed 16 June 2013.

17. Stenographic Protocol, 98th Meeting of the *Nationalrat* of the Austrian Republic,

March 22, 2011. http://www.parlament.gv.at/PAKT/VHG/XXIV/NRSITZ/NRSITZ _00098/fname_213224.pdf. Accessed June 16, 2013.
18. See http://www.bka.gv.at/site/cob_47370/currentpage_3/7674/default.aspx. Accessed June 16, 2013.
19. BGBl No. 510/1994; entered into force on January 1, 1995; for an account of the lawmaking process, see, for example, Grabner (1999).
20. In 1987, the German *Enquete Kommission* of the German Bundestag on Chances and Risks of Genetic Engineering published its final report, which caused broad controversy concerning both content and process. http://hdl.handle.net/10068/124174. Accessed June 16, 2013.
21. For the text of the referendum, see http://www.parlament.gv.at/PAKT/VHG/XX/I /I_00715/fname_139588.pdf. Accessed June 16, 2013.
22. Stenographic Protocol, 69th Meeting of the *Nationalrat* of the Austrian Republic, April 10, 1997. http://www.parlament.gv.at/PAKT/VHG/XX/NRSITZ/NRSITZ_00069/fname _114127.pdf. Accessed June 16, 2013.
23. Stenographic Protocol, 116th Meeting of the *Nationalrat* of the Austrian Republic, April 16–17, 1998. http://www.parlinkom.gv.at/PAKT/VHG/XX/NRSITZ/NRSITZ _00116/index.shtml. Accessed June 16, 2013.
24. For the declaration of the Federal Chancellor, see http://www.bka.gv.at/site/cob _47370/currentpage_3/7674/default.aspx. Accessed June 16, 2013.
25. Stenographic Protocol, 98th Meeting of the *Nationalrat* of the Austrian Republic, March 22, 2011. http://www.parlament.gv.at/PAKT/VHG/XXIV/NRSITZ/NRSITZ _00098/fname_213224.pdf, p.26. Accessed June 16, 2013.

References

Adam, Barbara, and Chris Groves. 2007. *Future Matters. Action, Knowledge, Ethics*. Leiden: Brill.
ANAP. 2010. "Austrian Nanotechnology Action Plan. Federal Ministry of Agriculture, Forestry, Environment and Water Management." http://www.nanoinitiative.at/1560 _EN.pdf.
Anderson, Benedict. (1983) 2006. *Imagined Communities: Reflections on the Origin and Spread of Nationalism*. London: Verso.
Austrian Chamber of Commerce. 1985. "Umwelt-Technik-Freiheit—Teilaufgaben mit Vorrang: Kernenergie." *Conturen*.
Brown, Nick, and Mike Michael. 2003. "A Sociology of Expectations: Retrospecting Prospects and Prospecting Retrospects." *Technology Analysis and Strategic Management* 15(1):3–18.
Callon, Michel, Pierre Lascoumes, and Yannick Barthe. 2009. *Acting in an Uncertain World: An Essay on Technical Democracy*. Cambridge, MA: MIT Press.
Calvino, Italo. 1972. *Invisible Cities*. San Diego: A Harvest Book.
David, Kenneth, and Paul B. Thompson. 2008. *What Can Nanotechnology Learn from Biotechnology*. Burlington, MA: Academic Press.
de Cillia, Rudolf, and Ruth Wodak, eds. 2009. *Gedenken im "Gedankenjahr," Zur diskursiven Konstruktion österreichischer Identitäten im Jubiläumsjahr 2005*. Wien: Studienverlag.
Distelberger, Teresa. 2009. "'Immer wieder steht Österreich befreit am Balkon'— Rekontextualisierungen der 'Balkonszene' und des Ausspruchs 'Österreich ist frei!'

im 'Gedankenjahr' 2005." Pp. 101–33 in de Cillia and Wodak, *Gedenken im "Gedankenjahr."*

Felt, Ulrike. 2010. "Leben in Nanowelten: Zur Ko-Produktion von Nano and Gesellschaft." Pp. 19–37 in *Technologisierung gesellschaftlicher Zukünfte: Nanotechnologien in wissenschaftlicher, politischer und öffentlicher Praxis,* edited by Petra Lucht, Martina Erlemann, and Esther Ruiz Ben. Freiburg: Centaurus.

Felt, Ulrike, Maximilian Fochler, Astrid Mager, and Peter Winkler. 2008. "Visions and Versions of Governing Biomedicine: Narratives on Power Structures, Decision-Making and Public Participation in the Field of Biomedical Technology in the Austrian Context." *Social Studies of Science* 38(2): 233–55.

Felt, Ulrike, Maximilian Fochler, and Peter Winkler. 2010. "Coming to Terms with Biomedical Technologies in Different Technopolitical Cultures. A Comparative Analysis of Focus Groups on Organ Transplantation and Genetic Testing in Austria, France, and the Netherlands." *Science, Technology, and Human Values* 35(4): 525–53.

Felt, Ulrike, et al. 2007. *Taking European Knowledge Society Seriously. Report of the Expert Group on Science and Governance to the Science, Economy and Society Directorate, Directorate-General for Research, European Commission.* Brussels: European Commission.

Felt, Ulrike, Simone Schumann, Claudia Schwarz, and Michael Strassnig. 2014. "Technology of Imagination. A Card-Based Public Engagement Method for Debating Emerging Technologies." *Qualitative Research* 14(2): 233–51.

Garde-Hansen, Joanne. 2011. *Media and Memory.* Edinburgh: Edinburgh University Press.

Gottweis, Herbert. 1997. "Neue soziale Bewegungen in Österreich." Pp. 342–58 in *Handbuch des politischen Systems Österreichs. Die Zweite Republik,* edited by Herbert Dachs et al. Wien: Manz.

Grabner, Petra. 1999. *Technik, Politik und Gesellschaft: Eine Untersuchung am Beispiel des österreichischen Gentechnikgesetzes.* Frankfurt am Main, Wien: Peter Lang.

Hacking, Ian. 1992. "World-Making by Kind-Making: Child Abuse for Example." Pp. 180–238 in *How Classification Works: Nelson Goodman among the Social Sciences,* edited by Mary Douglas and David L. Hull. Edinburgh: Edinburgh University Press.

Hecht, Gabrielle. 1998. *The Radiance of France. Nuclear Power and National Identity after World War 2.* Cambridge, MA: MIT Press.

Hobsbawm, Eric, and Terence Ranger. 1983. *The Invention of Tradition.* Cambridge: Cambridge University Press.

Hubauer, Anton. 2006. *Das Atomkraftwerk Zwentendorf—Berichterstattung in Öl-Journalsendungen.* Österreichische Mediathek. www.mediathek.at/downloadplatform /file/source/1159122. Accessed September 15, 2010.

Jasanoff, Sheila, ed. 2004. *States of Knowledge. The Coproduction of Science and Social Order.* London: Routledge.

———. 2005. *Designs on Nature: Science and Democracy in Europe and the United States.* Princeton, NJ: Princeton University Press.

———. 2011. "The Politics of Public Reason." Pp. 11–32 in *The Politics of Knowledge,* edited by P. Baert and F.D. Rubio. Abingdon: Routledge.

Law, John. 2004. *After Method: Mess in Social Science Research.* New York/London: Routledge.

Natter, Bernhard. 1987. "Die "Bürger" versus die "Mächtigen." Populistischer Protest an den Beispielen Zwentendorf und Hainburg." Pp. 151–170 in *Populismus in Österreich,* edited by Anton Pelinka. Wien: Edition Junius.

Nelkin, Dorothy. 1995. "Forms of Intrusion: Comparing Resistance to Information Technology and Biotechnology in the USA." Pp. 378–90 in *Resistance to New Technology:*

Nuclear Power, Information Technology and Biotechnology, edited by Martin Bauer. Cambridge, UK: Cambridge University Press.

Nenning, Günther. *Profil,* November 7, 1978.

Nowotny, Helga. 1979. *Kernenergie: Gefahr oder Notwendigkeit.* Frankfurt am Main: Suhrkamp.

———. 1980. "The Role of the Experts in Developing Public Policy: The Austrian Debate on Nuclear Power." *Science, Technology, and Human Values* 5: 10–18.

Radkau, Joachim. 1995. "Learning from Chernobyl for the Fight against Genetics? Stages and Stimuli of German Protest Movements—A Comparative Synopsis." Pp. 335–55 in *Resistance to New Technology: Nuclear Power, Information Technology and Biotechnology,* edited by Martin Bauer. Cambridge, UK: Cambridge University Press.

Seifert, Franz. 2003. *Gentechnik—Öffentlichkeit—Demokratie. Der österreichische Gentechnik-Konflikt im internationalen Kontext.* Wien: Profil Verlag.

Torgersen, Helge. 2002. "Austria and the Transatlantic Agricultural Biotechnology Divide." *Science Communication* 24(2):173–83.

Winner, Langdon. 1986. "Do Artifacts Have Politics?" *Daedalus* 109(1): 121–36.

Wodak, Ruth, Rudolf de Cillia, Martin Reisig, and Karin Liebhart. 2009. *The Discursive Construction of National Identity.* Edinburgh: Edinburgh University Press.

Wynne, Brian. 1991. "Knowledges in Context." *Science, Technology and Human Values* 16(1): 111–21.

———. 1992. "Misunderstood Misunderstanding: Social Identities and Public Uptake of Science." *Public Understanding of Science* 1(1): 281–304.

SIX

Remembering the Future: Science, Law, and the Legacy of Asilomar

J. BENJAMIN HURLBUT

Progress, far from consisting in change, depends on retentiveness . . . Those who cannot remember the past are condemned to repeat it.

—George Santayana (1980)

Introduction

In February 1975, a group of the world's leading molecular biologists gathered in a large hall at the Asilomar conference center in Pacific Grove, California. They were there at the behest of the US National Academy of Sciences to assess the risks associated with recombinant DNA (rDNA) technology and to establish guidelines to govern research in this area. Their intent was to lift a year-old voluntary moratorium on rDNA research by producing a consensus on how to proceed safely. The organizing committee hoped that the guidelines would simultaneously limit biohazard risks and forestall regulatory intervention. The meeting was an expression of scientific solidarity, but also of control—of the authority of a scientific community to constitute itself, to predict possible futures, and to define responsibilities for them in the present.

The Asilomar meeting is often remembered as a historic event that established a foundation for scientific self-regulation in an unknown and potentially dangerous domain. It also marked a crucial early moment in the development of biotechnology; biotechnology's many scientific, technological, and economic achievements, the story goes, trace their genealogies back to Asilomar (Berg 2008, 2001; Schapiro and Capron 2001; Dworkin 1978; Fredrickson 2001b; Singer 2001). Asilomar is remembered as charting a course between the Scylla of technological risk and the Charybdis of over-regulation. It is said to have resolved the scientific and political uncertainties around rDNA technology through an act of prudence by the scientific community. Most importantly, Asilomar is remembered as a great success story in managing public alarm that, had it not been contained, would have inhibited scientific research and the emergence of biotechnology. It is credited with having engendered public trust, thereby leaving science and the technological futures it creates largely unmolested by state intervention.

Asilomar persists in memory as a precedent for making sense of present problems. A 2002 commentary in *Science* on human cloning noted, "Twenty-five years ago, when the future of recombinant DNA technology was at stake, hope prevailed over fear, and reasoned debate over sensationalism. We must do our utmost to ensure that history repeats itself in the debate now before us" (Feldbaum 2002, 975). A 2004 *Nature* article on synthetic biology asked, "Is it now time for another Asilomar?" (Ball 2004). A reflection on the Human Genome Project's ELSI program recalled Asilomar's instrumental role in "resolv[ing] an emerging science policy crisis" and described ELSI as "the heir to the scientific legacy of Asilomar" (Sharp, Yudell, and Wilson 2004). The 2012 controversy over publicly disclosing details of studies that made the H5N1 flu virus potentially transmissible between humans was, in the words of the National Science Advisory Board for Biosecurity, "an Asilomar-type moment"—in which a risk controversy could be resolved through a temporary moratorium "until [the research community] can develop guidance for the safe and responsible conduct of such research" and "with little detrimental effect on scientific progress" (Berns et al. 2012, 661).

Yet approaches to the governance of emerging technology have changed dramatically since 1975, as have the epistemic, institutional, economic, and political worlds in which they are enacted. Why, then, does Asilomar continue to capture the imagination? Why is it so frequently invoked and given such precedential weight? This essay explores how Asilomar persists in American deliberation on the governance of science and technology as a site of memory (Nora 1989), particularly in domains of "emerging technology." Remarkably, the Asilomar story persists even around technoscientific projects that bear little resemblance to rDNA. I argue that Asilomar is invoked because it crystallizes a widely shared imaginary of science and law—an imaginary of "governable emergence"—wherein not only is science imagined as an engine of change, but law is cast as always trailing behind and thus reactive to and potentially inhibitory of scientific progress (i.e., the "law lag"). Asilomar-in-memory, as I call it, perpetuates this imaginary by grounding it in a historical precedent.

Building on the insight that technological futures and social orders, jointly imagined, are key sites in the fabrication of modern forms of power (Jasanoff, this volume), I examine the role of memory in the practices of reenactment that perpetuate and sustain one particularly powerful socio-technical imaginary. I show that imaginations of the right allocations of responsibility in the governance of technoscience are unreflexively reinscribed via analogical matching of present to past, thereby reproducing and rendering durable particular forms of power and authority. This analysis explicitly takes up the themes of time and change developed in Jasanoff's introductory essay. As Asilomar became Asilomar-in-memory, it came to operate as a figure with which to legitimate acts of recollection and reenactment. In short, by exploring the dyad of memory and prediction in imaginations of (technological) novelty, I elucidate some forms of "retentiveness" (to borrow Santayana's term as quoted above) that are engendered by a particularly powerful—and widely shared—imaginary of progress. Asilomar-in-memory is one apparatus for sustaining an imaginary that shapes what lessons go (un)learned, what allocations of power go (un)interrogated, and what repetitions of history are thereby sustained.

Rather than hunt for power cloaked in knowledge or expose interests within institutions, the concept of sociotechnical imaginaries highlights the role of collective imagination in the making of presents and futures and thus in the structuration of power (Jasanoff, this volume). Sociotechnical imaginaries push past microhistorical attention to contingency while remaining historically situated. In this sense, the concept highlights an important interplay between structure and agency. It illuminates the role of imagination in the production and realization of social, political, and technological futures, exposing the extent to which imagination is a locus of collective agency, even as it generates and sustains structural constraints.

Asilomar-in-memory crystallizes and reinscribes three dimensions of an imaginary of governable emergence. First, it privileges technoscience as the source of novelty, and thus as a force of historical change and a cause of social reordering. This account ascribes priority of agency to science. Technology emerges independently of the social orderings around it: science acts, while society reacts. Second, in imagining technoscience as the driving force behind sociotechnical change, it constructs and ascribes agency to a "scientific community" that claims the competency (and responsibility) to generate and adequately characterize novelty and to decree what forms of novelty warrant societal attention. Third, in giving scientists this role, Asilomar-in-memory offers a specific, programmatic vision of the right allocations of responsibility between institutions of governance. If technology is seen as the site of social emergence, governance becomes a matter of ontological discernment: of asking, "what is new here?" and thus, "what can reasonably be predicted given existing knowledge?" Framed as epistemic matters—that is, as problems of properly assessing the risks of novel technological constructions—problems of governance become questions for experts. The "scientific community" thus acquires a gatekeeping role, based not on some principle of scientific autonomy or purity but on an imagination that science is *the* institution most capable of governing technological emergence. Normative questions of what is at stake, what is the public good, and who has the authority to define benefits and harms are thereby rendered subsidiary to expert assessments of novelty. The construction of society as necessarily reactive renders normative questions secondary to scientific assessments of the possible. This division of labor is institutionalized in the role of contemporary bioethics, with its focus on downstream consequences and its preoccupation with the question of whether a given technological domain is sufficiently novel to engender new normative problems (Parens, Johnston, and Moses 2008; see, e.g., Burley and Harris 2002; Marcus 2004).

By constructing scientific experts as the most capable predictors and, thus, governors of futures, Asilomar-in-memory in effect positions novelty and change as external to and independent of wider social supervision. This obscures the extent to which novelty is itself constructed in sociotechnical terms: the measure of novelty is its potential to impact upon or revolutionize society. This erasure rests in turn on a concomitant imagination of society: on the one hand the (in)ability of its mechanisms of governance to assimilate new technologies and contain their potential for harm and on the other, society's (in)ability to react reasonably and responsibly, to abstain from interfering except where risks genuinely demand intervention and thus to allow good technologies to emerge unimpeded via the market into the world. Thus, Asilomar-in-memory implicitly delegates to science the authority to construct—and constrain—the public imagination of what counts as legitimate and valuable progress.

Asilomar-in-memory crystallizes and reinscribes an imagination of law[1] as well as science. Corollary to the notion of science as the site of emergent novelty is that of law as laggard, primed to react only when circumstances demand (Jasanoff 2007, 1995a; Testa 2011). Indeed, Asilomar-in-memory privileges an imagination of restraint, suggesting that the law's response should be limited by a prior scientific judgment that a given form of novelty has "standing" as a potential object of legal action. Law ought not act until it is in a position to *react* to scientific judgments. Thus it authorizes scientific experts as de facto common law adjudicators of the technological future: we may not demand precautionary legal action until science determines that risks are imminent. In short, the notion of law lag is an expression of the imaginary of governable technological emergence. Law inevitably lags, and must lag, if science is to be free to generate novelty.

By naturalizing a narrative of technological emergence and constructing law as only reactive, Asilomar-in-memory also privileges authoritative imaginations of the future that are offered in the technical idiom of risk—"the moral statements of scientized society" (Beck 1992, 179). Thus, questions of whether and how technology serves the public good are rendered subsidiary to narrowly technical assessments of the potential for harm (Winner 1986; Gottweis 1998; see also Jasanoff 1995b). Importantly, however, this privileging of a risk discourse entails a corollary taking-for-granted of the benefits of science for society: benefits are naturalized through the notion that technology emerges under its own momentum along a linear trajectory from laboratory to market.

In what follows, I trace the crystallization of the sociotechnical imaginary of governable technological emergence and of laggard law in the figure of Asilomar-in-memory. In the first part of the chapter, I recount key elements of the history of the rDNA controversy and examine how that history is remembered and invoked as precedent. In subsequent sections, I analyze two moments—Congressional deliberations over human cloning and the deliberations of the US Presidential Commission for the Study of Bioethical Issues on synthetic biology—that exemplify how Asilomar-in-memory is invoked— and with what consequences—in moments of political uncertainty.

Making Memories

In the early 1970s, new techniques were developed for cutting and recombining sections of DNA, opening up tremendous possibilities for experimentation and future industrial applications. It became possible for the first time to move genes between evolutionarily distant species. This raised worries that hazardous biological constructions might be produced. The 1975 Asilomar meeting on rDNA was convened to assess these worries and to plot a course for lifting a voluntary research moratorium adopted the year before (Krimsky 1984; Wright 1994).

The meeting was by invitation, and its focus was narrow by design. The organizing committee identified select molecular biologists who were already working with rDNA techniques or were likely to in the future. A few nonscientists were also invited, including several lawyers and a handful of journalists. The organizers intended the meeting to produce concrete assessments of risk and a consensus on precautionary best practices (Berg 2001,

2008; Singer 2001). This was a challenging task. At the meeting, many expressed reservations about any systematic regulation of scientific research, voluntary or otherwise. However, over the course of three days, consensus grew that adopting a credible strategy of self-regulation for containment of biohazards within the lab could also contain the risk of public opposition and legislative action (Krimsky 1984; Wright 1994). The guidelines for physical and biological containment produced at the meeting ensured that any uncertainty about the hazards of rDNA constructs would be offset by the certainty that they would not escape the lab (Berg et al. 1975).

The focus on containment reduced uncertainty to a technical problem that could be managed through researchers' knowledge, skill, and prudence. By confining problems to the laboratory setting and solutions to the competencies of scientists, the Asilomar guidelines in effect excluded other forms of input, expert and nonexpert alike. Given that the objects of concern were circumscribed to the laboratory, input from other experts (e.g., ecologists) and even potentially affected publics was not seen as necessary. For instance, a firm commitment was made to prohibiting field release of genetically engineered organisms. It was, in the words of then National Institutes of Health (NIH) Director Donald Fredrickson, "an important rule to reduce anxiety in the first several years of use of the technology" (Fredrickson 2001a, 40). If engineered products could be prevented from entering the wider world, so too could questions of how to govern them.[2]

Yet this was a departure from already established framings. Since the late 1960s, ethical and social implications of genetic modification in humans had figured centrally in discussions about genetic engineering among leading scientists, theologians, elected officials, and others (Wolstenholme 1963; Evans 2002; Benya 2012). Although such concerns did figure in its early discussions, the planning committee made the strategic decision to set them aside (Berg 2001; Krimsky 1984). At the opening of the Asilomar meeting, committee member David Baltimore described social issues as "peripheral," leading "to complicated questions of what's right and what's wrong—complicated questions of political motivation . . . ," and suggested this was not the right time to discuss them (quoted in Wright 1994, 149).

Another widespread concern before Asilomar was the possible use of rDNA technology as a tool of biological warfare. Given that the unintentional production of novel infectious agents in rDNA experiments was a prime concern for Berg and others, it was not much of a stretch to imagine that the same technologies might be used intentionally to produce such agents for military or terrorist purposes. Nevertheless, the committee set biowarfare concerns aside, characterizing them as matters of national security rather than of scientific responsibility (Wright 1994), thereby in effect rejecting responsibility for consequences that the scientific community itself did not imagine as plausible hazards arising from their work.

Excluding these dimensions had three important consequences. First, by conceptually circumscribing risky activities to the laboratories in which dangerous entities might be generated, the organizers denied publics who had no place in these laboratories any role in the initial deliberations. Second, this narrow focus helped to constitute the meeting participants as *the* scientific community competent to assess and manage risk and thus to be responsible for governing rDNA research. Third, in claiming the capacity to assess the risky futures of rDNA techniques, the Asilomar conferees claimed the ability to foresee future benefits as well. The meeting thus shifted *all* imaginations of the future of rDNA, both hazardous and beneficent, into a technical register. Futures became contained in the molecules themselves, and governing these potential futures required those doing the governance to be able to "read" the molecules on everyone else's behalf (Jasanoff 2005; Gottweis 1998).

The scientists gathered at Asilomar described rDNA not merely as a scientifically promising research tool but as a likely source of technological goods (Dworkin 1978; Berg et al. 1975; Wright 1994). They could (and did) claim that rDNA would transform agriculture, that synthetic therapeutics would be produced in high volumes at low cost, and that myriad other industrial applications and economic and social benefits would accrue from it. Thus the Asilomar participants simultaneously constituted themselves as the community most capable of seeing rDNA's potential—a claim that in turn depended on the notion that novelty is generated by laboratories and can be adequately characterized at the level of molecules—and took upon themselves the authority to discern progress and thereby to regulate on behalf of the public good.

The Asilomar conference produced consensus recommendations for "containment" of biohazards (Berg et al. 1975), which NIH took up and implemented in somewhat modified form through the Recombinant DNA Advisory Committee (RAC). Yet this was not the end of the story. In 1976 and 1977, some members of Congress launched efforts to produce rDNA regulations that included more input from potentially affected communities. One leading critic of Asilomar was Senator Edward Kennedy (D-MA), who initially rejected the legitimacy of the Asilomar guidelines because, in his view, the public had been inappropriately excluded: "The factors under consideration [at Asilomar] extend far beyond [the scientists'] technical competence. In fact they were making public policy. And they were making it in private" (quoted in Culliton 1975, 1188).

Created by NIH in 1974, the RAC was the centerpiece of government oversight and the scientific community's preferred mechanism for governing research. The RAC was, in effect, a committee of experts[3] whose principal role was to evaluate the risks of proposed research protocols. But to Kennedy, the RAC was little more than a mechanism for implementing guidelines that had been written by an interest group without public input. One of Kennedy's proposals for bringing in the public was to create a nonscientific governing body addressing "social and ethical issues" to work in parallel with the RAC. This body was to be modeled on the National Commission for the Protection of Human Subjects of Behavioral and Biomedical Research, created by the 1974 National Research Act in the wake of scandals over ethically questionable biomedical research, foremost among which was the infamous Tuskegee syphilis study.

However, Kennedy's proposal met strong opposition from Asilomar scientists, who argued that public involvement would invent, not solve, problems. Their reason was that, although social and ethical problems indeed belonged to the public, science had yet to produce anything that warranted public (re)action. As Donald Fredrickson put it, "this was 1977 and by no means were there available sufficient facts to give much grist for the philosopher's mill . . . the burden of the RAC was not for amateurs" (Fredrickson 2001b, 141).

Indeed, it was precisely this claim—that governability depends on first grasping the scientifically correct imagination of possible futures—that diffused and ultimately silenced Kennedy's dissent. Kennedy's picture of a partisan science in need of public oversight was undercut by the competing picture of an unknowledgeable and unruly public threatening premature control. His assertion that public policy making had been captured by the private interests of scientists was countered by an imagination that the good government of technological emergence depends on anticipating and containing risk—not only the risk of science to society, but likewise (and perhaps even more) of society to science. In April 1977, molecular biologist Roy Curtiss wrote to Donald Fredrickson: "I am . . . extremely concerned that, based on fear, ignorance and misinformation, we are about to embark on over-regulation of an area of science and scientific activities" (Curtiss 1977). Later that year, 137 scientists attending the Gordon Conference on Nucleic Acids signed an open letter to Congress likewise predicting that "the benefits of recombinant DNA research will be denied to society by unnecessarily restrictive legislation" (Gilbert 1977). Widely circulated, these letters helped tip the political scales toward an imagination of society, rather than science, as the primary source of risk (Wright 1994). Articulated in the register of expert prediction, the future imagined at Asilomar therefore came to serve as a declaration of the appropriate forms of control through which the potencies of rDNA research could be adequately discerned, governed, and set free. Legislative efforts were dropped, and by 1979 the rDNA controversy had all but faded from memory.

Asilomar-in-Memory

In 1977, at the height of political uncertainty over rDNA regulation, NIH director Fredrickson lamented that the public dialogue around rDNA had given "rise to dangerous overreaction and exploitation, which gravely obstructed the subsequent course" (Fredrickson 1979, 151). Two decades later his recollection was different. "From the awesome promise of Asilomar has come a new science, a new medicine, and a new industry of genomics and proteomics." The engagement of "science, public, government, and industry with questions about the meaning of the social compact" initiated at Asilomar was "an entry point into the future" (Fredrickson 2001b). This recollection is typical. According to Harold Varmus, also a former director of the NIH, "the importance of the achievement [of Asilomar] cannot be overemphasized." Without it, "the world of biomedical science would be very different today" (Varmus 2009, 149).

Remembered as the beginning of a golden age for science, Asilomar is also a precedent for the governance of emerging technologies. It is "eponymous for the discussion of scientific policy issues" and "a landmark in the evolution of social awareness and the assumption of responsibility in the biomedical research community." It is a "model for the discussion and resolution of difficult scientific issues" whose "proof of principle" is its historical legacy: the "blossoming of the field of molecular genetics" (Schechter and Perlman 2001). As early as 1980, Maxine Singer maintained that Asilomar and its successes were not merely expressions of good science but also of good democracy. The scientific accomplishments of molecular biologists "had unprecedented support from enlightened societies and governments. It has been a joint venture, and we should celebrate together" (Singer 1980, 1317). Looking back, Paul Berg also suggested that the most important accomplishment of Asilomar lay in the containment not of physical but of political risks; the future was achieved by "gaining the public's trust" (Berg 2001). In this sense it marked "the beginning of an exceptional era for science and for the public discussion of science policy" (Berg 2008). Former NIH director Harold Varmus has suggested that one of the lessons to draw from Asilomar is that "active and open engagement with an anxious public can produce successful outcomes" (Varmus 2009, 149). Thus the settlement that in the 1970s was constructed as a nonpolitical matter of containing physical risk is remembered (and celebrated) as a political achievement. The right ordering of politics is seen as having followed from the proper management of science's material productions (see also Jasanoff 1995b).

In these retellings, Asilomar is said to have quelled public anxiety and overreaction, although Asilomar was followed in fact by multiple years of controversy in which federal, state, and municipal governments threatened more stringent regulation of research. By taking preemptive action at Asilomar, the dominant story goes, scientists contained the risks of an anxiety-driven public response, supplying instead a scientifically realistic and, as it happens, calming imagination of the future. In this respect, it is also a story about restraining politics and law. Regulations would have "unwittingly creat[ed] restrictions and reiterations and, above all, delay" (Fredrickson, 2001b, 173). Berg, commenting on science policy controversies of the twenty-first century, has argued, "if you wait too long . . . the sides become entrenched like with GMO" (quoted in Markoff 2009). There is, he suggests, "A lesson in Asilomar for all of science: the best way to respond to concerns created by emerging knowledge or early-stage technologies is for scientists from publicly-funded institutions to find common cause with the wider public about the best way to regulate—as early as possible" (Berg 2008).

Indeed there have been a number of attempts to repeat history through reenactments of Asilomar, sometimes right down to the physical staging, in which Asilomar's retold plotlines are virtually always kept intact: self-regulation by the scientific community contains pubic anxiety and engenders (legislative) restraint, thereby preparing the way for an orderly unfolding of a beneficent technological future. In 2009, the Association for the Advancement of Artificial Intelligence organized a meeting at Asilomar on governance of artificial intelligence that was modeled on the 1975 meeting. It was undertaken not in response to any clear and present technological danger but to an emerging "*perception* of urgency by non-experts" (AAAI Presidential Panel on Long-Term AI Futures 2009, emphasis in the original). Its aim was to contain the risk of unruly public imagination that might foreclose technological emergence as imagined by scientists. As with the 1975 meeting, "proactive reflection" would "ensure the best outcomes for AI research, enabling society to reap the maximal benefits of AI advances" (AAAI Presidential Panel on Long-Term AI Futures 2009).

In March 2010, the Climate Response Fund (CRF) organized a meeting on "responsible conduct of research on climate engineering" (i.e., geoengineering) at Asilomar. Noting that the 1975 meeting had "set a precedent for discussion of science for which risks are associated with research," CRF president Margaret Leinen reported that history was repeating itself, if in a quite different scientific context: "[H]eld just 7 months ago, [the Asilomar conference] has already helped to significantly expand the scope and breadth of international discussion and has prompted greater and deeper thinking about research on climate engineering" (Asilomar Scientific Organizing Committee 2010).

These invocations and reenactments of Asilomar looked past the specific uncertainties associated with rDNA, substituting a wide range of different technologies posing different mixes of risks and benefits. The sources of innovation, although diverse, were all treated as "emerging technologies"— that is, as domains of novelty whose futures were yet to be authoritatively imagined with allocations of responsibility thereby delineated. These reenactments reinscribed Asilomar-in-memory as an imaginary of governable emergence. Yet, like Asilomar itself, the reenactments would not have been effective in shaping governance if American society did not buy into the story. Asilomar-in-memory is compelling because the imaginary it crystallizes—of technoscience as a primary driver of historical change and of a social world that is always playing catchup—coheres with a specifically American civic epistemology that privileges formally grounded "sound science" as a foundation for public knowledge and public reason giving (Jasanoff 2005). As I demonstrate in the remainder of this chapter, this much is evidenced in the ways an imaginary of governable emergence is invoked not only by scientists in attempts to manage publics but by policy makers in constructions of their own democratic responsibilities vis-à-vis science and the public.

Cloned Controversy

The imaginary of governable emergence figured centrally in one particularly fraught policy domain—US congressional debates over human cloning. Even on an issue as ethically complex as cloning, the notion that society cannot shape but only react to (and potentially inhibit) technological change profoundly informed public debate. It constrained the public moral imaginations that informed deliberation and the capacity of the law (and, by proxy, of the public) to shape technological futures.

In February 1997, Congress reacted rapidly to the announcement of the cloning of Dolly the sheep, holding hearings and drafting legislation. In the hearings, cloning was described as an utterly unprecedented technology, profoundly destabilizing the moral as well as the biomedical status quo. Comparisons were made to rDNA, to in vitro fertilization, to the splitting of the atom, and to the Copernican revolution. For Senator Bill Frist (R-TN), cloning "challenged our imaginations." Ian Wilmut, Dolly's creator, had, in the words of Senator Edward Kennedy (D-MA), "broken the biological equivalent of the sound barrier" (US Senate, Committee on Labor and Human Resources 1997), and Senator Tom Harkin (D-IA) predicted that it "held untold benefits for humankind."

Yet, the significance of the achievement was less clear from a moral standpoint. To Senator Frist, cloning held the potential for "both good and evil." Senator Jeffords declared, "this research at once completely fascinates me and scares me to death." Senator Kit Bond, sponsor of legislation to criminalize human cloning, was less equivocal: "Human cloning is something that we as a society cannot and should not tolerate" (US Senate, Committee on Labor and Human Resources 1997).

Differences of opinion notwithstanding, lawmakers seemed to agree tacitly that government cannot shape but can only react to technoscientific change. While some were convinced that congressional action was urgently needed to prevent cloning from running amok, others doubted that research could be controlled even if legislators took action. Echoing the US Supreme Court's assertion in *Diamond v. Chakrabarty* that law cannot hold back science "any more than Canute could command the tides" (*Diamond v. Chakrabarty* 447 US 303, 1980), Senator Harkin declared with conviction that "the march of science" could not be restrained by congressional meddling: "What utter, utter nonsense to think that somehow we are going to hold up our hands and say 'Stop'!" (US Senate, Committee on Labor and Human Resources 1997).

The hearings repeatedly expressed the notion that science is unpredictable and driven by its own internal energy and that democratic institutions necessarily lag behind. Science committee chairman, Senator Jim Sensenbrenner (D-WI), observed, "in the area of cloning embryos, it is obvious that science is ahead of both the law, morals and ethics." The role of Congress, he asserted, must be to reconnect them so that "science can carry forward" (US House, Committee on Science 1997).

Asilomar surfaced as an example of a moment when lawmaking would have been counterproductive. In testimony before Congress, NIH director Harold Varmus repeatedly pointed to the case of rDNA in the 1970s as a moment in which swift congressional action could have destroyed a nascent field. Stressing the inhibitory effects of law, he used Asilomar to demonstrate the virtues of congressional restraint in the governance of science: "Legislation and science frequently do not mix very well." The lesson of Asilomar for lawmaking? Congress often made its greatest contributions to science by doing nothing, he said. "Much deliberation was given to the question raised by the cloning of DNA . . . the consequence of not having legislation to prevent such research is directly linked to the fact that we now have an extremely vibrant and benefit-generating biotechnology industry in this country" (US Senate, Committee on Labor and Human Resources 1997).

Congressman George Brown (D-CA) agreed, but he emphasized that Asilomar had succeeded in protecting democracy as well as science. He noted that the same congressional committee that was now addressing cloning had held hearings on rDNA legislation almost exactly two decades earlier. Building on Asilomar, these hearings had allowed Congress to contribute to "a broad educational process in society" that drew together legislators, the research community, and the general public, allowing them to "leave the initial ignorance and anxiety behind, giving way to a process of education and rational discussion" (U.S. House, Committee on Science 1997). By moving slowly and trusting science to lead, the committee had guaranteed that legislative action would be informed by the results of innovation, allowing reasoned deliberation to emerge in the interim. Brown called for similar restraint on human cloning.

Speaking as a representative of the Biotechnology Industry Organization, Michael West of Advanced Cell Technology, a Massachusetts-based biotech company engaged in stem cell research, argued that voluntary scientific moratoria were the most effective mechanism for containing risks while also allowing for science to progress and for benefits—both technological and democratic—to accrue. The legacy of Asilomar had proved that the "power of responsible, voluntary restraint" by scientists can produce not only health-giving technologies but "informed public debate" (US Senate, Committee on Labor and Human Resources 1997).

Brown, Varmus, and others argued that the initial public response to a novel technology is always unreasonable because the implications of technology can no more be anticipated by laypersons than the technology itself. Therefore, in order for policy makers and the public to fulfill their democratic duties, they must rely on scientific authority to predict accurately which risks are plausible, which are governable, and which regulatory rules are likely to inhibit progress. This debate cast Congress as allied with science as an educator of public reason rather than as a publicly authorized norms maker with jurisdiction over science. In short, the hearings built upon and propagated the imaginary of governable emergence.

Here again Asilomar-in-memory operated as a key figure. As a precedent, it demonstrated not only that scientific self-regulation is a path to a beneficent future but also the necessity of reining in the normative public imagination.[4] It framed public anxiety as endangering the public good—the beneficent technological future that would emerge but for unwarranted meddling. It was Congress's responsibility, in its capacity as the people's representative and in its processes of public reasoning, to defer to this authorized, and authorizing, imagination of science. Legislative deliberation was thereby reimagined as an instrument to align populist thinking to the model of rationality espoused by the nation's elite scientists.

As in earlier debates, the discourse of risk permeating the cloning discussions included the risks of society for science. This discourse weighed morally laden public imaginations against science-based predictions. Moral uncertainty was translated into discrete and manageable packages of risk with associated strategies of containment. Good governance required that the public, through its representatives, assimilate and defer to authoritative imaginations of the future in order to channel its reactions most constructively. For instance, R. Alta Charo, a lawyer-bioethicist and a member of the National Bioethics Advisory Commission, told a Senate subcommittee, "these kinds of . . . discussions about *the scientific advances that are at risk* if we were to ban all such forms of research *could be more important than any kind of legislation* you ultimately come up with, because it *will help us to understand what it is we are balancing* and make a reasoned choice" (US Senate, Committee on Labor and Human Resources 1997, emphasis added).

The National Bioethics Advisory Commission, tasked by President Clinton with assessing the ethical dimensions of human cloning, also shifted moral uncertainty into the register of risk. The commission concluded that, given the state of the art, cloning was unethical because it posed excessive physical risks to human subjects (NBAC 1997). It recommended a temporary moratorium (as with rDNA) on reproductive applications of human cloning. This conclusion transmuted ethical uncertainty into incomplete scientific knowledge and encouraged scientists to better predict the likely outcomes of such an experiment. The commission declared that "time is an ally . . . allowing for the accrual of further data from animal experimentation, enabling an assessment of the prospective safety and efficacy of the procedure in humans, as well as granting a period of fuller national debate on ethical and social concerns" (NBAC 1997, iii). The logic was that in safety lies the way to permissibility, that publics with more information naturally move toward greater rationality, and that the best pathway to naturalization of novelty is public restraint. Pointing to Asilomar, NBAC suggested that a moratorium would be more flexible and thus more effective than "government intrusion into the freedom of scientific inquiry via legislative fiat" (NBAC 1997, 96).[5]

In all of these imaginings, governance tended to get articulated in the language of risk containment—the likelihood that a possible harm will come to pass and restraint as the means necessary to prevent the bad while allowing the good to emerge unimpeded. At the same time, the transmutation of moral uncertainties (and dangers) into the idiom of risk in effect excluded other moral framings—for example, moral disgust, distrust of scientific authority, playing God—through an ostensibly prenormative reality check, in an epistemological framework with no space for moral interrogation.

Governability in Practice

The plotlines of Asilomar-in-memory are replayed not only in narratives of risk and benefit but in the casts of characters who are called upon to assume responsibilities of governance and notions of the ways in which governance should be carried out. In crystallizing an imaginary of governable emergence, Asilomar concomitantly shapes ideas of the right distributions of agency and responsibility between science and law in practices of governance. In this section I focus on these dynamics by examining the deliberations on synthetic biology (synbio) of the Presidential Commission for the Study of Bioethical Issues (hereafter PCBI).

In 2009, President Obama responded to the production of the "synthetic organism" *Mycoplasma mycoides* JCVI-syn1.0 by calling upon his newly formed Presidential Bioethics Commission to study synthetic biology. The commission went a step further, offering a set of guiding principles for the governance of emerging technology more generally (PCBI 2010). Asilomar was constantly in the background of the commission's deliberations. As in earlier bioethical decision making, the discussion frequently circled back to the question, "What is new here?" Had all relevant concerns already been addressed in the 1970s, and were existing regulatory approaches adequate for controlling synthetic biology? The commission by and large answered the question of adequacy in the affirmative, not because it found no material difference between rDNA and synbio, but because synbio represented a similar case of governable emergence.

The question of novelty was foundational for the commission. It predicated the assessment of the project of synthetic biology on an assessment of its technological products. Referencing "scientific evidence," it declared that the newly created synthetic organism "does not amount to creating life as

either a scientific or a moral matter." This judgment rendered ethical thinking unnecessary. The commission restricted the scope of appropriate moral
imagination by declaring that the likelihood of such a scenario "remains
remote for the foreseeable future"; "what remains realistic is the expectation
that over time research in synthetic biology may lead to new products for
clean energy, pollution control . . . vaccines, and other medicines" (PCBI
2010, 3). As with the cases discussed above, the commission drew the continuum of the past up against predicted futures to assess whether there was
a form of novelty that warranted reaction. However, its purpose was as much
to assess its own (and the wider public's) role in governance as to determine
whether there was something new that needed governing. The commission
described its own mandate as identifying "appropriate ethical boundaries
to maximize public benefits and minimize risk" (PCBI 2010, 4).[6] In short,
in treating emergence as the work of science, and governance as limited to
containing risk, the commission positioned itself as necessarily reactive, at
once denying itself any place in the making of technological futures and
tacitly designating that as the role (and responsibility) of technoscience.

This move is reflected in the commission's recommendations, the centerpiece of which is "prudent vigilance"—processes for "assessing likely benefits along with assessing safety and security risks . . . as technologies develop
and diffuse into public and private sectors" (PCBI 2010, 4). Put simply, prudent vigilance is a posture that the scientific community is asked to assume
in order to serve as the first line of defense against technology-derived risks.
Prudent vigilance entails, in the commission's words, "enhanced watchfulness" by those best positioned to watch, namely, researchers themselves.
Given the early stage of technological development and diffusion, the commission suggested that the synbio scientific community was best positioned
to assume the tasks of governance—implying, at the same time, that scientific communities are best able to foretell both the social and the technological future. In short, the commission asserted that governance can and
should be the responsibility of scientists until technologies have sufficiently
emerged—or until a future in which their imminence becomes imaginable.

Predictably, the commission invoked Asilomar as precedent:

Individual scientists were among the first to raise concerns about the possible
risks posed by synthetic biology research. . . . The willingness and initiative of
the scientific community to engage in this level of introspection is both reassuring and essential. Similar to researchers in the early years of recombinant
DNA research in the mid-1970s, those closest to this emerging field have exercised caution. While self-governance is not a sufficient means to mitigate all

risks, it is *likely an effective way to control many of the risks associated with emerging technologies, including synthetic biology, particularly at this early stage.* Individual scientists and students typically are *the first to notice the laboratory door ajar.* (PCBI 2010, 129, emphasis added)

The commission's metaphor of the laboratory door ajar evokes the containment measures first articulated at Asilomar. But here it informs not only a judgment of the role of science, but a corollary judgment about the commission's own role and the forms of public scrutiny and democratic judgment it represents. Only after technologies have emerged and threaten to cross the threshold are other mechanisms of governance empowered to kick in. In this way, the commission also reaffirmed the linear model of innovation that many observers have rejected as empirically erroneous and unhelpful for policy (Stokes 1997; Nowotny, Scott, and Gibbons 2001). But here the linear model figures in an imaginary not merely of emerging technology but of governable emergence—that is, of the practices of democracy appropriate to technological innovation. The imagination of linear progress—from science to society, lab to market—is, therefore, not merely a function of how technology is seen to "happen" but of the responsibilities of creativity and restraint that science assumes on behalf of (and to the exclusion of) law in serving the dual role of fabricating and governing technological emergence.

Linearity figures centrally in the sociotechnical imaginary that Asilomar-in-memory inscribes: orderly (i.e., governable) emergence unfolds in a predictable way—from science to technology, present to future, laboratory to world. As I have shown, this imaginary is two-sided: the imagination of technoscientific emergence entails a corollary imagination of law. However, as the synthetic biology case demonstrates, these imaginations not only describe understandings of agency but actively produce them, by allocating responsibility for governance. Indeed it is precisely this dynamic of co-production (Jasanoff 2004)—in which visions of technoscientific progress are always already tied up with visions of governance—that the theoretical framework of sociotechnical imaginaries developed in this volume underscores. Here I wish to draw attention specifically to the ways sociotechnical imaginaries can shape the organization of practices of governance, through constructions of the scientific community, of law, and of the democratic public—and of their respective competencies in tasks of deliberation and governance.

In delegating governance responsibilities to the prudently vigilant scientist, the commission also set tacit limits on the role of law. The principles of governance that the commission enumerated include, for example, regulatory parsimony: "only as much oversight as is truly necessary to ensure justice, fairness, security and safety while pursuing the public good." This formulation limits the law's role in governance: "undue restriction" by the law risks "inhibit[ing] the distribution of new benefits" and may even "[prevent] researchers from developing effective safeguards" (PCBI 2010, 28).

Although the commission defers extensively to scientific self-regulation, as encoded by Asilomar-in-memory, its vision of governance does not passively defer to scientists' stated views on the subject. Indeed, in recommending prudent vigilance, the commission actively overruled the concerns of leading scientists that synthetic biology might provide tools for would-be bioterrorists. For instance, a 2007 report issued by the J. Craig Venter Institute and authored by leading figures in synthetic biology rejected the notion that the scientific community could be made responsible for ensuring biosafety and biosecurity. It did so by flatly denying that the precedent of Asilomar was applicable to synthetic biology: "There have been suggestions that synthetic genomics needs 'another Asilomar.' But Asilomar was an exercise in self-governance: the community determined and imposed on itself those procedures needed to ensure safety. Bioterrorists, by definition, are not willing to accept the norms of the research community" (Garfinkel et al. 2007, 17). As the molecular biologists did at Asilomar, so Venter and his colleagues engaged in boundary work to demarcate plausible civilian uses of synbio from illegitimate uses by terrorists, but here the purpose was to shed, rather than assume, responsibility for self-governance.

The commission did not buy this reasoning. It responded to the scientists' concerns by arguing that, when moored to an accurate picture of the state of technology, the bioterror scenario was at present implausible. It is "not yet possible" to create a disease-forming pathogen without the mainstream scientific community's technical and financial resources. But since everything of potential significance necessarily happens "at the laboratory level," the commission did not absolve research scientists of regulatory responsibility. Rather, it recommended "responsible stewardship" by the scientific community, built on a "culture of responsibility," as an adequate safeguard. To demonstrate that such a culture was the right path to the future, the commission too pointed to Asilomar: "The scientists who participated at Asilomar recognized that the uncertain nature of the risks associated with their efforts demanded that they act cautiously and with utmost attention to the public interest . . . [They] developed a shared culture of responsibility to assure safe conduct of research in the largely uncharted world of genetic engineering. In the 35 years since Asilomar, the then-nascent field of genetic engineering research has flourished" (PCBI 2010, 144). Thus the commission not only endorsed the self-regulatory capacity of the scientific community but furthermore asserted, against the protests of leading scientists in the field, that such a community in fact exists.[7] In the process, the bioethics body conjured up its own Asilomar-in-memory, shifting responsibility to an empowered, but imagined, scientific community. Thus the commission performed its oversight function by declaring science capable of overseeing itself. With this move, it simultaneously insulated (market-driven) engines of technological progress from excessive public scrutiny and constructed heterogeneous scientists as a singular scientific community, responsible for its own creations. It offered a vision of science consistent with the imaginary of governable emergence: science is at once autonomous and in the service of society. Science is rightfully autonomous by virtue of its instrumental role: as a source of technological emergence and also as that institution most competent to anticipate emerging technological futures and to ensure their governability as they move from lab to market. Thus in imagining science as an endless frontier of innovation and a cornucopia for society, the commission also constructed a figure of science to which responsibilities of governance can be delegated—even over the objections of scientists.[8]

The commission also offered a corollary construction of the public. True to the popular (although substantially discredited) notion of an ignorance-driven public that policy grounded in "openness" and "engagement" must first educate (Irwin and Wynne 1996; Wynne 2002), the commission positioned itself as standing just beyond the threshold of the laboratory door to mediate between scientific production and public response. For instance, the commission recommended a "publicly accessible fact-checking mechanism for [publicly circulating claims about] prominent advances in biotechnology" to facilitate reasoned deliberation and improve "public perception and acceptance of emerging technologies" (PCBI 2010, 4).

As examples of the sorts of discourse that such fact-checking would hold in check, the commission offered "playing God" and "creating life." Quite apart from the absurdity of fact-checking whether something amounts to "playing God," this recommendation affirms that expert fact-checkers should have the authority to decide when debate should move into the public sphere—in other words, when the novelty of an "is" claim is sufficiently secure to warrant democratic reflection on what "ought" to be done about it. More significantly, however, it makes the mediation between science's "is" and society's "ought" a task for expert bioethics, and it does so in the name of protecting the integrity of both science and democracy.

The commission's foregrounding of the role of public deliberation bears the fingerprints of its chairwoman, political theorist Amy Gutmann (see, e.g., Gutmann and Thompson 1998). The commission asserted that "an inclusive process of deliberation, informed by relevant facts and sensitive to ethical concerns, promotes an atmosphere for debate and decision making that looks for common ground . . . and seeks to cultivate mutual respect" (PCBI 2010, 5). The "principles" outlined in the report highlight "justice and fairness"—the pillars of Rawlsian liberal democracy. Given this inclusive and participatory vision of governance, it is remarkable that the threshold for authorized public deliberation is so clearly vested in a prudently vigilant scientific elite.[9] But this allocation of responsibility is entirely coherent within the US sociotechnical imaginary of governable emergence and the constructions of law and science that sustain it.

An irony of history nicely demonstrates this coherence. In 2010 the PCBI served precisely the role that Senator Kennedy had proposed for a similar body in 1976—one composed of a majority of nonscientists assessing ethical, legal, and social issues associated with an anxiety-provoking emerging technology in order to ensure that public values anticipate (Guston and Sarewitz 2002) and help to guide technological development. Indeed, the PCBI explicitly noted that it had a "rare and exceptional opportunity to be forward looking instead of reactive" (PCBI 2010, 3); and with its emphasis on "deliberation," the commission affirmed the kind of openness and inclusion that Kennedy thought was missing in the 1970s. Yet where Kennedy had rejected Asilomar, the commission invoked it, but as an affirmation, rather than a negation, of deliberative democracy. Articulating its own plotline of Asilomar-in-memory, the commission exercised its "opportunity to be forward looking" by looking backward. Recollecting imagined postures of governance that had opened the way to a governable future, it reproduced them: conjuring up a scientific community to be made responsible for "pursuing the public good" and restraining democracy to "only as much oversight as it truly necessary" (PCBI 2010, 28).

Thus the synbio case again underscores that the sociotechnical imaginary of governable emergence is simultaneously an imaginary of the forms of public reason appropriate to the governance of science. It is society's democratic agency that must be modulated lest its (over)reactions encumber science and inhibit the very dynamics of technological emergence from which society stands to benefit. The commission endorses the imagination of a public in need of control, and it demonstrates how the silencing of public unease—even to the point of suppressing expression in the news media and public fora—is held without irony to be consistent with democratic imperatives. Far from seeing scientific authority as playing second fiddle to participatory democracy, the commission called upon science to ensure that democracy functions well. That an imagination of the governability of emerging technologies can so deeply inform basic ideas and practices of democracy—and that these prior assumptions can escape the notice of a commission chaired by one of America's most respected theorists of deliberative democracy—merits reflection.

Conclusion

In a pivotal moment in the development of biotechnology, a group of scientists gathered at Asilomar to offer a vision of the future that lay latent within rDNA. I have argued that Asilomar was pivotal in another way: it produced a powerful, enduring narrative about the relationship between science and technology and the forms of governance appropriate to technological change.

Asilomar-in-memory has become a means of telling a specific story about science, politics, and law, one that is retold—and reenacted—over and over again. It is a story that never wears thin because it resonates with the constitutional commitment to the separation of science and the state in American political culture (Jasanoff 2005). It crystallizes a powerful sociotechnical imaginary—and the civic epistemology that sustains it—in historical memory. In this chapter, I have highlighted the work that is required to build and maintain such an imaginary, the role of memory in that effort, and the manners and spaces of collective recollection through which the imaginary is reenacted and reinscribed in the fabrication of power.

On one level Asilomar-in-memory is a story about the education of the public imagination. In this story, the governance of science became a public problem not through an organic, Deweyian emergence of an interested public (Dewey 1991) but through science's accounts of what is new and what uncertainties merit public deliberation. Importantly, what authorizes this posture is an imagination of the subservience of law to science that is widely shared by elite representatives of the very publics whose imaginations scientists seem so reluctant to set free. Most fundamentally, then, Asilomar-in-memory reflects an imaginary of governability that demands delegation of responsibility to science by law.

On another level, Asilomar-in-memory operates like a legal precedent for establishing the respective responsibilities of science and law. It privileges the scientific community's predictions of risk and imaginations of the good above those of the public at large. It anchors a vision of the governability of technological emergence—and corollary constructions of responsibility and control—in an imagination of governance as the containment of risk

(Jasanoff and Kim 2009). Furthermore, it authorizes science to measure new and as yet ungoverned potentialities against known forms of novelty, thereby delimiting uncertainty to defining strategies for the containment of discernible risks.

This imagination of science as a wellspring of (always governable) innovation entails a construction of the law as intrinsically reactive, lagging, predisposed toward holding things back, and ignorantly failing to appreciate science's true potential to define—and pursue—the good. This imaginary of law lag is predicated on the linearity of technological emergence—novelty emerges into, not out of, a world that it lawfully reshapes. This imagination is a mechanism for delegating power. It requires democratic deference to the agenda-setting authority of science in governance: should society presume to intervene before science says what is really new, the results will inevitably run afoul of progress. Yet in each episode discussed above, the technology to which society was prematurely reacting was as yet only a figment of imagination—an anticipated but unrealized future. Indeed, it is the very work of imagining such futures that brings them into being, thereby creating a semblance of inevitability where agency and habits of deference are at work.

More than just a story about the governance of technological progress, Asilomar-in-memory draws together some ubiquitous features of late modernity—uncertainty, power, knowledge, technology, and rapid, destabilizing change—and renders them coherent, orderly, and controllable. It is a simple fable for a complex age, one that promises predictability when the future is uncertain and renders uncertainty governable without friction. At the first Asilomar meeting, the disconcerting power of a scientific revolution seemed to be tamed in the space of a few days. So recalled, Asilomar-in-memory portends similarly gratifying futures, with risks checked and benefits harnessed through successive periods of authorized imagination, with entry cards for deliberation suitably controlled and allocated.

In this chapter, I have suggested that memory is a powerful instrument for regulating the dynamics of imagination. Imaginations of the future entail the selective remembering, retelling, and reenactment of the past, reweaving threads of memory into the fabrics of power. Memorializing tames the past into recognizable plotlines and renders prior experience understandable and concrete, yet also durable and powerful, transmuting random events into de facto precedents for ordering society: for allocating responsibility and articulating imperatives of (in)action in the present. Sociotechnical imaginaries crystallized in memory are mechanisms of social retentiveness. In this sense, and contra Santayana, those who employ memory as an instrument of imagination may, in remembering the past, find themselves condemned to repeat it.

Notes

1. Here I use "law" to refer generally to forms of governance that are seen to engender lawfulness, whether via formally legislative or judicial mechanisms or via forms of regulation and oversight that are authorized by law or assume law-like institutional and procedural forms.

2. In yet another performance of the imagined priority of scientific judgment over legal process, the rule was unceremoniously dropped by the RAC a few years later, although not without a successful reassertion of legislative and judicial authority under the National Environmental Policy Act (Jasanoff 1995, 151–9).

3. The RAC had a small number of seats reserved for nonexpert "members of the public."

4. The justification for this move is the notion that public wariness of science and technology derives from public ignorance—the so called knowledge deficit model of the public (Wynne 1993). But the notion of public deficiency that I am highlighting here entails more than a mere assertion of ignorance-driven misunderstanding. Rather, it is fundamentally rooted in a normative evaluation of public reasoning that is in turn grounded in a theory of democracy. The claim is not simply that people don't know, but rather that value-laden evaluation that is not constrained by a scientifically authorized assessment of plausibility is inadequate to the task of legitimately democratic governance.

5. The notion that reproductive cloning is impermissible because physical risks outweigh benefits has become a binding norm that has largely come to displace moral debate and alternative regulatory imaginations. According to one influential view, reproductive cloning ought to be conceptually (and legally) separated from non-reproductive applications of the technology so that moral debate—and practices of governance—correspond to a realistic picture of scientific practice (National Research Council 2002). Interestingly, this is an instance where scientists have actively called upon policy makers to codify a scientific norm in law—to ban reproductive cloning. This is entirely in keeping with the imaginary of governance that this paper explicates. First, the law codifies a scientifically authorized imagination of the public good by containing the risks of technological misuse. And second, the law sustains science's preferred boundary work, maintaining the imagination of the "scientific community" as univocal and authoritative by rendering anyone who violates its taboos not merely a dissident or apostate but a criminal.

6. A more accurate formulation would be "to maximize *predicted* public benefits and minimize *predicted* risk," but the elision of the normative work of prediction goes with the territory.

7. In this respect, the constructions of science as a locus of epistemic authority, as a source of technological novelty, and as a self-regulating social community are unified in the imaginary of governable emergence. In order to ascribe agency to science (and reactivity to law), the virtues of knowledge, creativity and virtuous self-governance must belong to science.

8. In this sense, Asilomar-in-memory manages the tension between the competing vi-

sions of the role of science in politics that were offered by the architects of postwar American science (Dennis, this volume). Whereas in Dennis's account science was made subject to political control, in my account it appears autonomous. But in fact this putative autonomy is less than pure. It derives from the instrumental role that science is made to play in delivering technological innovation to the market, a role that science invites upon itself as it declares itself capable of generating governable technological futures. This certainly reflects Vannevar Bush's vision of a pure science and the corollary notion of linear progress from science to technology that justified it. But it also reflects Price's vision of a well-governed and instrumental science, overseen by due technical expertise and policed to ensure that it serves public needs. Thus insofar as scientific autonomy is deemed prerequisite for producing progress, science is also made responsible for regulating itself, even where it would rather refuse the responsibility. In this Faustian bargain, science maintains independence by continually constructing society as necessarily dependent upon it and thereby claiming authorities (and responsibilities) of governance. Scientific freedom comes at the price of delivering the futures that the state and society see fit to demand.

9. Or, perhaps better, "prudent vigilantes."

References

AAAI Presidential Panel on Long-Term AI Futures. 2009. "Interim Report from the Panel Chairs." http://www.aaai.org/Organization/Panel/panel-note.pdf. Accessed January 23, 2012.

Asilomar Scientific Organizing Committee. 2010. "The Asilomar Conference Recommendations on Principles for Research into Climate Engineering Techniques." Climate Institute. http://www.climateresponsefund.org/images/Conference/finalfinalreport.pdf. Accessed January 23, 2012.

Ball, Philip. 2004. "Synthetic Biology: Starting from Scratch." *Nature* 431 (7009): 624–26.

Beck, Ulrich. 1992. *Risk Society: Towards a New Modernity*. London: SAGE. Publications.

Benya, Frazier. 2012. "Biomedical Advances Confront Society: Congressional Hearings and the Development of Bioethics, 1960–1975." PhD diss., University of Minnesota, 2012.

Berg, Paul. 2001. "Reflections on Asilomar 2 at Asilomar 3." *Perspectives in Biology and Medicine* 44 (2): 183.

———. 2008. "Asilomar 1975: DNA Modification Secured." *Nature* 455 (7211): 290.

Berg, Paul, David Baltimore, Sydney Brenner, Richard O. Roblin, and Maxine F. Singer. 1975. "Summary Statement of the Asilomar Conference on Recombinant DNA Molecules." *Proceedings of the National Academy of Sciences of the United States of America* 72(6): 1981–4.

Berns, Kenneth I., Arturo Casadevall, Murray L. Cohen, Susan A. Ehrlich, Lynn W. Enquist, J. Patrick Fitch, David R. Franz, et al. 2012. "Adaptations of Avian Flu Virus Are a Cause for Concern." *Science* 335 (6069): 660–61.

Burley, Justine, and John Harris. 2002. *A Companion to Genethics*. Malden: Blackwell.

Culliton, Barbara J. 1975. "Kennedy: Pushing for More Public Input in Research." *Science* 188: 1187–9.

Curtiss, Roy. 1977. Letter to Donald Fredrickson, Director, National Institutes of Health. April 17. http://profiles.nlm.nih.gov/DJ/Views/Exhibit/documents/regulation.html. Accessed November 4, 2010.

Dewey, John. 1991. *The Public and Its Problems*. Athens: Swallow Press.

Dworkin, Roger B. 1978. "Science, Society, and the Expert Town Meeting: Some Comments on Asilomar." *Southern California Law Review* 51(6): 1471.

Evans, John H. 2002. *Playing God?: Human Genetic Engineering and the Rationalization of Public Bioethical Debate*. Chicago: University of Chicago Press.

Feldbaum, Carl. 2002. "Some History Should Be Repeated." *Science* 295 (5557): 975.

Fredrickson, Donald. 1979. "A History of the Recombinant DNA Guidelines in the United States." In *Recombinant DNA and Genetic Experimentation: Proceedings of a Conference on Recombinant DNA*, edited by J. Morgan and W.J. Whelan. New York: Pergamon Press.

Fredrickson, Donald S. 2001a. *The Recombinant DNA Controversy: A Memoir: Science, Politics, and the Public Interest, 1974–1981*. Washington, DC: ASM Press.

Fredrickson, Donald. 2001b. "The First Twenty-Five Years after Asilomar." *Perspectives in Biology and Medicine* 44 (2): 70–182.

Garfinkel, Michele, Drew Endy, Gerald Epstein, and Robert Friedman. 2007. "Synthetic Genomics: Options for Governance." J. Craig Venter Institute, The Center for Strategic and International Studies, MIT.

Gilbert, Walter. 1977. "Recombinant DNA Research: Government Regulation." *Science* 197 (4300): 208.

Gottweis, Herbert. 1998. *Governing Molecules: The Discursive Politics of Genetic Engineering in Europe and the United States*. Cambridge, MA: MIT Press.

Guston, David H., and Daniel Sarewitz. 2002. "Real-time Technology Assessment." *Technology in Society* 24 (1–2): 93–109.

Gutmann, Amy, and Dennis Thompson. 1998. *Democracy and Disagreement*. Cambridge, MA: Belknap Press.

Irwin, Alan, and Brian Wynne. 1996. *Misunderstanding Science*. New York: Cambridge University Press.

Jasanoff, Sheila. 1995a. *Science at the Bar: Law, Science, and Technology in America*. Cambridge, MA: Harvard University Press.

———. 1995b. "Product, Process, or Programme: Three Cultures and the Regulation of Biotechnology." In *Resistance to New Technology*, edited by M. Bauer. Cambridge: Cambridge University Press.

———. 2004. *States of Knowledge: The Co-Production of Science and Social Order*. New York: Routledge.

———. 2005. *Designs on Nature: Science and Democracy in Europe and the United States*. Princeton, NJ: Princeton University Press.

———. 2007. "Making Order: Law and Science in Action." Pp. 761–86 in *The Handbook of Science and Technology Studies*, edited by Edward J. Hackett, Olga Amsterdamska, Michael Lynch, and Judy Wajcman. Cambridge, MA: MIT Press.

Jasanoff, Sheila, and Sang-Hyun Kim. 2009. "Containing the Atom: Sociotechnical Imaginaries and Nuclear Power in the United States and South Korea." *Minerva* 47(2): 119–46.

Krimsky, Sheldon. 1984. *Genetic Alchemy: The Social History of the Recombinant DNA Controversy*. Cambridge, MA: MIT Press.

Marcus, Steven J., ed. 2004. *Neuroethics: Mapping the Field*. New York: Dana Press.

Markoff, John. 2009. "Scientists Worry Machines May Outsmart Man." *New York Times*, July 26, sec. Science. http://www.nytimes.com/2009/07/26/science/26robot.html. Accessed December 7, 2010.

National Research Council. 2002. *Scientific and Medical Aspects of Human Reproductive Cloning*. Washington, DC: National Academies Press.

NBAC. 1997. *Cloning Human Beings: Report and Recommendations of the National Bioethics Advisory Commission.* Rockland, MD.

Nora, Pierre. 1989. "Between Memory and History: Les Lieux de Mémoire." *Representations* 26(April 1): 7–24.

Nowotny, Helga, Peter Scott, and Michael Gibbons. 2001. *Re-Thinking Science: Knowledge and the Public in an Age of Uncertainty.* Malden: Blackwell.

Parens, Erik, Josephine Johnston, and Jacob Moses. 2008. "Ethics: Do We Need 'Synthetic Bioethics'?" *Science* 321 (5895): 1449.

Presidential Commission on Bioethical Issues in Biomedical Research. 2010. "New Directions: The Ethics of Synthetic Biology and Emerging Technologies." Washington DC. http://bioethics.gov/cms/sites/default/files/PCSBI-Synthetic-Biology-Report-12.16.10_0.pdf. Accessed January 13, 2011.

Santayana, George. 1980 [1905]. *Reason in Common Sense.* New York: Dover.

Schapiro, Renie, and Aleander M Capron. 2001. "Remember Asilomar?" *Perspectives in Biology and Medicine* 44(2): 162.

Schechter, Alan N., and Robert L. Perlman. 2001. "Editors' Introduction to the Special Issue on the 25th Anniversary of the Asilomar Conference." *Perspectives in Biology and Medicine* 44(2): 159–61.

Sharp, Richard R., Michael A. Yudell, and Samuel H. Wilson. 2004. "Science and Society: Shaping Science Policy in the Age of Genomics." *Nature Reviews Genetics* 5(4): 311–15.

Singer, Maxine. 1980. "Recombinant DNA Revisited." *Science* 209 (4463): 1317.

———. 2001. "What Did the Asilomar Exercise Accomplish, What Did It Leave Undone?" *Perspectives in Biology and Medicine* 44(2): 186.

Stokes, Donald E. 1997. *Pasteur's Quadrant: Basic Science and Technological Innovation.* Washington DC: Brookings Institution Press.

Testa, Giuseppe. 2011. "More Than Just a Nucleus: Cloning and the Alignment of Scientific and Political Rationalities." In *Reframing Rights: Bioconstitutionalism in the Genetic Age,* edited by Sheila Jasanoff. Cambridge, MA: MIT Press.

US House, Committee on Science. 1997. *Biotechnology and the Ethics of Cloning: How Far Should We Go?* March 5. HRG-1997-SCI-0008. Text in LexisNexis, Congressional Hearings Digital Collection.

US Senate, Committee on Labor and Human Resources. 1997. *Scientific Discoveries in Cloning: Challenges for Public Policy.* March 12. HRG-1997-LHR-0005. Text in LexisNexis, Congressional Hearings Digital Collection.

Varmus, Harold. 2009. *The Art and Politics of Science.* New York: W.W. Norton & Co.

Winner, Langdon. 1986. "On Not Hitting the Tar-Baby." Pp. 138–54 in *The Whale and the Reactor: A Search for Limits in an Age of High Technology.* Chicago: University of Chicago Press.

Wolstenholme, Gordon. 1963. *Man and His Future.* Boston: Little Brown & Co.

Wright, Susan. 1994. *Molecular Politics: Developing American and British Regulatory Policy for Genetic Engineering, 1972–1982.* Chicago: University of Chicago Press.

Wynne, Brian. 1993. "Public Uptake of Science: A Case for Institutional Reflexivity." *Public Understanding of Science* 2(4): 321–37.

———. 2002. "Risk and Environment as Legitimatory Discourses of Technology: Reflexivity Inside Out?" *Current Sociology* 50(3): 459–77.

SEVEN

Social Movements and Contested Sociotechnical Imaginaries in South Korea

SANG-HYUN KIM

Introduction

The literature on Korean history and society long shied away from viewing science and technology and related policies as objects of social and political analysis. When attention was drawn to science and technology, it focused almost exclusively on their role in the rapid industrialization of postcolonial South Korea (see, e.g., Branscomb and Choi 1996). More recently, the tide has turned, and there is a growing body of social science literature on public controversies over science and technology issues in South Korea. But with few exceptions these studies have tended to confine analysis to the interests, values, and strategies of the major social actors in the controversies (see, e.g., Kang and Jang 2013; P.S. Kim 2012). There is no doubt that it is important to study how key actors frame their issues and how they mobilize resources and support for their positions. The basic assumption of this chapter, however, is that it is also necessary to go beyond such undertakings and to explore the underlying visions of technoscience and social order that guide and shape the very thoughts, reasoning, and actions of the actors involved.

As Sheila Jasanoff and I, along with numerous others, have pointed out, public debates concerning the development and use of science and technology are likely to be informed by distinctive visions of the right relations among science, technology, the state, and society (Jasanoff 2005; Jasanoff and Kim 2009, 2013; Kim 2014; see also Hecht 1998; Mizuno 2009; Prakash 1999). Those visions, in turn, embed and are embedded in the processes through which the meanings, roles, and purposes of science and technology become closely intertwined with broader conceptions of national identity, history, and the future. Such coproductions of science, technology, and nationhood, though by no means static or permanent, can endure over a considerable period of time, encompassing the landscape, scope, and nature of disputes related to science and technology in a given nation. The concept of "sociotechnical imaginaries" is particularly useful to capture the dynamics and durability of these processes. As defined in the introduction to this volume, sociotechnical imaginaries refer to "collectively held, institutionally stabilized, and publicly performed visions of desirable futures, animated by shared understandings of forms of social life and social order attainable through, and supportive of, advances in science and technology" (for an earlier definition, see Jasanoff and Kim 2009, 2013).

In this chapter, I employ the concept of sociotechnical imaginaries as an interpretive framework for understanding and analyzing South Korean debates on science and technology issues in a wider social and political context. The analysis focuses on three cases that sparked intense conflicts nationwide—specifically, the active pursuit of nuclear power–centered energy policy, the regulation of the ethics and safety of biotechnology, and the import of US beef potentially contaminated with bovine spongiform encephalopathy (BSE), commonly known as mad cow disease. The chapter first briefly describes the ways in which South Korea's visions of science and technology have become interwoven with enduring projects of nation building and how the resulting sociotechnical imaginary shaped the initial formulation of the state's policies and actions in each case. It then examines the responses of South Korean social movements to these developments and shows how the disputes they generated were correspondingly entangled with alternative imaginaries concerning the current state and future of the Korean nation.[1]

As discussed in the sections below, in contesting the state's efforts to develop and utilize science and technology, social movement activists not only challenged the official visions of development and national interests but also questioned the proper role and place of science and technology in society. These activists were occasionally able to force the government to retreat from—or at least delay—its original policies and plans. Yet it proved very difficult for them to dethrone the dominant sociotechnical imaginary that viewed science and technology primarily as a form of power and as instruments to serve state-led national development. By positing a stark binary opposition between the hope of catching up with more "advanced" nations through economic growth and industrial competitiveness on the one hand and the fear of falling behind with stagnation and backwardness on the other, the official imaginary left little room for radically different visions of how to connect science and technology productively with nationhood.

Science, Technology, and the Imagination of Modern Korea

Western science and technology was introduced to Korea in the seventeenth century, but it was not until the mid-to-late nineteenth century that they became inextricably identified with the state's project of national development. Faced with growing encroachment from Western powers, many intellectuals in East Asia looked to Japan as a potential model for their nations' modernization and independence. While largely conforming to Western ideals of progress and development, Japan's Meiji Restoration seemed especially effective in appropriating modern institutions, knowledges, and technologies as an essential means of regaining sovereignty and strengthening national power. Impressed by Japan's apparent success, and with an even greater sense of urgency to ensure national survival against powerful Others, Korean elites increasingly came to view modernization as a state-led strategy of achieving a "rich nation and strong army" (*puguk kangbyŏng*, 富國强兵; Huh 2006). That instrumental vision of modernization was bolstered by a statist and collectivist form of Social Darwinism, which was popular in East Asia at the time, as well as by Korea's long tradition of Confucian statecraft founded on the centralization of administrative power. Integral to this vision was the task of acquiring and mastering Western science and technology to build up the nation's economic and military power and thereby to gain a higher position in the hierarchy of the world's civilizations as Koreans saw it.

This imaginary of science and technology for national empowerment remained in force even after Korea became a Japanese colony in 1910. The harsh decades of Japanese rule (1910–1945) disrupted Korea's attempts to modernize on its own terms. But as part of Japanese empire building, a range of measures were taken to accelerate industrialization and modernization in Korea. The colonial state significantly extended the networks of railways, telecommunications, and electricity and promoted various light and heavy industries such as textiles, chemicals, metals, and hydroelectric power (Eckert 1996). In addition, technical education and training were expanded, accompanied by the establishment of industrial testing laboratories. As a consequence, modern science and technology were introduced on a scale that Korea had never encountered before. Although Koreans were rarely admitted to advanced scientific and engineering training at universities (G.B. Kim 2005), wartime needs after the 1930s tended to give them more access to technical knowledge, skills, and resources. Japan's military mobilization also disseminated nationalist discourses of science and technology, as exemplified in catchphrases such as "serving the nation through science" (*kagaku hōkoku*, 科學報國) and "nation building through technology" (*kijutsu rik-*

koku, 技術立國; Mizuno 2009). These initiatives and experiences reinforced a cultural understanding that the primary role of science and technology should be to serve national development as defined by state authorities. The sociotechnical imaginary of modernization forged during colonial rule, in effect, very much resembled that adopted earlier by Korean nationalist elites already impressed by the Japanese model. Despite their resistance to colonial rule, these elites enthusiastically advocated a similar vision of science and technology for development, except that the nation they wished to develop was a future independent Korea rather than a subject of the Japanese empire.

The entanglement of an instrumental view of science and technology with nationalism, statism, and developmentalism repeatedly appeared in the writings of Korean nationalist leaders in the late nineteenth century and the colonial period. From Yu Kil-Chun (1856–1914), an influential scholar-politician, and Yi Kwang Su (1892–1950), a famous writer-journalist, to Kim Yong-Gwan (1897–1967), an advocate of invention, they all insisted that Koreans should domesticate modern science and technology to empower their nation and to assert its position in the global power hierarchy (S.-H. Kim 2012). In postcolonial South Korea, those views became more deeply entrenched under Park Chung Hee's military regime (1961–1979). In pursuing its aggressive campaign to modernize the country, the Park regime introduced a series of new policies and institutions that aimed to directly link the promotion of science and technology with national economic growth. The commencement of the Five-Year Science and Technology Promotion Plans (1962), the setting up of the Economy and Science Council (chaired by the president; 1963), the enactment of the Science and Technology Promotion Act (1966), and the creation of the Ministry of Science and Technology (1967) were but a few early examples of these efforts. The sociotechnical imaginary consolidated during this period was well reflected in the regime's slogans: "nation building through science" (*kwahak ipkuk,* 科學立國) and "technological self-reliance" (*kisul charip,* 技術自立; Ministry of Science and Technology 1976).

Ironically, South Korea's efforts to enhance its scientific and technological capacity were at first heavily dependent on foreign technical aid, especially from the United States. Yet, if South Korea borrowed US know-how, there was no corresponding buy-in to the idea of science as an autonomous, self-regulating domain of basic research. Spearheaded by wartime presidential adviser Vannevar Bush (see Dennis, this volume), the vision of a "social contract for science"—which sought to disentangle science and technology from excessive politics and state intervention—became influential in postwar America. Reliance on US technical assistance, however, did little to import this vision to South Korea. The Park regime used foreign aid as a strategic tool to achieve the nation's technological self-reliance. For its part, the United States was happy to support state-led nationalist development strategies in Asia—along with a centrally coordinated mobilization of science and technology—as a possible means of containing the spread of communism in the region (Rostow 1960).

Eventually, South Korea's rush to development came under sharp criticism from within, by student activists, intellectuals, religious leaders, and trade unions. The top-down policies advocated by the Park regime and the following military governments entailed a variety of social and political problems, including urban and rural poverty and the suppression of civil and labor rights. Opponents criticized these policies as driven by a destructive "(economic) growth-first" ideology, and (like dissidents in Suharto's Indonesia, see Moon, Barker, this volume) they viewed the state's imaginary of development as sacrificing justice and democracy in favor of the interests of the establishment and big business (Korea Democracy Foundation 2008). And yet the developmentalist thrust of the military regimes was widely accepted by the South Korean public. Several groups of intellectuals who had staunchly opposed authoritarian rule—for instance, those involved in *Sasanggye* (思想界, 'World of Thought'), one of South Korea's most influential magazines in the 1960s—also seemed to share the official vision of nationalist developmentalism (B.-H. Kim 2003). While it continued to be debated whether the military regimes' policies would lead to a genuinely national development, the basic idea of strengthening the Korean nation through rapid economic growth was not effectively challenged. Even more rarely questioned was the constitutive role of science and technology in South Korea's path to modernization.

Social Movements and the Challenges of Science and Technology

Some dissident intellectuals—religious progressives, in particular—did voice concerns about the dehumanizing and alienating aspects of modern science and technology in the late 1960s and 1970s (Lee 2007). It took some time, however, before the logic of science and technology for national development came to be more thoroughly problematized. Protests from the left were not initially a source of critique. During the early to late 1980s, many student activists and progressive intellectuals turned to Marxism as the core ideological basis of the antimilitary dictatorship movement (Shin 1995).[2] Their orthodox and dogmatic interpretations of Marxism, however, led to a yet another version of the scientism and technological determinism that were already prevalent in South Korea. Although highly critical of the development policies of the military regime, social movement activists of this period by and large subscribed to the conventional view that science and technology were politically neutral and essential vehicles for national development. For these activists—among whom left-wing nationalists formed the largest group—the most pressing issue regarding science and technology was South Korea's technological dependence on the United States and other foreign powers, exacerbated by its neocolonial capitalist political economy.

Only after the democratic transition unfolded through the late 1980s to the mid-1990s, did a new trend surface. The opening up of political space for a newly expressive civil society not only revitalized traditional activism such as labor, farmers', women's, and human rights movements but also stimulated the rise of new social movements focusing on issues such as women's health (as opposed to more traditional women's rights) and the environment. The public increasingly claimed an active and participatory role as citizens. Different ideas of "nationhood" and "development"—with greater dissent and grassroots autonomy incorporated into them—began to be more clearly articulated, undermining the predominant conception of "national development" in South Korea. A group of new social movement activists extended this critical current into contesting the state-led initiatives in science and technology. They were particularly concerned that the developmentalist drive toward risky technologies—for example, nuclear power and biotechnology—posed threats to the public interest and democracy and would ultimately block South Korea's progress toward becoming a democratic nation founded on the values of social justice, equality, participation, and sustainability (Jasanoff and Kim 2009; S.-H. Kim 2014). Gradually, the once unquestioned preeminence of science and technology as instruments of nation building, and their presumed links with the desirable future of the Korean people, came under political scrutiny.

Three of the activist challenges to the government's policies concerning science and technology in post-military rule South Korea stand out as especially significant—namely, contestations over nuclear power–centered energy policy, the regulation of biotechnology, and the importation of US beef and the presumed risk of BSE. The anti-nuclear power movement is the earliest and most notable example of South Korean activists' confrontation with science, technology, and the state. The nongovernmental organization (NGO) coalition to strengthen the regulation of biotechnology differed from previous activist efforts in that it grappled more explicitly with the issue of how to govern science and technology ethically in a democratic society. The campaign against the government's decision to import US beef represented yet a different political dynamic, in which leading activists themselves inclined to a technocratic and scientistic approach in advancing their claims. In all three cases, however, we see a persistent anxiety about possible obstacles to South Korean economic viability and competitiveness, which to some extent blunted the force of the popular critique.

Opposition to Nuclear Power

South Korea began to build commercial nuclear power plants as early as 1970. By the mid-to-late 1980s, nine nuclear power plants were in operation and dozens more were scheduled for construction (Korean Nuclear Society 2010). Nuclear safety issues did not yet receive much attention, either from the media or the general public. This was partly because the military regimes censored and suppressed dissident voices, but it was also because, as noted above, the South Korean developmental state was able to garner mass consent for its technoeconomic policies (Jang and Lee 2006). The government succeeded in representing the failure to achieve rapid economic growth and to join the league of advanced industrial countries as the most serious risk for the Korean nation. The vision of nuclear power as a crucial instrument to avoid this risk—by helping the nation move away from its already heavy dependence on foreign oil and meet the rising industrial energy demand—therefore appealed to a large portion of the South Korean public. The inscriptions in the monuments at the Kori-1 nuclear power plant and the Korea Atomic Research Institute—written by Park Chung Hee and his successor Chun Doo Hwan, respectively—neatly summarize this vision: "Torchlight for National Restoration" and "Atomic Power Is National Power." In the nation's emerging sociotechnical imaginary, nuclear safety was conceived as a subordinate technical problem that could be readily fixed by further research and development investment. The anti-dictatorship social movements, for their part, were preoccupied with the issues of human rights, economic justice, and political democracy and did not see nuclear safety as urgent or as tied to those more immediate and fundamental social problems.

Doubts slowly grew as the first generation of South Korea's environmental activists raised the expansion of nuclear power plants as one of the most urgent and important issues to be addressed (Korean Pollution Research Institute 1987). These activists took the risk of nuclear energy very seriously and strove to publicize its potential environmental and health hazards. Interestingly, their criticism focused as much on the state's promotion of risky technologies in the name of national development as on specific safety issues. In other words, the movement against nuclear power was simultaneously a rejection of a specific technology and a struggle against the dominant imaginary of developmental nationalism. Anti-nuclear activists initially framed nuclear power—far from being a vehicle for authentic national development—as a threat to the motherland imposed by neocolonial, dependent capitalism. The military regimes' commitment to nuclear energy was seen as driven by multinational corporations, such as Westinghouse, and only serving the interests of the ruling establishment and foreign powers (the United States, in particular; KPRI 1987). From the early 1990s onward, this dependency argument began to lose its influence, as the government's efforts to develop South Korea's own standardized reactor and plant design made steady progress (Korean Nuclear Society 2010). Activist groups then shifted from the critique of dependent capitalism to that of a runaway state out of democratic control, reframing their resistance as a battle against the legacy of authoritarian developmentalism. The conflicts over nuclear power thus continued to be shaped by the broader political struggle over the direction and pace of national development.

The 1990s witnessed a number of fierce local disputes over the construction of new nuclear power plants and the disposal of radioactive waste (Park 1995). The anti-nuclear power movement protested against the government's technocratic approach to assessing and managing the risks of nuclear energy and called for more transparency and democratic participation in policy making. The campaign had wide public appeal, especially for local residents near the contested areas. The ensuing controversies forced the government to postpone or even cancel some of its original plans and to introduce new regulatory measures for nuclear safety. For instance, in the mid-1990s, the Ministry of Science and Technology issued the Nuclear Safety Policy Statement, incorporating the five regulatory principles of independence, openness, clarity, efficiency, and reliability, and established the Nuclear Safety Commission to oversee the licensing and regulation of nuclear power plants (Korea Institute of Nuclear Safety 2000). But these changes had little bearing on the official vision of making South Korea one of the leading nuclear nations. Nor did growing public distrust of the government's ability and intention to strictly regulate nuclear safety derail the state's commitment to nuclear power. The imaginary of national development through advanced science and technology still pervaded South Korean society. This helped the government contain criticism and maintain its strongly pro-nuclear energy policy.

A case in point was the more recent controversy over the siting of a radioactive waste repository. In 2004, a year after the Wido—an islet off the Buan

County—was nominated as a potential site for a low- and intermediate-level (LILW) facility with an interim storage of high-level waste (spent nuclear fuel), residents in the county self-organized a referendum. The result was a landslide rejection of the siting proposal, even though the government had promised to build a multimillion dollar electron-positron accelerator in Buan as part of an economic compensation package (Kim and Cho 2004). The referendum was not legally binding, but it was a big political defeat for the government. In response, the government actively sought to relink national development goals with the desire for regional development by reiterating that the disposal of radioactive waste was vital to national interests, and by enacting a special act providing economic development incentives— far more than the construction of an accelerator—to the siting region.[3] In 2005, residents in the Gyeongju, Pohang, and other cities voted by a large majority to host a LILW repository (E.J. Kim 2011). The story indicates that, for many South Korean citizens, the physical risks of nuclear energy were secondary to the lure of advanced technologies. Environmental and safety risks, in particular, were constantly balanced against the social, political, and economic risks of failing to develop (or of falling behind) nationally as well as regionally. Anti-nuclear activists were unable to offer a compelling alternative vision to challenge this imaginary of atoms for national development. Even Japan's Fukushima disaster, while revitalizing local and national anti-nuclear movements, did not lead to changes in the existing policy to promote and expand nuclear energy (Korea Hydro and Nuclear Power Co. 2013); some even viewed Fukushima as an opportunity for South Korea to further strengthen its position among the advanced nuclear nations.

Regulation of Biotechnology

South Korea's research and development into new biotechnology—based on the use of recombinant DNA and other novel techniques—increased rapidly from the early 1980s, with the introduction of the Genetic Engineering Promotion Act (renamed the Biotechnology Promotion Act in 1995; Shin 2009). As in the case of nuclear power, the South Korean state and its followers envisaged biotechnology as a promising tool for the nation's industrial development and framed both its benefits and risks in terms of developmental goals. Perhaps not surprisingly, although the South Korean state was strongly interventionist in some respects, with an expansive machinery of regulation, biosafety and bioethical issues were left aside as secondary problems. Policy documents and the writings of government officials and scientists repeatedly expressed concern that regulation in these areas would impede the development and commercialization of biotechnology and thereby hinder "national development" (S.-H. Kim 2014). In the late 1990s, partly influenced by the publicity surrounding the sheep Dolly cloned at Edinburgh's Roslin Institute, a series of domestic biotechnological events attracted considerable media and public attention, ranging from the development of genetically modified rice and other crops, to the cloning of a dairy cow named Young-Long, to the use of and attempts to create human embryos for medical research and treatments (B. Kim 2005). Those events were welcomed by many as a sign of South Korea's advanced biotech capabilities. Yet they all occurred in the virtual absence of regulation. Even some ardent supporters of the government's policy were worried that such a situation might send a negative message to the international scientific and business communities about South Korea's ability to effectively and responsibly govern advances in biotechnology and the life sciences.

These unchecked developments were harshly criticized by a group of environmentalists, feminists, and other progressive NGO activists who were concerned about the social, ethical, health, and environmental implications of biotechnology (Lee 1999).[4] Under pressure, the government also began to suggest that more rigid biosafety and bioethics regulations would be introduced but only through the amendment of the Biotechnology Promotion Act and largely because noncompliance with international regulatory standards could frustrate South Korea's ambition to become an advanced biotech nation. For NGO activists, the idea of adding certain biosafety and bioethics clauses to the Biotechnology Promotion Act was utterly unacceptable. Nine civic NGO groups—including two of the country's largest environmental organizations—immediately launched a coalition campaign, the Alliance for Biosafety and Bioethics (ABB), advocating for new legislation to regulate the safety and ethical aspects of biotechnology—both agrifood and medical (Lee 1999). Activists in the ABB were heterogeneous, representing diverse interests, objectives, and priorities. They were nevertheless commonly opposed to the ideology of developmentalism underpinning the imperative of national economic growth over and above other public concerns, which they and their predecessors had fought hard against during the military regimes. Like nuclear power, biotechnology came to be seen as a symbol of that very hated ideology, embodying a state-led alliance among science, technology, and commercial interests.

Rejecting the popular conception that biotechnology is an apolitical instrument of supposedly unproblematic national development, ABB activists contended that, unless tightly controlled, the development and commercialization of biotechnology would endanger the public interest—including protection of human rights, social justice, equality, public health and safety, ecological sustainability, and other civic values—and would ultimately thwart the realization of genuine democracy in South Korea (Han and Kim 2000). There were some environmental and feminist activists who believed that the manipulation of genes and life processes is an inherently bad and dangerous technology, against nature and human dignity, and should thus be opposed outright. Overall, however, the NGO coalition managed to broaden the scope of the debate to encompass larger questions such as how to construct a more equal, more democratic, and more sustainable sociopolitical and economic order and how science and technology should be governed to achieve that goal. Through their campaign to impose strict social controls on biotechnology and to mandate public participation in regulatory decision making, NGO activists did not just interrogate various safety and ethical risks related to biotechnology. Concurrently, they also disputed the state's claim of national unity, the primacy of economic development, and the adequacy of formal political democracy. In sum, they called into question the presumed role and place of science and technology in a democratic society.

As the controversy over Hwang Woo Suk's human embryonic stem cell (hESC) research demonstrated, however, the imaginary of science and technology for national development was powerful and resilient (S.-H. Kim 2014). Hwang and his coworkers' supposed breakthroughs in hESC research—the derivation of stem cells from cloned human embryos and the subsequent creation of patient-tailored stem cells—were hailed as a milestone by all mainstream political parties, as well as by the government, industry, the scientific community, and the media. The Hwang team's experimental results were published—both times as cover stories—in the internationally renowned journal *Science* (Hwang et al. 2004, 2005). Enthusiastic supporters of Hwang regarded these achievements as exemplifying the development of world-class, indigenous technology that would enable South Korea to catch up and compete with more advanced industrial nations. They attacked NGO critiques of South Korea's rush to hESC research as "unpatriotic" acts against "national interests" (Kang, Kim, and Han 2006). Even after it was revealed that Hwang's research team had violated ethical codes for egg procurement and deliberately fabricated scientific data, many South Koreans remained sympathetic to him (*Nocut News* 2006). With the prosecution of Hwang for fraudulent misuse of funds, South Korea's interest in and support for hESC research waned, but other areas of biomedical research—for instance, adult stem cell research—continued to be approached in the same way (Paik 2012). NGO activists faced formidable difficulties in disseminating their vision that science and technology should be controlled to reflect a public interest defined from below, with attention to issues of social, gender, and environmental justice; in this respect, they failed to overthrow South Korea's deeply entrenched official sociotechnical imaginary.

Import of US Beef and BSE Risk

In April 2008, South Korea's incoming conservative president, Lee Myung-Bak, announced, as part of the Korea–United States Free Trade Agreement (KORUS FTA), the lifting of a ban on the import of US beef aged over thirty months (Yonhap News 2008). The ban had been in place since the first discovery of BSE in US cattle in 2003, identified in an adult dairy cow from the state of Washington. Prior to this, South Korea was the third largest market for US beef exports (Jurenas and Manyin 2010). The announcement prompted a series of massive protests and candlelight vigils that brought more than a million citizens out onto the streets of Seoul and other major cities between May and August 2008—the biggest political rallies since the 1987 June Uprising that had forced the military regime to concede the demand for constitutional reform. Although the public outcry about the BSE risk of imported US beef was at the heart of the rallies, the agenda quickly expanded to include a range of wider social and political issues, evincing widespread lack of confidence in the Lee regime. Many protesters were frustrated that President Lee did not have courage to stand up against US economic pressure. Others suspected that it was just the beginning of the conservative counteroffensive to reverse the democratic political reforms introduced by the two centrist liberal governments during the preceding ten years. Progressive social movement activists went further, portraying the decision as a result of the intensification of the neoliberal policies that had already been initiated by the previous governments (Hong 2008; Lee et al. 2010).

Conflicts over how to assess and regulate the risk of BSE contamination in US beef imports played a pivotal role in the unfolding of the candlelight demonstrations. The controversy revolved mainly around the definition of specified risk materials (SRMs) in beef and the appropriate cutoff age for determining its boundary (Ha 2012; E.-S. Kim 2012; J. Kim 2013). The Lee government's initial agreement with the United States permitted the import of both boneless and bone-in beef from cattle aged over thirty months, which were banned for trade before April 2008 on the grounds that they were generally thought to be more vulnerable to BSE infection. Adopting a new, looser definition of SRMs and new criteria for their removal—based on the guidelines of the World Organisation of Animal Health (OIE)—the Ministry for Food, Agriculture, Forestry and Fisheries also allowed the import of brain, eyes, spinal cord, and skull from cattle younger than thirty months, which, again, were previously banned. Furthermore, it was agreed that quarantine inspections should be conducted on the basis of random sampling rather than on all imported US cattle. This nontrivial shift in the assessment and regulation of BSE risk provoked strenuous opposition from civil society; more than a thousand NGOs formed the People's Council against Mad Cow Disease and called for an immediate withdrawal of the agreement (PCAMCD 2008a).

The decision to resume the import of US beef was, for these NGO groups, nothing less than a humiliating surrender to US economic interests at the expense of South Korea's national sovereignty and the health and safety of its citizens. Activists in the People's Council severely criticized the Lee government's change of regulatory criteria and procedures—and the OIE guidelines that it relied upon—as scientifically flawed and politically motivated (PCAMCD 2008b). The government ministries and mainstream scientific organizations, on the other hand, insisted that new policy measures—such as the adoption of OIE standards and the choice of random sampling inspections—were scientifically sound (Korean Academy of Science and Technology 2009; Ministry for Food, Agriculture, Forestry and Fisheries and Ministry for Health, Welfare and Family Affairs 2008). The real problem, they argued, was that the BSE risk of US beef had been grossly exaggerated by activist propaganda and the public's scientific illiteracy, causing much unnecessary turmoil and jeopardizing South Korea's opportunity to further national interests through the KORUS FTA. While reaching opposite conclusions on the safety issue, therefore, both sides of the debate shared the view that the misuse or politicization of science was threatening the future of the nation.

Similarly to the cases of nuclear power and biotechnology, South Korea's policy on US beef was embedded in the imaginary of developmental nationalism. The Lee government and its supporters did not simply ignore the potential risk of US beef. Rather they feared that unfounded doubts about the BSE risk could damage the KORUS FTA, which they perceived as a necessary step for South Korea to enhance its competitiveness and catch up with more advanced nations in an increasingly globalized world economy. With mounting pressure from civil society, the government had to renegotiate the agreement with the United States and, in the end, temporarily banned the import of US beef aged over thirty months and of brain, eyes, spinal cord, and skull from younger cattle again.[5] But it was maintained that science should serve the nation not only by building indigenous technological capability but also by quelling public disquiet that could undermine the nation's global competitiveness.

The major difference with the other two cases lay in the ways in which progressive activists responded to the government's policy. Some of the activists involved did cast doubt on the taken-for-granted relationship between technical experts and national development, thereby challenging the established sociotechnical imaginary. They charged that the government's technically oriented approach inappropriately narrowed the scope of its assessment and management of the BSE risk and demanded more stringent precautionary measures that would openly acknowledge uncertainty and the importance of social and political judgments. On the whole, however, the majority of NGOs were more inclined toward proving that the decision to resume the import of US beef was scientifically indefensible. This was primarily because, unlike in other cases, activists could obtain competent expert assistance. A number of progressive-leaning veterinary doctors and scientists, medical doctors, and public health specialists viewed the government's and the OIE's criteria for the definition, removal, and surveillance of SRMs as scientifically unsound and politically manipulated to serve US interests (J. Kim 2013). With the help of these experts, the anti–United States beef campaign sought to redraw and defend the distinctions between fact and value and between science and politics (PCAMCD 2008b). The question of the proper role of science in society consequently was framed by both sides in this case in the traditional terms of protecting science from political interference.

Conclusion: Contesting Sociotechnical Imaginaries—Possibilities and Difficulties

As discussed above, South Korea's engagement with science and technology issues has been profoundly shaped by a sociotechnical imaginary that defined the risks and benefits of science and technology in society predominantly in terms of implications for the future prosperity and empowerment of the nation. Deeply immersed in this imaginary, political elites, the bureaucracy, the scientific community, and industry believed that even a high degree of environmental, health, and safety risks could be tolerated, or left to be tackled at a later stage, if the rapid development and utilization of science and technology could help counter what they saw as the bigger risk—failure to ensure sustained national development. The state's official discourse repeatedly emphasized that the South Korean public, as dutiful and responsible national subjects, should embrace rapid technoeconomic advances as the nation's primary objective. As dramatically demonstrated by the fervent support for Hwang Woo Suk and his hESC research, a considerable section of the general public also willingly supported the task of building a technologically advanced nation and seemed prepared to accept the lax control of risks and even violations of international ethical norms as the price that latecomer nations had to pay for rapid technoeconomic development (S.-H. Kim 2014).

The challenges to the state's advocacy of developmental nationalism by student activists, trade unionists, and other progressive groups did not initially extend to the critical examination of science and technology issues. It was with the upsurge of new social movements since the late 1980s that hitherto unexplored relationships among science, technology, and the desirable future of the nation gradually began to be articulated and problematized. Progressive NGO activists, including environmentalists and feminists, launched numerous local and nationwide campaigns against the state-led projects of technoeconomic development, protesting South Korea's developmentalist approaches to science and technology as unjust, anti-democratic, and ecologically destructive. In several cases, they succeeded in disrupting the implementation of specific state policies: they delayed the construction of new nuclear power plants, revoked the original plan to build a radioactive waste repository, forced the strengthening of biosafety and bioethics regulations, and pressured the government to renegotiate the conditions of importing beef with the United States. More significantly, these struggles highlighted South Korea's pro-development, technocratic governance of science and technology—and of their potential risks—as serious threats to a rising alternative imaginary of the public interest and participatory democracy (Jasanoff and Kim 2009; S.-H. Kim 2014). From the viewpoint of progressive activists, at stake was whether, in the face of seemingly unstoppable developmentalism, South Korea could rebuild itself as a nation truly committed to the values of social justice, equality, participation, and sustainability.

Yet decades of protests and resistance by social movement activists have produced mixed results for the politics of science and technology in South Korea. The actions and criticisms of progressive NGOs did not readily translate into an effective across-the-board challenge to the prevailing sociotechnical imaginary. Social movement activists were not able to construct a coherent, alternative vision of technoscience and social order. Even when they tried, it proved extremely difficult to articulate and convey their vision to the public and to society at large. The difficulty was not solely due to the lack of human, technical, and financial resources that they could mobilize. South Korea's democratic political space, although continuously expanding, was still strongly constrained and mediated by the imaginary of developmental nationalism, in which the imperative of securing technoeconomic sovereignty figures as a key constitutive element of national identity. Not only conservative descendants of the military regimes but also many of the liberals who fought against the military dictatorship shared the basic logic of science and technology holding the keys to national development. The successful siting of a LILW repository in Gyeongju city and the persistence of biomedical policies focusing on the nation's indigenous technological capability even after the notoriety of the Hwang Woo Suk scandal were but a few examples attesting to the power of the dominant sociotechnical imaginary (E.J. Kim 2011; Paik 2012).

This should not be taken to imply that South Korea lacked—or had relatively weak—counterhegemonic, bottom-up forces. The history of South Korea can be rewritten as a history of resistance from below: the autocratic Rhee Syngman regime was brought down by the popular April 19 revolution of 1960; the assassination of Park Chung Hee by his own chief of intelligence in 1979 was precipitated by a prolonged series of mass protests against the military dictatorship; hundreds of pro-democracy protesters were killed in the Kwangju Uprising against a new military junta led by Chun Doo Hwan in May 1980; and Chun was eventually forced to step down in 1987 by massive rallies involving millions of citizens nationwide (see Bowman, Chen, this volume for comparisons with Rwanda and China). Such experiences and memories did not simply vanish from South Korean society but paved the way for numerous community-oriented, participatory, grassroots interventions similar to those of Hasan Poerbo and Onno Purbo in Indonesia (see Barker, Moon, this volume). These attempts to counter top-down developmentalism encompassed diverse areas of life and society and were neither trivial nor merely experimental. Many of them were, in fact, incorporated into the governance of Seoul—the capital city of South Korea—when Park Won-Soon, a lifelong civil rights lawyer and the founder of the People's Solidarity for Participatory Democracy, was elected as mayor in 2011 (Lee 2013).[6]

For many South Koreans, however, the more powerful experiences and memories were those of the rapid industrial transformation of South Korea—from one of the poorest countries in Asia in the 1950s; to a fast developing country with an average gross domestic product growth rate of 8–10 percent throughout the 1960s to the 1980s, and the twenty-ninth Organisation for Economic Cooperation and Development member country in 1996; to the world's fifteenth largest economy in 2008. The growing middle class, and even those who were less privileged, felt they had benefited from, and were proud of, these achievements. The liberal political group, which had been active in the anti-dictatorship struggle and came to power between 1997 and 2007, and its supporters were no exception. They separated "industrialization" from "democratization" and portrayed them as the twin symbols of the strength of the Korean nation. Often, the latter had to be justified as a more effective way to facilitate the former. Although the governments of Kim Dae Jung and Roh Moo Hyun implemented a number of new political reforms to enhance procedural democracy, South Korea's imaginary of developmental nationalism remained largely intact.

Successes in applying science and technology for national empowerment—for example, the export of South Korea's own standard nuclear power plant model or Hwang Woo Suk's alleged creation of patient-tailored stem cells via somatic cell nuclear transfer—effectively performed this imaginary, perhaps even more powerfully than did Saturn 5 in Ezrahi's analysis of the American state's instrumental use of technology (see Jasanoff, Introduction, this volume). It was no coincidence that Kim Dae Jung and Roh Moo Hyun repeatedly praised the promotion of science and technology for industrialization as one of the most important contributions the preceding military regimes had made to the Korean nation and declared that they would follow suit. Progressive activists in South Korea were thus in a much tougher position than their counterparts in Europe (see Felt's analysis of the Austrian case, this volume). Their mobilization to democratize development itself, and accordingly to reshape the relations among science, technology, the state, and society, essentially required the deconstruction and reconstruction of the very foundations of the Korean national identity. This was a task beyond their capacity. Moreover, anti-communist sentiments propagated by the Cold War were still alive and well in South Korea, which made it doubly difficult for progressive social groups to disseminate their criticism of capitalist development.

In the absence of concrete alternative imaginaries, hard-won, participatory experiments to engage NGO representatives and lay citizens in relevant policy making had only limited success. The impasse persuaded some activists that pitting democracy against technocracy was not enough: instead, they needed their own independent expert authority to evaluate, criticize, or legitimize policy decisions. This invited scientism and technocracy back into the thinking and practices of social movement activists, further destabilizing South Korean civil society's already precarious efforts to redraw the balance between factual claims and values and so to democratize technical decision making. For instance, the critique of the government's handling of the BSE risk put forward by the People's Council tended to reinforce—rather than resist—the traditional view of expert authority that earlier activists had worked hard to problematize in controversies over nuclear power or biotechnology. The council's defense of the public's right to oppose the import of US beef was itself an exercise in counter-expertise; as such, it paradoxically limited and undervalued the power of citizens to know, and intervene in, issues traditionally dominated by state bureaucrats and technical experts. The role of science in society imagined by anti–United States beef activists neither conformed to nor contradicted the official imaginary of developmental nationalism. The technocratic nature of their reasoning, however, failed to advance an alternative sociotechnical vision with more equalized access to epistemic authority and political influence.

Within South Korea, public controversies on science and technology have been, and still are, framed by the dichotomy of authoritarian rule versus political liberalization, or of material versus post-material values. In the light of the Hwang scandal, more attention began to be paid to the role of nationalism in scientific controversies, but only as one external factor distorting the political neutrality of science and technology (Kim et al. 2006). However, a closer look at South Korean disputes over the development and use of science and technology reveals that their political dynamics are more complex than some accounts seem to suggest. Across different techno-economic initiatives, proponents were firmly committed to the vision of developmental nationalism that imagined science and technology as essential supports for South Korea's standing as a powerful industrial nation. Conversely, progressive NGO activists believed that they were engaged in a broader struggle to protect South Korea's as a just, sustainable, and democratic nation from a pro-development alliance among science, technology, the state, and corporate power.

As I have attempted to show in this chapter, the concept of sociotechnical imaginary helps to elucidate these multifaceted processes of coproducing science, technology, and nationhood. The study of sociotechnical imaginaries also has normative implications. Having experienced a series of costly controversies, South Korea devoted greater effort to improving the governance of science and technology. But even that discussion quickly turned technical, focusing on evaluating the effectiveness of a range of mechanisms for public engagement (Yoon 2013). In the meantime, the developmentalist sociotechnical imaginary remained a powerful presence, seeming to become even stronger with the election of Park Geun-Hye—a leader of the conservative New Frontier Party and daughter of Park Chung Hee—as president in

2012. Without critically examining the nature and workings of that imaginary, attempts to fundamentally reform the governance of science and technology seem doomed to fall short. To make progress, we will need more detailed analyses—along the lines of Felt's and Hurlbut's studies in this volume—to understand better how South Korea's dominant sociotechnical imaginary is sustained through practice and performance by the state, the media, industry, and even parts of civil society.

Notes

1. This chapter is based on the research funded by the National Science Foundation (NSF Award No. SES-0724133) and subsequently supported by the National Research Foundation of Korea (NRF-2008-361-A00005). I thank Sheila Jasanoff for her constant and invaluable guidance, advice, and encouragement.

2. In South Korea, the term "progressive" has historically been associated with various traditions of radicalism including Marxism and has been distinguished from "liberal." More recently, however, its meaning has been broadened to include more diverse ideological orientations.

3. The new act also specified that high-level radioactive waste would not be stored in the facility. It should be noted, however, that the risk difference between low-/intermediate-level and high-level waste was not perceived as a major issue by residents in the Buan County and did not play a central role in the referendum (Rho 2006).

4. These activists were well aware of the activities of NGOs opposing new biotechnology abroad—for example, Greenpeace, GeneWatch UK, the Council for Responsible Genetics, the Rural Advancement Foundation International (later, the ETC Group), Our Bodies Ourselves, and others.

5. In fact, the South Korean government did not enforce a legally binding ban on the import of US beef, but made a "voluntary export restraint" agreement with the United States. Accordingly, brain, eyes, spinal cord, and skull were still not recognized as SRMs.

6. The People's Solidarity for Participatory Democracy is one of the largest and most influential progressive NGOs in South Korea. Park was elected as an independent candidate supported by a liberal-progressive coalition that included not only the major opposition Democratic Party but also progressive/left-wing political parties and many social movement activists. It is noteworthy that, as of 2012, the annual budget for the city of Seoul was 21.8 trillion won (20 billion USD), nearly 7 percent of South Korea's total budget.

References

Branscomb, Lewis M., and Young-Hwan Choi, eds. 1996. *Korea at the Turning Point: Innovation-Based Strategies for Development*. Westport, CT: Praeger.

Eckert, Carter. 1996. "Total War, Industrialization, and Social Change in Late Colonial Korea." Pp. 3–39 in *The Japanese Wartime Empire, 1931–1945*, edited by Peter Duus et al. Princeton, NJ: Princeton University Press.

Ha, Dae-Cheong. 2012. *Globalization of Risk, Risk of Globalization: Mad Cow Disease Controversy in Korea*. PhD diss., Program in History and Philosophy of Science, Seoul National University (in Korean).

Han, Jae-Kak, and Byungsoo Kim. 2000. "Technological Development in Bio-Industry: Civic NGOs' View." *Science and Technology Policy* 125: 51–61 (in Korean).

Hecht, Gabrielle. 1998. *The Radiance of France: Nuclear Power and National Identity after World War II*. Cambridge, MA: MIT Press.

Hong, Seongtae. 2008. "The Candle Assembly and Democracy." *Economy and Society: Journal of the Korean Critical Sociological Association* 80: 10–39 (in Korean).

Huh, Donghyun. 2006. "The Korean Courtiers' Observation Mission's Views on Meiji Japan and Projects of Modern State Building." *Korean Studies* 29: 30–54.

Hwang, Woo Suk et al. 2004. "Evidence of a Pluripotent Human Embryonic Stem Cell Line derived from a Cloned Blastocyst." *Science* 303(5664): 1669–1674.

Hwang, Woo Suk et al. 2005. "Patient-specific Embryonic Stem Cells derived from Human SCNT Blastocysts." *Science* 308(5729): 1777–1783.

Jang, Moon Seok, and Sang-Rok Lee, eds. 2006. *Reading Dictatorship at the Border of Modernity: Mass Dictatorship and the Park Chung Hee Regime*. Seoul: Green-bi (in Korean).

Jasanoff, Sheila. 2005. *Designs on Nature: Science and Democracy in Europe and the United States*. Princeton, NJ: Princeton University Press.

Jasanoff, Sheila, and Sang-Hyun Kim. 2009. "Containing the Atom: Sociotechnical Imaginaries and Nuclear Power in the United States and South Korea." *Minerva* 47(2): 119–46.

———. 2013. "Sociotechnical Imaginaries and National Energy Policies." *Science as Culture* 22(2): 189–96.

Jurenas, Remy, and Mark E. Manyin. 2010. *US-South Korea Beef Dispute: Issues and Status. CRS Report for Congress RL34528*. Washington, DC: Congressional Research Service.

Kang, Minha, and Jiho Jang. 2013. "NIMBY or NIABY? Who Defines a Policy Problem and Why: Analysis of Framing in Radioactive Waste Disposal Facility Placement in South Korea." *Asia Pacific Viewpoint* 54(1): 49–60.

Kang, Yang-Gu, Byoungsoo Kim, and Jae-Kak Han. 2006. *Silence and Fanaticism*. Seoul: Humanitas (in Korean).

Kim, Bo-Hyeon. 2003. "*Sasanggye*'s Theory of Economic Development, How Different Was It from That of the Park Chung-Hee Regime?: Developmentalism against Developmentalism." *Political Critique* 10: 345–80 (in Korean).

Kim, Chul-Kyoo, and Sung-Ik Cho. 2004. "The Structure and Dynamics of Social Conflict around Nuclear Waste Facility: Focusing on Buan Struggle." *Economy and Society: Journal of the Korean Critical Sociological Association* 63: 12–39 (in Korean).

Kim, Eun Ju. 2011. "Gyeongju Radioactive Waste Disposal Facility and the Life of Local Residents: Competing Discourses and Change of 'Locality.'" *Journal of Regional Studies* 19(3): 21–46 (in Korean).

Kim, Eun-Sung. 2012. "Technocratic Precautionary Principle: Korean Risk Governance of Mad Cow Disease." *Journal of Risk Research* 15(9): 1075–100.

Kim, Geun Bae. 2005. *The Emergence of Modern Korean Scientific and Technical Manpower*. Seoul: Munhak kwa Chisŏngsa (in Korean).

Kim, Jongyoung. 2013. "Construction of Oppositional Knowledge: Hybrid Expert Alliance and Development of Oppositional Logics in the 2008 Candlelight Movement." *Korean Journal of Sociology* 45(1): 109–53 (in Korean).

Kim, Pan Suk. 2012. "Advocacy Coalitions and Policy Change: The Case of South Korea's Saemangeum Project." *Administration and Society* 44(6 Suppl): 85S–103S.

Kim, Sang-Hyun. 2014. "The Politics of Human Embryonic Stem Cell Research in South Korea: Contesting National Sociotechnical Imaginaries." *Science as Culture* 23(3): 293–319.

——. 2012. "Historical Genealogy of South Korea's Sociotechnical Imaginaries." Paper presented to the International Workshop: Intersections with Science, Medicine, and Technology in Korea, Binghamton University, October 20, 2012.

Kim, Se-Kyun, Kab-Soo Choi, and Seongtae Hong, eds. 2006. *The Hwang Woo Suk Affair and Korean Society*. Seoul: Nanam (in Korean).

Korea Democracy Foundation. 2008. *The History of Democratization Movement in Korea*, vols. 1–3. Seoul: Dolbegae (in Korean).

Korea Hydro and Nuclear Power Co. 2013. White Paper on Nuclear Power. Seoul: Ministry of Knowledge Economy (in Korean).

Korea Institute of Nuclear Safety. 2000. *Ten-Year History of the Korea Institute of Nuclear Safety*. Daejeon: Korea Institute of Nuclear Safety (in Korean).

Korea Pollution Research Institute. 1987. *Pollution Research No. 16: Nuclear Power and the Korean Peninsula*. Seoul: Korea Pollution Research Institute (in Korean).

Korean Academy of Science and Technology. 2009. *Voice of the Korean Academy of Science and Technology* No. 1 (in Korean).

Korean Nuclear Society. 2010. *Korean Nuclear Power: A Fifty-Year History*. Seoul: Korean Nuclear Society (in Korean).

Lee, Hye-Kyung. 1999. "Biotechnology and Civic Movements." Pp. 322–48 in *The Paradox of Progress*, edited by the Center for Democracy in Science and Technology. Seoul: Dangdae (in Korean).

Lee, Sang Rok. 2007. "Critical Intellectuals' Perspectives on the Modernization of South Korea during Park Chung-Hee's Regime." *Critical Studies on Modern Korean History* 18: 215–51 (in Korean).

Lee, Seung-Ook, Sook-Jin Kim, and Joel Wainwright. 2010. "Mad Cow Militancy: Neoliberal Hegemony and Social Resistance in South Korea." *Political Geography* 29: 359–69.

Lee, Young-Hee. 2013. "An Evaluation of Participatory City Planning of Seoul: Focused on Seoul Plan." *Economy and Society: Journal of the Korean Critical Sociological Association* 98: 106–33 (in Korean).

Ministry for Food, Agriculture, Forestry and Fisheries and Ministry for Health, Welfare and Family Affairs. 2008. *Mad Cow Disease Scare: Ten Questions and Answers* (in Korean).

Ministry of Science and Technology. 1976. *Science and Technology Annual* (in Korean).

Mizuno, Hiromi. 2009. *Science for the Empire: Scientific Nationalism in Modern Japan*. Palo Alto, CA: Stanford University Press.

Nocut News. 2006. 70% of the Public Want to Give Hwang a Second Chance. January 16 (in Korean).

Paik, Young Gyung. 2012. "Stem Cell for the Present: Reconfiguration of Stem Cell Research, Ethics and Bio-Industry in South Korea after the Hwang Scandal." *Journal of Science and Technology Studies* 12(1): 185–207.

Park, Jae-Mook. 1995. *Locally Based Anti-Nuclear Movements and Citizen's Participation: A Comparative Analysis on Four Cases of Anti-Nuclear Facilities Movements*. PhD diss., Department of Sociology, Seoul National University (in Korean).

People's Council against Mad Cow Disease (PCAMCD). 2008a. *Press Release: Launch of People's Council against Mad Cow Disease* (in Korean).

People's Council against Mad Cow Disease (PCAMCD). 2008b. *Proceedings: Forum on the Scientific Truth about Mad Cow Disease and the Korean Society's Countermeasures* (in Korean).

Prakash, Gyan. 1999. *Another Reason: Science and the Imagination of Modern India*. Princeton, NJ: Princeton University Press.

Rho, Jin-Chul. 2006. "Risk-Communication around Deciding the Place for Radioactive Waste Disposal Site, and Self-Decision." *Economy and Society: Journal of the Korean Critical Sociological Association* 71: 102–25 (in Korean).

Rostow, Walter W. 1960. *The Stages of Economic Growth: A Non-Communist Manifesto*. Cambridge: Cambridge University Press.

Shin, Gi-Wook. 1995. "Marxism, Anti-Americanism, and Democracy in South Korea: An Examination of Nationalist Intellectual Discourse." *Positions: East Asia Cultures Critique* 3 (2): 510–36.

Shin, Hyang-Sook. 2009. "The Advent of Genetic Engineering Promotion Act in Korea the 1980s." *Korean Journal for the History of Science* 31 (2): 475–504 (in Korean).

Yonhap News. 2008. The Outcomes and Achievements of Korea-U.S. Summit Meeting. April 20 (in Korean).

Yoon, Su-Jin. 2013. *Science and Technology Policy and Citizen Participation*. Seoul: Korea Institute of S&T Evaluation and Planning.

EIGHT

Building from the Outside In: Sociotechnical Imaginaries and Civil Society in New Order Indonesia

SUZANNE MOON

Introduction

Sociotechnical imaginaries frequently gain their authority because they are backed by an institution capable of wielding considerable resources, usually the state. Considering civil society organizations, however, offers the opportunity to explore the role of sociotechnical imaginaries in opposing, challenging, or redirecting the priorities and privileges of a state-driven social order. To do so may require attention to works of dramatically smaller scope than nationwide technological initiatives, for the simple reason that civil society organizations may not be able to muster the resources to implement large projects or their work may be distributed across multiple small projects. Yet the smaller scale of individual technological projects should not blind us to their power to assert a materialized alternative to the status quo or to promote a collectively imagined form of social life and social order that is the hallmark of the sociotechnical imaginary (see, e.g., Barker; Felt, this volume). Individual activities may coalesce into movements that sustain themselves because of, rather than despite, their distributed nature.

As sociotechnical imaginaries are taken up by civil society organizations, sometimes modified or bolstered by outside ideas yet still retaining the core vision that allows them to be recognizable, they can both rhetorically assert desirable social futures and demonstrate the material possibility of such through technological projects. In the case of a society like Indonesia's New Order, in which conventional political action was highly constrained by authoritarian leadership and economics tightly tied to political order, technical projects became a way of insisting that practical and meaningful alternatives to the status quo, both economic and political, were available.

Studying sociotechnical imaginaries may therefore help us to properly read certain forms of civil action and to consider the material ways that civil society actors carry a message of reform through society.

This chapter explores the operation of sociotechnical imaginaries as forms of critique in postcolonial Indonesia, especially those directed against the economic and political logic of Suharto's New Order. As such, they articulated not just alternative technical arrangements but a challenge to arrangements of power and influence within the New Order developmental state. These critiques are explored through Hasan Poerbo's community-based development projects in the city of Bandung during the late 1970s and 1980s. Poerbo, an architect and professor at the Institute of Technology in Bandung (ITB), became deeply involved in a number of urban projects in the city that aimed to produce greater economic equity by including frequently marginalized people into technological activities fundamental to the smooth functioning of their neighborhoods and cities. In contrast to the New Order linkage of political control with the centralized operation of technological development, Poerbo's proposals and projects distributed both control and important sections of technological work away from centers of power. Poerbo's work engaged a sociotechnical imaginary also seen in the cooperative movement promoted and encouraged by Muhammad Hatta, a respected politician and economist, whose thinking about economics, technology, justice, and civic responsibility became a core resource for political activists opposed to the New Order (Noer 2002).

Projects like Poerbo's, operating at a local scale with consequences that may be difficult to fully trace, are usually dwarfed in discussions of technological development by the powerful and clearly discernable effects of Suharto's New Order, with its dramatic uptick in industrial production and its centralized organization. Edward Aspinall characterized many nongovernmental organizations (NGOs) organized to pursue community development as a political refuge or retreat from Suharto's repressive authoritarianism, when overt political action was no longer possible (Aspinall 2005, 95–56).

For Aspinall economically oriented projects sidelined "real" politics (such as the formation of parties and study groups or public demonstrations) because they too easily legitimated the New Order by helping to resolve problems of poverty without clearly rejecting the policies that helped to sustain those problems. In 1992, William Liddle had a broader take on the value of civic activism, arguing that the international orientation of activist groups, as well as their ability to "cut across or combine in new ways the old ethnic, religious and class cleavages of Indonesian society," made them a notable force to support democratic transition; time has shown that observation to be true (Liddle 1992, 459). But what of the practical, technological, and economic projects that many activists engaged in? Were they truly as politically empty as Aspinall implies? Neither Liddle nor Aspinall offers much insight into how these more bottom-up projects contributed to the political environment of the New Order.

Although many NGOs and foreign-funded antipoverty projects did indeed operate in cooperation with the government, a close look at the sociotechnical logic of their work reveals a political project with more teeth than is suggested by the existing literature. A sociotechnical imaginary linking technical and economic activity with ideas about citizenship and justice developed through the influence of Muhammad Hatta in the earliest years of Indonesia's postcolonial period and became over time a blueprint for the coproduction of a social and technical order that challenged New Order thinking in important ways.

Imagining a Cooperative Economy

The sociotechnical imaginaries at work in the 1970s and 1980s in Indonesia are best comprehended by looking to earlier Indonesian history, particularly the influence of Muhammad Hatta, who articulated a vision of postcolonial Indonesia whose endurance is crucial to this story (Hatta 1972). Since the 1920s and 1930s, colonial Indonesia's political parties divided on many issues, including economics (Shiraishi 1990; Legge 1972, 109–29). Yet they shared an understanding that Indonesians suffered primarily because of exploitative Dutch economic practices, which were themselves facilitated by the lack of political rights and respect accorded to Indonesians (Elson 1984; Shiraishi 1990; Moon 2007). Anticolonial activism was therefore always political-economic, intertwining issues of citizenship, political representation, and economic activity and participation in society (Boeke 1946; Hatta 1972).

After the revolution in 1945, the question of how to construct a viable and just alternative to the colonial past was heavily contested, as became evident in the crafting of the Indonesian constitution (Constitution of the Republic of Indonesia 1945; McIntyre 2005; Ricklefs 2008). Sukarno, an activist who would later become president, produced a compromise that the Islamic, communist, democratic, and social democratic political groups could accept. The constitution, based on the agreed-on principles of the Pancasila,[1] called for a democratic state in which deliberation among representatives would produce consensus (Legge 1972, 181–239). The constitution further called for state ownership of industries vital to public welfare, with the rest of the economy based as much as possible on cooperative principles—a capitalist system that nonetheless embraced collective action on the part of ordinary people.

Muhammad Hatta, a prominent revolutionary himself, was ideally placed to help this diverse group reach consensus. A devout Muslim, he nevertheless supported a secular state; he believed that individual rather than state enterprise should be the basis of economic growth, and he saw in cooperatives the best combination of collective action and reward for individual initiative (Hatta 1954a). Hatta spent the rest of his life tirelessly campaigning via speeches on radio and in person, in education programs, and in his writings, to convince Indonesians to form cooperatives, to teach them the principles of cooperative capitalism, and to link economic and technical organization to social justice and responsible citizenship in postcolonial society.

Hatta's cooperative principles emerged from the anticolonial commitment to widely shared economic prosperity as the foundation of a just society. Although framed in terms of economics, the colony's technical organization was frequently invoked to explain the workings of its injustice. Dutch businesses employed expensive, imported machines, while few Indonesians could afford to do the same. Indonesian businesses were more likely to operate artisanally in fields like batik making that relied on simple tools and skilled labor. Popular thinking aligned the Dutch with foreign, capital-intensive industries and Indonesians with small and labor-intensive craft production. Whatever the accuracy of such stereotypes, they show that worries about economic equity were implicitly also concerns about the sociotechnical character of Indonesian society.

Over time, Hatta produced a powerful vision to help Indonesians see an alternative to colonial economics. In 1934, after being released from prison where he had been interned for anticolonial agitation, he expressed disillusionment with politics as usual and the infighting among Indonesia's anticolonial political parties. He turned to economics rather than political ideology to bring about Indonesian independence. He dreamed of creating an autonomous space for indigenous economic activity within the colony, one in which Indonesian technical abilities and creative energies would be freed from dependence on the Dutch. Writing in *Daulat Ra'jat* ('the sovereignty of the people'), a newspaper he coedited with Sutan Sjahrir, Hatta proposed that Indonesians create model communities outside the centers of Dutch control on Java, in which Indonesians could disengage from traditional exploitative relationships while they built a new society—both socially and technologically (Hatta 1934). The first group of emigrants would collaborate to build homes, break ground for cultivation, and plant crops. The next wave of immigrants would join in home building, agriculture, and other essential work for the young village. Each generation of migrants would prepare for the next, building the schools, hospitals, and other elements that would collectively transform a settlement into a functioning community.

Hatta imagined seeding such communities throughout the colony, creating an independent Indonesia not from rhetoric, and ideological violence, but from collective sociotechnical initiatives designed to produce stability, justice, and equitable opportunities. Indonesians, he said, would "grow from their own strength" (Hatta 1934). The material dimension was essential. The collective work involved in physically constructing the community and its array of productive technological activities created social coherence. Such technological activity was a form of civic engagement that had both political and economic ends. The combination of individual initiative with group responsibility became the foundation of Hatta's later advocacy for Indonesian cooperatives.

Hatta's communities never came to fruition. A few days after the article was published, he was arrested and sent to Boven Digul, an internment camp in West Papua (Mrázek 2009). Hatta wrote with particular eloquence, but his ideas were not entirely idiosyncratic. Other activists called for similar, if less utopian, autonomous economic action as the basis for a peaceful yet effective response to colonialism (Moon 2007). These notions of anticolonial action were ultimately rendered moot in the face of Japanese invasion and the subsequent war for independence. But Hatta kept alive the idea of building civic and economic virtues in self-governing cooperatives as an appealing path toward a just postcolonial society.

Hatta was able to win a constitutional commitment to cooperatives after Indonesian independence because it promised a truly inclusive economy built on the principle of "family spirit" and "gotong-royong" or community cooperation, appealing across political divisions and resonating culturally (Kahin 1952; Sutter 1959, 238–9). Ordinary Indonesians embraced cooperatives, which multiplied dramatically in the 1950s, yet few leading politicians did more than pay lip service to them. Sukarno treated cooperatives more as an adjunct than as foundational to the economy itself. But in promoting and teaching about cooperatives, Hatta articulated for a wide audience a vision that emphasized the importance of the cooperative activities of ordinary "small people," both for the spread of prosperity and for the creation of an ethos of individual action and collective responsibility that served both economic and political needs (Hatta 1952, 33).[2] "It is only in co-operatives that solidarity and individuality can come to flower in conditions of harmonious relationship. By continually keeping alive this solidarity and individuality, the co-operative movement instills in the human breast a feeling of social responsibility" (Hatta 1952, 32). Cooperatives operated not just as pragmatic systems of production but as technologies of social justice, creating a habit of justice and civic responsibility in the fabric of daily life. Just society therefore would flow not only from the policies and directives of government down but also from the bottom up, through the individual and collective economic behaviors of ordinary people.

Hatta defined the cooperative as a site of emergence for proper economic and civic behavior. Participation in cooperatives was therefore a more generative kind of economic participation than working in a conventional business. Cooperatives encouraged what Indonesians call *swakarya*—autoactivity or initiative—because they allowed capital-poor Indonesians to pool their resources and skills and establish their own goals and methods of work, rather than having those dictated by someone else.[3] The cooperative ideal called for members actively involved in decision making, giving scope for and rewarding initiative and creativity. Participation built not only innovative material outcomes but self-respect, helping Indonesians to conquer their colonially induced "inferiority complex" (Hatta 1952, 31; 1955). Hatta's project was sociotechnical because he insisted that one had to simultaneously build infrastructure and self-confidence, industry and social harmony, in order to have widely shared prosperity. If dependency rather than autoactivity came to the fore, no matter how many factories they might build, Indonesians were in danger of being "a nation of coolies and a coolie among nations" (Hatta and Yasni 1981, 84).

Although Hatta backed credit and purchasing cooperatives as useful economic tools, his larger goals required production cooperatives, which would collectively direct technological and economic development in ways that responded to Indonesian goals and values. Hatta hoped to transform the nation from a producer of mainly raw materials to a producer of finished goods, but his framework notably left decision making to cooperatives themselves. Industrialization might go faster by importing factories and factory managers from abroad, but doing so would simply reinvent colonial systems of inequity. Hatta argued that industries grown from small-scale cooperatives were superior because they would transition from home-based production to larger scale, factory organization only as individuals consensually "in the spirit of the family" deemed it appropriate to do so. Hatta's was in some sense a democratic vision of industrialization in that ordinary people would shape the outcome of Indonesia's industrial change by their own actions and according to their own values (Hatta 1952, 53–67). In turn, the habits of cooperation and mutual care, and the daily practices of consensus building, would translate to the political realm, as people behaved in public life in ways reflecting the lessons of the cooperatives. As drivers of both prosperity and social responsibility, cooperatives were, for Hatta, "a good element for strengthening economy and morality" (Hatta 1952, 31).

The technological outcome of Hatta's cooperative industrialization was purposely underspecified; yet, that it would be newly technological was central to his thinking. The shape of the technological future—whether improved artisanal methods, mass production technologies, local inventions, or foreign imports—would be determined not by a small group of elites but by the collective engagement of many individuals Technological change would emerge not as a forced march toward a predetermined goal but as an exercise in democratic trail blazing (Hatta 1954a). In *Djalan Keekonomi dan Kooperasi*, the expansion of batik cooperatives from artisanal to industrial-style production, exemplified a home-grown success story (Hatta 1954a, 145–53). When multiplied across many cooperatives, this work would drive the Indonesian economy away from its dependent position, effectively and logically controlling industrial expansion from below: "The Indonesian community should form a living organism, a living machinery of wheels within wheels, in which each wheel causes other wheels to turn. . . . It is the aim of democracy that each section of humanity, large or small, take care of its own interests with a feeling of responsibility to all. Each group should be able to take the initiative in providing the common good on the basis of one for all and all for one" (Hatta 1952, 31).

Indonesia under Sukarno's leadership saw neither the social harmony nor the economic growth Hatta hoped for. In the 1950s, as Indonesia struggled with inflation that resulted from wartime disruptions and falling prices for key exports, intense political partisanship developed both at the level of high politics and in the lives of ordinary people, with powerful antipathies developing between members of Muslim parties and Sukarno's Communist Party (Hasyim 2007, 16–17). Sukarno egged along these divisions, hoping to gain more power for the presidency than the original constitution allowed.[4] As his biographer J.D. Legge (1972, 276) argued, Sukarno was "manipulating these divisions in order to advance his political preferences rather than make a genuine attempt to conciliate opponents." Yet even Sukarno could not control the consequences of his actions. His "Guided Democracy," introduced in 1957 to calm political divisiveness by eliminating the election of parliamentary representatives and substituting an appointed cabinet, failed to repair the deepening fissures within society

(Legge 1972, 279–310). Hatta lamented this angry politicization, including the politicization of cooperatives. He resigned from the vice-presidency in protest, winning the respect of many Indonesians.

In contrast to Hatta, Sukarno focused on building large industries and gave little attention to cooperatives. Sukarno viewed economic equity as satisfied by guaranteeing that business in Indonesia was run by Indonesians. He borrowed extravagantly to build the economy without foreign investment, exacerbating the declining economic situation. By 1965, inflation had reached 600 percent per year, and Sukarno's political gamesmanship had produced a culture of corruption and deep-seated public anger. By the mid-1960s, the country was mired in economic failure and political crisis (Legge 1972; Elson, 2001).

The New Order: New Visions of Political Order and Industrial Development

Sukarno's presidency ended in 1965, after the mysterious assassination of five generals, which was blamed on Sukarno's Partai Kommunis Indonesia (PKI).[5] The deep polarization of society exploded in the anticommunist massacre of 500,000 people that took place at the behest of then General Suharto, who took control of government after the assassinations and who asked the military to "take care" of the "communist problem" (Elson 2001, 120–66). The political divisions that had festered under Sukarno erupted, producing a shattering violence of ideology and retribution. Suharto called his government the "New Order" to distinguish it from Sukarno's disreputable rule, responding to the debacles of 1965 by prioritizing political quiet and economic growth, pursuing both ends aggressively (Elson 2001).

Central to Suharto's strategy to restore order to the nation was his plan for economic development. As scholars have frequently noted, Indonesia's development efforts were as much about restoring political order as they were about improving economic conditions (Vatikiotis 1998; Elson 2001). Suharto encouraged Indonesians to focus on the work of development and to turn away from participation in politics outside of the three officially approved, and tightly controlled, political parties. In the late 1960s, this approach was not uncongenial for many Indonesians, who were tired of the chaos under Sukarno and horrified by the events of 1965. As Taufik Abdullah recalled in 2007, "Catchwords like 'development,' 'program,' 'economy,' 'democracy,' and social justice not only became the subjects of discussions, but also formed the mantra Indonesians believed would help heal social and political ills" (Abdullah 2007, xxi). Universities ran symposia and seminars suggesting new policies for the future, focusing on economics, not politics (Simposium Kebangkitan 1966; Symposium Pembangunan 1966; Perumusan Hasil Seminar 1967). Initially the intelligentsia happily focused on constructive solutions to Indonesia's economic problems rather than political criticism. "Gone were the days of pseudo-socialistic economy and the never-ending ritual of cursing neo-colonial and capitalist exploitation. The topic of the day was: How best to carry out the development programs?" (Abdullah 2007, xxi). This move defined civic engagement as economic engagement but stripped ordinary people of their political voice, rather than, as Hatta might have preferred, creating a public ethos of social responsibility. Nevertheless, development, or *pembangunan*, became the focus of national planning and identity making (Heryanto and Lutz 1988, 1–24; Budiman 1982).

The sociotechnical imaginary that informed official Indonesian development followed in general outline the modernization precepts common in international, especially American, academic circles (Gilman 2003). Following the counsel of a team of Berkeley-educated Indonesian economists (known as the "technocrats"), Suharto's first five-year plan aimed to increase the percentage of the economy devoted to industry as compared with agriculture (Departemen Penerangan Republik Indonesia 1969). Because Indonesia was nearly bankrupt, Suharto appealed to foreign investors to provide capital and expertise, wage labor opportunities for ordinary Indonesians, and skills training for a few higher level managers and engineers. Modernization theory promised that this industrial transformation would eventually reward all of society with higher living standards and, as time went by, a richer, more diverse supply of consumer goods with which to construct new ways of living. Indonesia's actual experience, if more mixed than the promises of modernization theory, certainly did have some of these attributes, especially for those who moved into the middle class (Heryanto 1999, 160–88).

The linkage that Suharto's government created between social control and economic growth defined the New Order (Elson 2001, 148–49). Suharto used economic patronage as a vital lever of political control. He dispensed access to lucrative business opportunities to those who followed his plans, but loans, licenses, and permits could always be taken away at the first sign of political disorderliness. As William Liddle put it, "the philosophy of the New Order, in the minds of its architects was a political order controlled from the top, rather than driven out of control from the bottom" (Liddle 1992). This political order was coproduced with a technical order; large, centralized industries offered fast economic results and served the project of quieting politics.

The New Order government and, for the largest and most lucrative projects, Suharto himself, were the obligatory passage points for those who wanted to be a part of Indonesia's *pembangunan*[6] (Liddle 1992, 278–81). Economic patronage not only tied civilians to the New Order, it also buttressed military support for Suharto's leadership. Because technocrat-authored budgets were too small to support the military, Suharto gave them instead the running of lucrative state enterprises from which they could draw off-budget revenue. Justified as part of the military's dual function (*dwifungsi*) of security and economic development, it allowed favored officers to operate these businesses with relative impunity, so long as they stayed within the boundaries Suharto set, and supported his leadership.[7] Both the oil industry and the sugar industries were run by trusted military generals who helped cement support for Suharto among army officers (Crouch 1978).

This approach had important consequences for the shape of technological change in Indonesia. Unlike the small-business cooperatives that proliferated in the 1950s and 1960s thanks to Hatta's inspiration, the superior accountability (to the New Order) of large, highly centralized organizations (whether Indonesian enterprises or partnerships with large foreign companies) worked in favor of a strongly top-down approach. Although they did not solely dominate Indonesia, a few large industrial conglomerates became miniempires. For example, Ibnu Sutowo, the former army general who ran Pertamina, Indonesia's national oil company, created a conglomerate that also included telecommunications, shipping, and resorts (Moon 2009). Multinational companies in fields like pharmaceuticals, textiles, and mining brought their industrial technologies to Indonesia. New Order officials publicly celebrated large and visible industries as the cornerstone of development, thus promoting their vision for the general public (Moon 2009).

Most importantly, this preference for large industries shaped participation in the project of development. Because patronage networks tended to filter all the way down through these organizations, the ability to be part of "development" depended on a person's access to these networks (Shiraishi 1997). And the power of patronage (i.e., its ability to control social behavior, whether through the pressure of kinship networks or through the more generalized threat of loss of access) helped to ensure that good political behavior reached even into the lower ranks of these enterprises (Suryakusuma 2004). To be sure, especially in the case of the largest organizations, an enormous number of people also gained both employment and skills training, which helped legitimize Suharto's development initiatives in the eyes of the wider public.

Patronage networks also geograpically shaped participation. Many companies established headquarters near Jakarta, easing their access to sources of power, as well as rumor and information about the inside workings of the New Order.[8] Residents of Indonesia outside of Java complained that industrialization was not spreading further across the archipelago (Wie 1988; Hill 1997). Yet even within Java, the combination of Jakarta-centric business operations and the working of patronage networks could produce an unexpected sense of alienation among ordinary people. In the city of Cilegon, for example, which in 1975 became the site for Indonesia's first state-operated steel mill, local residents believed that it was impossible to be employed there unless one was a "foreigner," that is, unless one had connections in Jakarta (Moon 2009, 273–74). That widely shared perception demonstrates how patronage networks in the New Order made development seem less an inclusive national enterprise than an exclusive project governed by insiders.

Critiquing the New Order: Reviving Hatta

By the early 1970s, many had become disillusioned with Suharto's vision of development. Intellectuals writing in scholarly and popular publications called for more attention to continued poverty, especially as it contrasted with the conspicuous consumption of a small yet privileged upper class (Arief 1977; Budiman 1982; Papanek 1976; Rahardjo 1985). Despite programs aimed to improve the circumstances for the poor in Indonesia, including state agencies to control supplies of staples like rice and health clinics, the benefits of development seemed to be accruing only to insiders, mocking the idea, still strongly felt by many Indonesians, that equality of economic opportunity ought to be the central goal of the postcolonial state. That such unhappiness was felt outside of intellectual circles is suggested by the outbreaks of violence aimed at foreigners in 1974 and Indonesian-Chinese in 1973 (who were perceived as enjoying special influence in the New Order; Elson 2001, 204–9).[9] Suharto responded both by reining in the real cause of the riots, elite political struggles, and conceding to public opinion by discouraging conspicuous consumption of foreign luxury items and building hospitals, schools, and mosques in poor areas around the country (Liddle 1992, 449–50; Masenas 1974). Such actions might have been welcome but failed to obscure the ways that the "small people" were left out of development as either actors or beneficiaries. Yet activists interested in economic equity and social justice faced grave personal risks if they engaged in conventional forms of political action, as Suharto cracked down on behaviors that threatened to ignite mass politics (Elson 2001; Aspinall 2005).

For critics of the New Order in this environment of disillusionment,

Hatta's grassroots philosophies of economic action, technological creativity, and civic development articulated a compelling alternative to politics as usual (LP3ES 1995). Hatta had continued offering educational programs and speeches about cooperatives after his resignation as vice-president. And despite the collapse of some cooperatives in 1965, Indonesians themselves, especially small-business people, formed them anew, collectively affirming the value of cooperatives even in a drastically changed political and economic environment (LP3ES 1981). Hatta also trained graduate students in economic development, especially the economic development of the islands outside of Java that the central government was frequently accused of ignoring. Just as he had in the early 1930s, Hatta turned away from the stalemates and failures of conventional political action and focused instead on the pragmatics of producing greater equity and material change on the ground, an approach he passed on to his students (Yasni 1968). Critics of the New Order, frustrated by the limits on political action and the direction of New Order development alike, found in Hatta's ideals a powerful critique of the emerging political and economic system. For example, when the NGO LP3ES (Lembaga Penelitian, Pendidikan dan Penerangan Ekonomi dan Sosial, or the Institute for Social and Economic Research, Education and Information) critically analyzed the state of Indonesia's cooperatives in the early 1970s, they pointedly compared the actions of contemporary cooperatives with Hatta's model, arguing that the state's heavy-handed regulation had effectively negated the ability of cooperatives to operate democratically and independently, rendering them ineffective for producing significant social development (Arifin and Nasution 1981; Djojohadikusomo 1981; LP3ES 1981; Swasono 1985). They argued for a return to Hatta's original vision, and implicitly his model of the independent (of the state) cooperative as an incubator of civic virtues.

Yet criticizing state policies was not enough. From the early seventies onward, the appeal of action, helping the disenfranchised instead of waiting for politics to sort themselves out, drove numerous groups and individuals to take up small projects on the ground. These projects focused on the structure of participation (that is, who would participate in what and under what conditions), the promotion of *swakarya* and individual responsibility, and the technological and economic activities that could make a difference in the lives of ordinary people. This activism drew heavily on both Hatta's vision and the on-the-ground working reality of cooperative enterprise. Their work gave that vision new life, creating an alternative sociotechnical imaginary in changed conditions. Although the dream of an economy built entirely on a cooperative basis was no longer realistic, the activists committed instead to defining a place in development for those ordinary people. In doing so, they made it clear that the New Order was leaving many behind, a stinging critique of its marginalization of vast numbers of people and the marginalization of the vision of the postcolonial state as a creator of widespread economic opportunities.

Because efforts were diffused across many groups, and visible sometimes more in plans than in actions, tracing the evolution of this imaginary can be difficult. Yet these ideals pop up again and again. For example, in the early seventies LP3ES worked with the Ford Foundation in the city of Cilegon, where the flagship Krakatau Steel plant was being constructed. Their project was to identify ways to help small craft workers and business people—carpenters, food stall operators, bricklayers, and the like—to benefit from the new steel industry. The plant itself would provide relatively few local jobs; LP3ES therefore worked to encourage creativity and initiative among ordinary business people so they could nevertheless gain something from this major industry (LP3ES 1975). It is easy to see how such actions could be understood as essentially collaborative with the New Order. Yet by emphasizing the need to create a place in the industrial economy, LP3ES underscored the ways industrial plans failed to actually spread benefits on their own terms.

Hasan Poerbo: Mass Housing and the Ethics of Participation

To see more clearly how Hatta's vision was rearticulated as a sociotechnical imaginary in this period, consider the work of Hasan Poerbo, an architect who spent his career developing projects that would help bring the poor and disenfranchised more effectively into the project of development. In March 1979, Poerbo, then a professor in the Department of Architecture at the ITB, became interested in the plans for the construction of mass housing for the city of Bandung. He published an article in the public affairs journal *Prisma*, published by LP3ES, that spelled out an alternative to the plans proposed by the government housing agency, PERUMNAS (Poerbo 1979, 3–13). The housing projects had been proposed in the wake of the unrest of the midseventies, constituting one of the New Order's responses to the growing disparities in wealth within Indonesia. The planners drew on European mass housing techniques to improve living conditions in the poorer sections of Indonesian cities, especially the crowded cities of Java. (Dunleavy 1981; Power 1997). As the name suggests, mass housing is constructed inexpensively and rapidly using industrial practices, including the use of standardized, mass-produced building components and designs, rationalized on-site building techniques, and specialized building equipment tailored to simplify the assembly of the elements. Mass housing reduces costs by shifting significant parts of the construction process from the building site to the factory.

The mass housing approach came to Indonesia through the United Nations, which had made the country a regional center for the study of mass housing in hot and humid climates (Larson 1958; Lembaga Penyelidikan Masalah Bangunan 1972). It had real appeal for developing nations: costs could be kept under control more easily than in traditional construction because of the smaller number of contracting firms involved, and locally sourcing basic materials in large amounts provided a ready market for domestic concrete and steel industries. New Order development plans looked for such linkages; the business model for the Krakatau steel plant, for example, relied on growing domestic demand (Rombe 1970).

Poerbo appreciated the inexpensive yet reasonable quality construction made possible by the mass housing approach as a boon for poor neighborhoods. But he objected to the exclusion of local workers' skill in building trades by the industrialized process of construction. For Poerbo, the issue of local technological participation in rebuilding these neighborhoods was at least as important as the distribution of new and improved infrastructure. The "informal sector" of skilled workers and small businesses included self-employed craftsmen and businesses run solely with family labor, working in established trades like brick making and bricklaying, carpentry, and tile making. Poerbo argued that mass housing projects should be modified to offer local employment to these groups, rather than disenfranchising them even further from the Indonesian economy (Poerbo 1979, 1980; Boenders and Poerbo 1981).

Poerbo therefore focused on how to integrate these informal workers into the mass housing model. The diversity of small, unlicensed builders created numerous challenges. The standards of quality they applied to their work were uneven and unpredictable. Many lacked training in building or business management. It was not unusual for small builders to simply disappear in the middle of a project if they encountered an intractable problem of technique or finance. Such unpredictability made the efficiencies and control of the mass housing model impossible to achieve. Those who did quality work reliably still faced difficulties. Because they lacked capital, they could not quickly purchase the required equipment or learn the specialized techniques involved.

Yet Poerbo insisted that these problems be tackled, not avoided. He criticized the "preoccupation with efficiency" that pushed the needs of the people's economy into the background (Poerbo 1979, 9). He noted that the Indonesian government had in the past invested in labor-intensive projects to provide jobs during difficult times; why not justify mass housing on a similar basis? Poerbo proposed to modify both the housing project and the practices of builders to form a compromise that instilled the industrial efficiencies of modern construction practices in small businesses and reoriented the efficiency-mindedness of PERUMNAS toward people-mindedness.

Poerbo first asked that modular building components be rethought to take better advantage of local materials and construction skills, mentioning a United Nations Industrial Development Organization project to improve the quality of traditional materials, to make them more readily usable in industrialized building (Poerbo 1979, 10). Getting the informal sector to adopt those materials and the industrial sector to use them posed difficulties. Poerbo suggested that informal sector builders form cooperatives that would in turn train workers in new technologies, materials, and modern business practices, helping them amass the skill and capital needed to be taken seriously by the industrial sector. The industrial sector itself should prioritize the use of, and learn how to successfully interact with, these cooperatives.

A comparison between Poerbo's plan and the logic of that followed by the government shows the former's wider significance. In the New Order, a project like this was meant not only to generate inexpensive infrastructure but also to function as a site of patronage, more easily accountable to networks of political and economic control than were countless smaller, traditional contractors. Consequently, the opportunity for technological participation was dictated by the networks and location of these centralized organizations. Companies located in Jakarta and those with connections, however distant, to important power brokers in the New Order would have the most to offer, keeping jobs and professional opportunities within their networks.

Poerbo's plan, in contrast, sought to maintain the benefits of industrial housing while shifting both employment and entrepreneurial growth to people local to the building sites. Both immediate work and the opportunity to gain technological skills would circulate wherever the housing was to be built, instead of in Jakarta. By prioritizing community participation, the geographical area of housing improvement became a site for deeper changes to technological and economic practices. Opportunities to participate in development would circulate outside the circles of influence in Jakarta. In a modest but not insignificant way, the proposal aimed to redirect the highly centralized logic of connections and geography that operated in the New Order, thus offering a plan for development that was inclusive and expansive. While there is nothing to suggest that Poerbo imagined such a project would revolutionize the political organization of the New Order, it is clear that his proposal would disrupt the circuits of power and privilege in favor of promoting wider participation. Had Poerbo's plan been adopted, development across the country would have become more than merely the sudden appearance of new infrastructure. It would have become an opportunity for active, local technological engagement with the making of that infrastructure, pulling control away from the center.

Reintegrating the Marginalized: A Sanitation Project in Bandung

One of the challenges facing Poerbo was the significant difference in the situation facing Indonesia by the early 1980s as compared with the early 1950s or even the 1930s when Hatta formulated his vision. Hatta had envisioned cooperatives as the foundation of economic development. By the 1980s, Indonesia was following a clear course of top-down, foreign investment–driven industrialization with cooperatives operating mainly at the margins. Did Hatta's thinking still have relevance in this environment? For Poerbo, members of LP3ES, and others, the issue had become the creation of a new economy built not on indigenous strength but on efforts to widen the base of participation in the industrial economy and the decentralizing of the benefits of that economy in the process. The sociotechnical order that was envisioned, one in which the promises of widespread economic opportunity were paired with the hope for responsible civic engagement, remained the same. But the vision for achieving that end had to change. Poerbo drew on internationally circulating discourses about appropriate technologies and participatory development to redeploy and justify Hatta's vision in this changed environment.

The intermediate or appropriate technology movement, inspired by E.F. Schumacher's seminal book *Small Is Beautiful*, questioned the ethical foundations of industrial society and the assumption that developing countries could improve their economies by adopting technologies of mass production (Schumacher 1973, 155–72; Dhakidae and Wikert 1977, 53–70). Schumacher instead endorsed technologies intermediate between craft and industrial organization because real increases in standards of living were more likely to be had from a "£1" development technology accessible to large numbers of un- or underemployed than from the "£1,000,000" technology that might produce more efficiently but employ fewer people. "Intermediate" and "appropriate" technologies were those that could be adopted by ordinary (often poor) people to increase their incomes and to supply basic needs less expensively. In this respect it harmonized well with Hatta's emphasis on encouraging technological change from the ground up because it directed technological attention in such a way as to make participation of the poor possible. Poerbo engaged in the international academic and development discourses on appropriate technologies, contributing, for example, to the University of Hawaii's East-West Center project on the use of appropriate technologies for the construction of low-cost housing (Poerbo and Kartahardja 1979, 67–87). Developing actual plans, techniques, and materials contested the idea that conventional approaches were inevitably the only practical ones.

For Hatta, cooperatives had always been about creating a new kind of individual, individually motivated and creative, yet responsibly engaged with the community. This was the foundation of a democratic society. The link between productive behavior and civic participation was also a theme in the literature of "participatory" development, which emerged in the 1970s and 1980s, to which Poerbo contributed (Poerbo 1995; Boenders and Poerbo 1981; Terrent and Poerbo 1989; De Soto 1989; Poerbo and De Soto 1992). Rather than imposing change from outside, supporters of participatory development sought to give voice and agency to the communities involved, bringing local individuals into planning and decision making in an effort to instill democratic mind-sets and produce more meaningful and enduring results. The participatory approach responded to the problems of those disenfranchised by industrialization who seemed to have lost the chance to speak for themselves; in urban areas this included especially the informal sector of workers living precarious lives of exploitation on the fringes of society without benefit of licenses, permits, or legal protection. Like Hatta's cooperatives, participatory development offered a concrete way to enable grassroots development of civic as well as material well-being.

The internationally circulating ideals of appropriate technologies and participatory development harmonized well with Hatta's ideas, even if the original hope of building an economy based entirely on a cooperative basis could no longer be entertained. By tapping into these other conversations about technological, social, and civic order, Poerbo brought international resources to bear on efforts to materialize alternative visions of development in Suharto's New Order

An urban sanitation project devised for Bandung by Poerbo and his collaborators in 1980 demonstrates how Poerbo materialized these various ideals. The project aimed to employ recycling technology and the work of local scavengers as an alternative to a conventional system of centralized trash collection. Poerbo led the project, which was cosponsored by the Centre for Environmental Studies at Bandung and the Institute of Social Studies in the Netherlands (Poerbo 1991, 60–69). The typical method for handling trash included the use of sanitation trucks to haul waste from households to a landfill on the edge of town. The justification for Poerbo's alternative design focused strongly on issues of environmental sustainability and cost containment. Not only were the costs of capital and operations high for ordinary trash collection, but any recycling done at the "back end" (at the site of the landfill) also called for expensive capital equipment. Poerbo argued that by mobilizing scavengers, who already sorted through city trash to find recyclable materials to sell, the city could recycle at much lower cost and reduce the size of the landfill. Yet his focus on cost should not blind us to the other work this project did.

Poerbo's plan would introduce numerous small sites for collecting recyclable goods, which he called Integrated Resource Recovery Modules. Each module provided areas for sorting and collecting recyclable materials and composting organic waste, a sales office, and services like a bath and toilet. Such modules, he argued, would allow scavengers to make a better living, while the city could spend less money on trash collection and landfills. His focus on scavengers and their place in Indonesian society is of special interest.

In many parts of Indonesia, scavengers had played an important but socially unrecognized role for years. They provided tons of scrap metal and plastics to Indonesian industries, although rarely were they well compensated. Middlemen, often quite wealthy, bought their scavenged material at low prices and sold them for much higher prices to local businesses. Entrepreneurial scavengers could not easily eliminate the middlemen, both because they could expect violent retribution and because of their own marginality in Indonesian urban life (Mangiang 1979). Living on the fringes of society, scavengers were frequently homeless, or squatters, and the middle class viewed them at best as public nuisances who made a mess during their scavenging activities and at worst as criminals, prostitutes, and drug addicts. Poerbo's plan aimed not merely to take advantage of scavengers but to offer them a safer, less marginal place in society. Scavengers would deal openly with the city rather than with middlemen and provide active input on how to improve technologies for collection or reclamation of recyclable materials (Poerbo 1991). Recycling modules would be placed to allow scavengers to reach them easily.

The consequences for the scavenging families who participated in the project went far beyond the technical organization of their previously informal work. Early on, Poerbo's team discovered that the scavengers themselves were less concerned with their precarious means of earning a living and more with their illegal status in society. Many lacked identity cards or the money to pay the fees to obtain them and therefore could not be legally married. Poerbo's team assisted thirty-five couples in marrying legally and, working with local NGOs, established an evening school for children, a relationship between the community and a local mosque, and a savings and loan cooperative. "Thus within three months," Poerbo remarked, "a group of scavenger families who had been trying to survive individually in a hostile environment had transformed themselves into a dynamic and creative community" (Poerbo 1991, 62). Poerbo's words and actions resonate with Hatta's belief in the power of cooperative techniques to transform both production and people.

As in the previous case, integrating marginalized people into society through recognized and productive technological work was an important goal for Poerbo. By making them part of Bandung's sanitation system, they became not merely accepted or recognized, but actually integral to the smooth material functioning of the city. Such a technologically systemic integration of people goes further than requests for or experiments in social tolerance or outreach because it makes them and their work necessary and desirable to others within society. Unlike their previous informal work that was ignored, this involvement would be socially explicit. Unlike charitable enterprises that provide goods to alleviate the problems of the poor, this project positioned the poor as necessary contributors to urban society.

How did such a project articulate with the larger politics of the New Order? Poerbo noted that the project became politically easier to do when Suharto made a famous speech in 1988 praising scavengers as a "self-reliant army" who found their own opportunities without looking for government handouts. This seemed to be a change from older New Order policies in which squatters were ruthlessly evicted and scavengers were assumed to be criminals. Poerbo's plan at least superficially harmonized with New Order leadership. Yet there is a world of difference between Suharto's position and that of Poerbo. Suharto praised scavengers but offered little else beyond, perhaps, less likelihood of being arrested. Poerbo on the other hand devised a concrete way to bring them more firmly into society, offering them citizenship in legal and technological terms. As a civil society action, it offered a different way of thinking about the continued existence of poverty in Indonesia by demonstrating an alternative to highly centralized technological organization and modeling the constructive relations that could connect work, social justice, and civic order.

Although it is difficult to establish a direct personal connection between Hasan Poerbo and Muhammad Hatta, there is a notable resonance between their priorities and their approach to improving Indonesian life. This resonance constitutes the sociotechnical imaginary that informed their work. Both Poerbo and Hatta sought to intertwine material change with change to the social and psychological conditions of Indonesians. They both attempted to undercut an industrial politics of elitism and patronage in favor of equitable opportunity that was nevertheless predicated on individual initiative and personal growth. Like Hatta, Poerbo saw the systematic exclusion of certain people from beneficial participation in the economy as a problem that required a solution. Meaningful inclusion in the economy was crucial and something to be planned and worked for, not merely hoped for. It was not enough to work as a low-level employee for a foreign corporation when Indonesians could pool their resources and become autonomous decision makers; it was not enough to simply tolerate scavengers if their technological work could instead be strongly integrated into the day-to-day functioning of cities. Like Hatta, Poerbo recommended the use of cooperatives as a vital tool to solve the problem of exclusion while making clear that the price of inclusion was the willingness to take initiative and learn new ways of doing things. Both encouraged an ethic of industrial change in which the capabilities and skills of people, and the coherence of communities were as crucial for the nation's development as the import of efficient, modern technologies.

Conclusion

Hasan Poerbo's efforts to devise alternatives to development as usual in the New Order offer us a way to consider the role of sociotechnical imaginaries when they operate without the support of state power and resources. To be sure, histories like this one can be difficult to trace. Projects inspired in different ways by Hatta's and Poerbo's imaginary might attract resources such as those provided by the United Nations, or international NGOs, but the very lack of a centralized actor or actors engaged in building institutional power around that imaginary makes it harder to discern. In this history, the power of the cooperative sociotechnical imaginary emerges from its persistence in the public imagination as an alternative, a viable path for the production of a more just and responsible social order for Indonesia. This persistence stems not just from Hatta's early appeal to a grassroots model of social and economic justice but to his insistence that material action could produce desired political and social outcomes as readily as conventional politics. Hatta's ideas spoke as effectively to the citizens of early postcolonial Indonesia who created cooperatives as they did to activists interested in appropriate technology projects in the New Order. By articulating technological plans that challenged the arrangements of power at the heart of the New Order, and materially demonstrating them within communities, technological activism was much more than mere populism. Projects both suggested economic possibilities outside the norm and, drawing on Hatta's thinking, built important forms of civic engagement by teaching individuals ways to engage positively with others for the common good.

Technological projects like Poerbo's operated within civil society as a form of political action in the New Order. By their very existence, they functioned as criticism—either by underscoring or clearly defining the needs going unmet or providing an unexpected opportunity to the marginalized. One might argue that any projects that helped the poor did tend to legitimate the New Order by operating under its imprimatur, and indeed Poerbo seemed far more interested in coming up with pragmatically helpful solutions to serious problems than alienating Indonesian leadership. However, by embedding a particular vision of participation and economic justice into the operation of his proposed technical systems, Poerbo undercut the sociotechnical logic of the New Order. Rather than rhetorically oppose Suharto's government, he made efforts to redirect its priorities in practice, making greater equality a material fact even as the state continued to operate more or less as usual. Just as Hatta had turned to pragmatic action when political activities no longer seemed to offer a viable response to the colonial state, so Poerbo and others like him turned to sociotechnical projects to produce a more just reality within the New Order.

Poerbo's projects exemplify the ways that Hatta's vision gained new relevance during the New Order as a critical sociotechnical imaginary. Suharto could not imagine a calm society that did not constrain flows of power between himself and a privileged elite dependent on his good will. The priorities of the New Order had dictated development, therefore, that proceeded from the needs and desires of insiders to state power, only slowly if at all incorporating those outside the networks of official patronage. Hatta, Poerbo, and others offered a glimpse of the alternative sociotechnical order that could result when development proceeded instead from the outside in.

Notes

1. The five principles of the *Pancasila* are in full as follows: Belief in the one and only God; just and civilized humanity; the unity of Indonesia; democracy guided by the

inner wisdom in the unanimity arising out of deliberations among representatives; and social justice for all of the people of Indonesia.

2. The source consulted was an English translation of Hatta's speech, provided by the Indonesian Ministry of Information.

3. The topic of *swakarya* came up frequently in the economic literature of the 1960s and continued to be of interest into the 1990s and beyond (Siregar 1969; Castles 1967; Dunham 1992).

4. From 1950 to 1955, control of Indonesia's parliamentary-style leadership changed five times, with the major divides in both religious and economic matters. Tragically reinforcing religious and economic divisions, he backed the Indonesian Communist party (the Partai Kommunis Indonesia, or PKI), in order to weaken the position of Islamic parties that called for foreign investment rather than Sukarno's preferred "Indonesianization."

5. There is still no clear explanation for the assassinations, although there are many theories (Elson 2001).

6. Later, Suharto's family also became deeply entrenched in this practice of dispensing patronage.

7. The concept of *dwifungsi* was developed at a seminar Suharto convened in 1966 with the Army (Elson 2001; Crouch 1978).

8. The region in question is often referred to as Jabotabek (Jakarta-Bogor-Tangerang-Bekasi).

9. In January 1974, what started as a small student protest (actually instigated, it was later discovered, by infighting among political elites) engaged the public imagination sufficiently to turn it into a large-scale riot.

References

Abdullah, Taufik. 2007. "The New Order: A Historical Reflection." In *Indonesia in the Soeharto Years: Issues, Incidents, and Images*, edited by John H. McGlynn and Hermawan Sulistyo. Jakarta: Lontar, in association with Ridge Book.

Arief, Sritua. 1977. *Indonesia: Growth, Income Disparity and Mass Poverty*. Sritua Arief Associates.

Arifin, Bustamil, and Muslimin Nasution. 1981. "Dialogue: Cooperatives and Government Participation." *Prisma* 23 (December): 22–26.

Aspinall, Edward. 2005. *Opposing Suharto: Compromise, Resistance, and Regime Change in Indonesia*. Stanford, CA: Stanford University Press.

Boeke, J.H. 1946. *The Evolution of the Netherlands Indies Economy*. New York: Netherlands and Netherlands Indies Council, Institute of Pacific Relations.

Boenders, Marinus, and Hasan Poerbo. 1981. *The Informal Sector Project in the Botabek Area: Inception Report*. Bandung: Centre for Environmental Studies.

Budiman, Arief. 1982. "The Pancasila Economy, Capitalism, and Socialism." *Prisma* 26 (December): 71–74.

Castles, Lance. 1967. *Religion, Politics, and Economic Behavior in Java: The Kudus Cigarette Industry*. New Haven; Detroit: Southeast Asia Studies Yale University.

Constitution of the Republic of Indonesia. 1945. (Undang-Undang Dasar Negara Republik Indonesia 1945). http://www.indonesia.go.id/id/files/UUD45/satunaskah.pdf.

Crouch, Harold A. 1978. *The Army and Politics in Indonesia*. Ithaca, NY: Cornell University Press.

Departemen Penerangan Republik Indonesia. 1969. *REPELITA: Rentjana Pembangunan Lima Tahun Republik Indonesia: 1969/1970–1973/1974.* Jakarta; Bandung: Departemen Penerangan.

De Soto, Hernando. 1989. *The Other Path: The Invisible Revolution in the Third World.* Harper Collins.

De Soto, Hernando, and Hasan Poerbo. 1992. *Masih Ada Jalan Lain: Revolusi Tersembunyi Negara Dunia Ketiga,* Jakarta: Yayasan Obor Indonesia.

Dhakidae, Daniel, and Jürgen D. Wikert. 1977. "People Really Matter: Two Exclusive *Prisma* Interviews with Margaret Mead and E.F. Schumacher." *Prisma* 7 (September): 53–70.

Djojohadikusumo, Sumitro. 1981. "The Role of the Civil Service Cooperative in the Indonesian Economy." *Prisma* 23 (December): 12–21.

Dunham, S. Ann. 1992. *Peasant Blacksmithing in Indonesia: Surviving against all Odds.* PhD, diss., University of Hawaii.

Dunleavy, Patrick. 1981. *The Politics of Mass Housing in Britain, 1945–1975: A Study of Corporate Power and Professional Influence in the Welfare State.* Oxford, New York: Clarendon Press, Oxford University Press.

Elson, R.E. 1984. *Javanese Peasants and the Colonial Sugar Industry: Impact and Change in an East Java Residency, 1830–1940.* Singapore, New York: Oxford University Press.

———. 2001. *Suharto: A Political Biography.* Cambridge, UK; Oakleigh, Victoria: Cambridge University Press.

Gilman, Nils. 2003. *Mandarins of the Future: Modernization Theory in Cold War America.* Baltimore: Johns Hopkins University Press.

Hatta, Muhammad. 1934. "Self-Help Dalam Emigratie." *Daulat Ra'jat.*

———. 1952 *The Co-operative Movement in Indonesia* (Speech delivered on the occasion of the Second Cooperatives Day, 12 July 1952). Djakarta: Ministry of Information.

———. 1954a. *Djalan Keekonomi Dan Kooperasi: Beberapa Fasal Ekonomi.* Djakarta: Perpustakaan Perguruan Kementerian.

———. 1955. *The Co-Operative Movement as an Institution for Teaching Auto-Activity and a High Economic Character: Opening Speech Delivered on Aug. 8, 1955 before the Third Seminar on Cooperatives Held in Bandung on the Initiative of ILO/FAO.* Djakarta: Ministry of Information.

———. 1972. *Portrait of a Patriot: Selected Writings.* The Hague, Paris: Mouton.

Hatta, Muhammad, and Z. Yasni. 1981. *Bung Hatta's Answers: Interviews.* Singapore: Gunung Agung.

Hasyim, Yusuf. 2007. "Killing Communists." In *Indonesia in the Soeharto Years: Issues, Incidents, and Images,* edited by John H McGlynn and Hermawan Sulistyo. Jakarta: Lontar, in association with Ridge Book.

Heryanto, Ariel. 1999. "The Years of Living Luxuriously: Identity Politics of Indonesia's New Rich." In *Culture and Privilege in Capitalist Asia,* edited by Michael Pinches. London: Routledge.

Heryanto, Ariel, and Nancy Lutz. 1988. "The Development of 'Development.'" *Indonesia* 46: 1–24.

Hill, Hal. 1997. *Indonesia's Industrial Transformation.* Singapore: Institute of Southeast Asian Studies.

Kahin, George McTurnan. 1952. *Nationalism and Revolution in Indonesia.* Ithaca, NY: Cornell University Press.

Larson, Carl Theodore. 1958. "The Regional Housing Centre, Bandung, Indonesia: A Summary Evaluation Report to the United Nations Technical Assistance Administration on

the Basis of Field Work Performed during the Period from July 1956 to June 1957."
Ann Arbor, MI.

Legge, J.D. 1972. *Sukarno: A Political Biography.* New York: Praeger.

Lembaga Penyelidikan Masalah Bangunan (Indonesia). 1972. *Housing and Urban Development in Indonesia.* Bandung.

Liddle, R. William. 1992. "Indonesia's Democratic Past and Future." *Comparative Politics* 24(4): 443–62.

LP3ES [Lembaga Penelitian, Pendidikan dan Penerangan Ekonomi dan Sosial]. 1975. *Seminar Pengembangan Usaha Swakarya Dan Swadaya Masyarakat Merak, Anyer, Dan Cilegon, (ADC).* Serang.

———. 1981. "Cooperatives: Search for Self-Reliance, a Field Report." *Prisma* 23 (December): 27–63.

———. 1995. *Pemikiran Pembangunan Bung Hatta.* Jakarta: Pustaka LP3ES.

Mangiang, Masminar. 1979. "The Economics of Scavenging." *Prisma* 5: 36–44.

Masenas, Clement. 1974. "The Tiger in Indonesia's Tank: Pertamina Plays a Vital Role." *New Standard,* December 21.

McIntyre, Angus. 2005. *The Indonesian Presidency: The Shift from Personal Toward Constitutional Rule.* Lanham: Rowman & Littlefield Publishers.

Moon, Suzanne. 2007. *Technology and Ethical Idealism: A History of Development in the Netherlands East Indies.* Leiden: CNWS Publications, 2007.

———. 2009. "Justice, Geography, and Steel: Technology and National Identity in Indonesian Industrialization." *Osiris* 24(1): 253–77.

Mrázek, Rudolph 2009. "Boven Digoel and Terezín: Camps at the Time of Triumphant Technology." *East Asian Science, Technology, and Society* 3(2): 287–314.

Noer, Deliar. 2002. *Mohammad Hatta: Hati Nurani Bangsa, 1902–1980.* Jakarta: Penerbitan Djambatan: Perwakilan KITLV (Koninklijk Instituut voor Taal-, Land- en Volkenkunde).

Papanek, Gustav. 1976. "The Poor of Jakarta." *Prisma* 3 (May): 33–43.

Perumusan Hasil Seminar Rentjana Pembangunan Pertanian Lima Tahun 1969–1973, Tgl. 1 S/D 5 Djuli 1967 Di Universitas Gadjah Mada Jogjakarta. 1967. Djakarta: Departemen Pertanian.

Poerbo, Hasan. 1979. "Mass Housing in Indonesia and Job Opportunities in the Informal Sector." *Prisma* 12 (March): 3–13.

———. 1980. *Partisipasi Masyarakat Dalam Pembangunan Kampung Terpadu: Kasus Babakan Surabaya Dan Cikutra, Bandung, Disampaikan Pada Rapat Kerja PSL Seluruh Indonesia, Jakarta, 13–15 Okt. 1980.* Bandung: PSLH.

———. 1991. "Urban Solid Waste Management in Bandung: Towards an Integrated Resource Recovery System." *Environment and Urbanization* 3 (1): 60–69.

———. 1995. *Working with People: Indonesian Experiences with Community-Based Development.* Toronto, Bandung [Indonesia]: Faculty of Environmental Studies, York University, Centre for Environmental Studies, Institute Technology of Bandung.

Poerbo, Hasan, and Albert Kartahardja. 1979. "Mass-Housing in Indonesia: In Search of New Solutions." Pp. 67–87 in *Low-Cost Housing Technology: An East-West Perspective,* edited by Louis J. Goodman. Oxford, New York: Pergamon Press.

Power, Anne. 1997. *Estates on the Edge: The Social Consequences of Mass Housing in Northern Europe.* New York: St. Martin's Press.

Rahardjo, M. Dawam. 1985. "The Role of the Community in Modernization." *Prisma* 36 (June): 3–7.

Ricklefs, M.C. 2008. *A History of Modern Indonesia Since C. 1200.* Stanford, CA: Stanford University Press.

Rombe, L., T. Ariwibowo, Manurung, et al. 1970. *Feasibility Study for Tjilegon Steel.* Jakarta, Indonesia.

Schumacher, E. F. 1973. *Small Is Beautiful: Economics as if People Mattered.* New York: Harper & Row.

Shiraishi, Saya. 1997. *Young Heroes: The Indonesian Family in Politics. Studies on Southeast Asia, No. 22.* Ithaca, NY: Cornell University, Southeast Asia Program.

Shiraishi, Takashi. 1990. *An Age in Motion: Popular Radicalism in Java, 1912–1926.* Ithaca, NY: Cornell University Press.

Simposium Kebangkitan Semangat Mendjeladjah Tracee Baru. 1966. Jakarta: Jajasan Badan Penerbit, Fakultas Ekonomi Universitas Indonesia.

Siregar, Arifin. 1969. "Indonesian Entrepreneurs." *Asian Survey* 9 (5): 343–58.

Suryakusuma, Julia I. 2004. *Sex, Power, and Nation: An Anthology of Writings, 1979–2003.* Jakarta, Indonesia: Metafor Publications.

Sutter, John Orval. 1959. *Indonesianisasi Politics in a Changing Economy, 1940–1955.* Ithaca, NY: Southeast Asia Program, Department of Far Eastern Studies, Cornell University.

Swasono, Sri-Edi. 1985. *Koperasi Di Dalam Orde Ekonomi Indonesia Mencari Bentuk, Posisi Dan Realitas.* Jakarta: Penerbit Universitas Indonesia.

Symposium Pembangunan Ekonomi Daerah. 1966. Medan: Universitas Sumatera Utara, Fakultas Ekonomi.

Terrent, James, and Hasan Poerbo. 1989. "Strengthening Community-Based Technology Management Systems".in *Doing Development: Government, NGOs and the Rural Poor in Asia.* Edited by Richard Holloway. London: Earthscan.

Vatikiotis, Michael. 1998. *Indonesian Politics under Suharto: The Rise and Fall of the New Order.* London, New York: Routledge.

Wie, Thee Kian. 1988. *Industrialisasi Indonesia: Analisis Dan Catatan Kritis.* Jakarta: Pustaka Sinar Harapan.

Yasni, Zainul. 1968. "Ekonomi swadaja: membangun daerah untuk kemakmuran bangsa: dengan case study, Sulawesi Selatan." PhD diss., Universitas Hasanuddin di Ujung Pandang.

NINE

Guerilla Engineers: The Internet and the Politics of Freedom in Indonesia

JOSHUA BARKER

Self-supported and self-funded, shoulder-to-shoulder, developing the nation and taking over other countries: that is what is happening in Indonesian information technology today. Don't expect World Bank or IMF loans. It is almost all resting on the strength of young people and ordinary Indonesian people, with hardly any help from the government much less from [political] parties; in fact, if anything the bureaucrats and police sweepings have made things more difficult. *Alhamdul-lilah*, the current from below has been able to build a nation between the cracks of the state's confiscations and anger, despite interrogations and imprisonments. The efforts at struggle are based on the simple ambition to see the Indonesian nation rely upon the strength of its minds, rather than its muscles. A nation able to triumph from its ability to think rather than to sweat and to shoot. (Purbo 2003b)[1]

In their recent book on the politics of Asian cyberspace, Deibert et al. (2012) argue that the idea that the Internet is destined to be a space of political freedom has now largely faded. Instead, they suggest, this idea is now best understood as a normative or rhetorical claim about what the Internet *should* be rather than what it is or what it will be. Underscoring this point, powerful actors like the former US Secretary of State, Hillary Clinton, and Google's leaders now find it necessary to regularly state their commitment to this normative ideal.

The embrace of this rhetoric by powerful US actors should not blind us to the fact that it is an ideal that has global reach and is embraced by a rather odd assortment of groups and individuals, often with very different backgrounds and agendas. If we are to understand the roots of this ideal—and its continuing power—we must examine the processes that constituted the association between the Internet and a politics of freedom in particular settings around the world. In this chapter, I examine the roots and character of this association as they developed in Indonesia in the 1990s and early 2000s. Specifically, I trace the activities and ideas of Onno Purbo, a man often called "the father of the Indonesian Internet."

Starting in the 1990s, Purbo and a small group of colleagues and students at the Institute of Technology in Bandung (ITB) were instrumental in establishing the first campus LAN and the first intercampus computer network linking together campuses and research centers in the cities of Jakarta and Bandung. They then went on to establish a dedicated satellite gateway for this network to the global Internet and to grow their intercampus network to include universities across the archipelago. While commercial Internet service providers (ISPs) have since become widespread and tens of millions of Indonesians now access the Internet via cell phones and wireless and fixed-line connections, for several crucial years this nationwide campus network was the main backbone for Internet communication in Indonesia. I will show that in the process of building this network, Purbo and his friends were also self-consciously building a sociotechnical imaginary that linked the technology to a politics of freedom. To some degree this was reflected in their jargon, as when they referred to themselves as "activists" (*aktivis*) using "guerilla" (*gerilya*) tactics in their "struggle" (*perjuangan*) to bring a "free net" to Indonesia. Such references were somewhat tongue in cheek, but they are suggestive of some of the roots of their sociotechnical imaginary. In part, their imaginary was an extension of a globalizing rhetoric about the Internet commons, but it was also inflected by the specific history of technology and nationalism in Indonesia and by ideas about bottom-up community development advocated by critics of the Indonesian government's development policies. Purbo himself was the son of development activist Hasan Poerbo, and he had been raised in an environment saturated with ideas about modernization from below.[2] Moreover, this imaginary was shaped by changes in Indonesia's political landscape during this period, as the country's longtime dictator, President Suharto, was eventually ousted under pressure from a student-led reform movement.

Besides its local inflections, what is also notable about the emergence of Indonesia's early Internet imaginary is that it developed largely independently of the institutions of state power. As Jasanoff and Kim (2009) have shown, when championed by governments, and supported by state institutions, sociotechnical imaginaries can acquire remarkable force. But what if they do not have these supports? The case of the Indonesian Internet is interesting because it shows how a sociotechnical imaginary can take shape— and acquire force—through the efforts of a bottom-up actor network that gradually gives shape to a broader media "counterpublic" (Warner 2002).

Beginnings of the Indonesian Internet

There is no other country in the world that has a 10+ years history of struggle building its information society, and has been able to build this information society into millions of people by 2003 from nil back in 1993. No other country in the world has built a large-scale system all on the basis of mutual self-support. [. . .] It's not surprising that many of those present at the World Summit on the Information Society exclaimed, 'Indonesia is inspirational.' (Purbo 2003a)

I first met Onno Purbo in 1995 at ITB. ITB is the oldest technical school in the country and its high temple of technology. Every incoming class of students is reminded that they represent the best that Indonesia has to offer. Faculty and graduates often seem to position themselves as high priests capable of mediating between the powerful and arcane world of new technologies and the mundane and familiar world of Indonesian underdevelopment. In the mid-1990s, ITB was the only place in Bandung where one could cheaply send and receive email. To get into the building where Purbo's research group was located I had to sign in with a security guard and state the purpose of my visit. Basoeki Suhardiman, a student who worked with Purbo, explained to me that they had learned long ago that it made sense to cultivate good relations with the security guard. They invited him up to learn how to use computers and always chatted with him when they passed his security post. Such security guards were sometimes plugged into the intelligence-gathering apparatus of Suharto's New Order regime, and it was best to keep them on your side. When Basoeki told me things like this he would speak in a hushed tone, and I often had trouble understanding him, not only because I had trouble hearing, but also because it all seemed overly melodramatic given that their activities in the lab were focused on technology not politics. I always felt that I was missing something. I had the same feeling some years later, after Suharto had been ousted and the military repression had been lifted, when I was discussing Purbo with a close friend of mine who works in a large telecommunications firm. He said, quite seriously, "be careful with Onno, he's dangerous." The tone and terms he used were somewhat anachronistic by this time, because they were drawn from the Suharto regime's discourse for branding opponents of the regime as dangerous communists, a label that could mean that those consorting with them might end up in prison or worse. Yet they still carried force. It was clear that Purbo somehow had come to represent a kind of politics that at least some people found threatening.

To understand what this politics is it is helpful to contrast the beginnings of the Internet with the beginnings of Indonesia's domestic satellite system, which was the last great communications innovation for the country prior to the advent of the Internet. As I have described elsewhere (Barker 2005), Indonesia was the first developing country to have its own domestic satellite system. The imaginary surrounding the satellite was from its inception all about national unity and, to a lesser degree, national development (a fact it shared with the Canadian satellite system it was modeled upon). While first put forward by academics at ITB, the idea of the satellite as a national unifier was quickly picked up by the authoritarian Suharto government, which turned it into a kind of state dogma for telecommunications. Under the authority of the state's monopoly telecommunications provider, the satellite would serve the explicit purpose of instilling official nationalist ideology across the archipelago. Suharto named the satellite Palapa, which referred to an oath taken by a precolonial Javanese king to unite the archipelago under his rule. At the inauguration of the satellite, Suharto dressed up in traditional garb, recited the Palapa oath, and turned the satellite on using a button in a replica of a Javanese ceremonial dagger. In the decades since, Indonesian children have been obliged to learn about the Palapa oath and its link to the satellite and national unity through school textbooks.

While the Palapa dogma exemplifies a sociotechnical imaginary imposed from the top down with the full force of state power, it was also an imaginary sustained by domestic capitalists vying for contracts to build out the ground segment of the system. These entrepreneurs used the Palapa ideology and arguments about national security to pressure bureaucrats and officials to buy more ground stations, particularly in border regions, where threats to national unity were feared to be greatest. The network they ended up building was thus one that reflected the deeply centralized imaginary promoted by the government and effectively hard wired its authoritarian politics into the terrestrial part of the system. There was nothing about freedom in the media ideology of the satellite, but it was a system that was from the outset all about politics.

The beginnings of the Internet, some two decades later, followed a very different trajectory of development. Rather than being coopted by the state for the purposes of social control, it emerged as a network of the people. Early Internet and proto-Internet development in Indonesia took place within loosely structured and sometimes overlapping communities of hobbyists—especially radio and computer hobbyists—and researchers working at the University of Indonesia (UI), ITB, and various government institutes like the National Agency for Assessment and Application of Technology (BPPT) and the National Institute of Aeronautics and Space.[3] In general terms, one can distinguish between two kinds of network development during the period up until 1993. On one hand, there were hobbyists who sought to link together computers into networks that could exchange data, messages, and small pieces of software. These efforts originated among computer hobbyists in the early 1980s, with technologies like bulletin board systems, which depended on linking computers together via the telephone network. Starting in the late 1980s, radio amateurs started connecting their radios to computers in order to establish wireless networks capable of transmitting data instead of just voice. On the other hand, there were a couple of much more ambitious and costly projects that aimed to connect Indonesian universities and research institutes to growing interuniversity networks in the United States and Germany. The first of these latter initiatives emerged at UI, and the second originated at BPPT, which was headed by B. J. Habibie.[4] Both were funded through moderately sized loans from the World Bank, and both envisioned a domestic network between universities and research institutes whose access to the globalizing Internet would be available through links at UI or BPPT, respectively. Habibie's ambition in particular was to replicate within Indonesia the kind of network that had emerged in the United States once the Internet had moved from the military into academia with support from the National Science Foundation.

Among researchers there was a great deal of enthusiasm for such international connections in part because many Indonesians studying in the United States, Canada, Europe, and Japan were involved in network development abroad. These expatriate students used the Internet to keep in touch with one another and to exchange technical know-how, software, and news. Graduates returning from abroad wanted to remain connected into these networks and to bring their colleagues and students in Indonesia into the growing conversation. The problems these people faced were the difficulty and high costs involved in connecting to UI or BPPT and, once the loans ran out, the high cost of connecting to the gateways abroad. Domestically, the phone system was under the monopoly of Telkom, a public corporation, which charged not by the call but by time used; and its rates were extremely expensive. Internationally, there was a duopoly, but the problem was the same. Just as importantly, telephone density was very low, and it could take years to get a line, and one often needed personal connections in Telkom to get one at all. (In the 1990s, the late Onghokham, an historian, was so thrilled about getting a phone line after years of waiting that he held a *slametan*, the kind of ritual Clifford Geertz [1960] wrote about that marks major life cycle events such as births, deaths, circumcisions, and marriages.)

Purbo first introduced the idea of the Internet to a broader public in

1990 when he published an article in the major daily newspaper *Kompas* (see Purbo 1990). At the time Purbo was pursuing his PhD in electrical engineering at the University of Waterloo in Canada. It was the first of many articles he would publish about the Internet over the next two decades in Indonesian newspapers. The article laid out in simple terms a concept for a low-cost network that would circumvent the problems of relying on the telephone network. He described a large-scale network of computers or "intelligent machines" that linked together the various LANs that already existed in research institutions, government, and industry. Such a network would have many benefits, most notably, email. Email had the advantage of speed, and it allowed people to send messages to multiple addressees, where one member's response would also be read by all members, thereby providing a group of people living far apart with a way to have an electronic discussion spanning months or even years "about anything from soccer to how to construct a computer" (Purbo 1990). He explained that many Indonesian students living abroad were already doing this, as were radio amateurs in Indonesia who were connecting to the global amateur packet radio network via satellite.[5]

In Purbo's view, the best way to expand the emergent Internet in Indonesia was to circumvent the monopoly-controlled wire infrastructure by using the kind of network that radio amateurs had been pioneering. And over the next few years a network of this sort did in fact come into being, connecting an expanding network of universities, government research institutions, and high-tech industry. Paguyuban Net ('Network Association'), as it came to be known, was most heavily concentrated in Bandung and Jakarta. Purbo's group at ITB was very active in developing this early wireless Internet, as well as a cable-based local area network (LAN) within the ITB campus.

The ITB group—known eventually as the Computer Network Research Group (CNRG)—had emerged out of ITB's Amateur Radio Club, where Purbo had long been active. CNRG was a bit of an unusual group for ITB, since it had no official status and no real hierarchy. It differed from other campus groups in that it had no hazing, no formal initiation, and the only real requirements for membership were that you had to be willing to spend long hours in the lab, often overnight, and you had to like *ngoprek*. *Ngoprek* consists of opening up the black box of technology, taking things apart, trying new components and new configurations, making things work in new ways, and then explaining, writing about, diagramming and taking pictures of what you have done so that others can replicate it. Or as one Indonesian website summarizes it, it means to take, dismantle, assemble, run, write, and share.[6] It is more or less what English-speaking radio amateurs refer to as "home brewing."

Ngoprek does not just mean a willingness to delve into the innards of software and circuitry, it also implies that one will participate in a broader public of fellow bricoleurs. In the case of amateur radio, this broader public builds upon the how-to books and magazines of print culture but extends beyond the limits of particular print languages through a shared culture of diagrams, hardware nomenclature, technical jargon, machine languages, and call signs; a shared market of electronic goods; and a shared practice of over-the-air interaction. Purbo consciously sought to cultivate such a culture, both for himself and for his students in CNRG. It was not enough to do so in the online listservs—which mainly focused on technology—that were the core content of the network. From early on Purbo concluded that in order to expand the circle of people involved in the network it was far more effective to go outside it. Both he and his students thus spent a lot of time giving talks, contributing articles to newspapers and magazines, and writing books.

Network expansion proved to be a complex political process, and nowhere was this truer than at ITB itself, where the LAN involved laying an Ethernet cable around campus. The college administration was not initially supportive of the project, and getting funds and permission from the bureaucrats in charge was simply too involved and time-consuming a process. Instead, the group used what one member told me in 1996 were stealthy "guerilla" tactics. It happened that the Electrical Engineering Department had labs in three corners of the campus, so initially the group laid wires that were ostensibly to connect their own labs. To do so, however, meant running their distinctive yellow cable through hallways and labs associated with other units, which sometimes gave rise to some territorial disputes, making negotiations necessary. As on any campus, there were complex politics around space, and in some cases CNRG was seen as an upstart since it was a nonstructural group and it was running its cable through other groups' space. At the same time, there was also the sensitive issue that CNRG was essentially a group of hobbyists, and although most of its members (though not all) were electrical engineers, most had no formal training in network technology or informatics. There was always a risk that they could be seen as mere novices who were meddling in a domain that could be seen, in ITB's rather conservative and hierarchical academic culture, as not rightfully theirs. CNRG dealt with these problems as best it could by always politely and respectfully "asking permission first" (*minta permisi dulu*) to the units they might encroach upon and by offering to provide offended parties with their own access to the network in return for their willingness to let the cable work go ahead. When units joined, they would have to provide their own cable and other infrastructure. In this way, as the physical network grew, so too did the actor network of those who were interested, knowledgeable, and committed to its development. For members of CNRG, the distinctive yellow cable of the ever-expanding network became a source of pride at having managed to wire the campus without any financial support from university authorities.

Throughout the process of building this network, Purbo was writing and publishing. The bulk of his writing focused not on his core research (integrated circuits) but on the nuts and bolts of what the Internet is, what the options are for network development, how one can use the Transfer Control Protocol and Internet Protocol suite protocol over packet radio, which institutions and individuals were contributing to network innovation, and how far network development had come. He encouraged his students to write books and contribute articles to magazines and taught them the most effective techniques to get technical concepts across to people who had little technical training: use analogies (e.g., email is like the postman), and don't use too much jargon. Purbo would then edit his underlings' books and help to ensure they got published.

In addition to their writings, members of CNRG were also obliged to give training, talks, and demonstrations to anyone who expressed an interest in joining the network. Within ITB, as other units heard about the network and saw the yellow cable, they would ask if they could gain access. Instead of doing the work themselves, CNRG members would provide training and assistance to representatives of the unit, who would then become the system administrators for that unit. When CNRG members traveled farther afield, to other universities in Bandung or in other cities, they would also give presentations. Often this meant giving a formal speech to the rector of the university, which would be followed by a couple of days of more informal technical demonstrations to students and faculty setting up the network. In this way, CNRG took the model they had developed at ITB and replicated it at other campuses, using wireless technologies first to connect campuses within the city of Bandung and then across the archipelago nation. This is how Indonesia's first nationwide campus network came into being.

Resources for the expansion of the network came from a variety of sources. At the beginning, older amateurs donated computers and radios. As the group started to develop a reputation, some funds flowed in from small commercial consultancies, which would be used to pay for equipment or divided up among the group. Units within ITB wanting to connect to the servers at CNRG had to pay the costs of the Ethernet cable needed to reach from their unit to CNRG, and they had to provide their own computers. Outside of the campus the same basic principles applied. CNRG would provide technical support and training but any necessary hardware had to be provided by the people who were going to use it. For a time CNRG worked with other groups within ITB (including the original instigator for the satellite system, Iskandar Alisyahbana) interested in building low-cost radio modems to supply the expanding network, but the price on commercially available modems dropped to a point that it made the local industry difficult to sustain.

The main challenge to network growth was overcoming the problem of access to the international Internet. As the Bandung network expanded, this problem became more acute because traffic on the system grew and the radio connections to BPPT and UI in Jakarta were very slow. On this front, CNRG depended on the largesse of senior faculty at ITB and on powerful alumni of the Electrical Engineering Department, to whom it had to go hat in hand. Iskandar Alisyahbana donated use of a ground station so that the group could try to establish access via an amateur satellite. Another senior faculty member, Samaun Samadikun, convinced a former student and higher-up in Telkom to donate use of a leased line (ISDN) between ITB and BPPT. And another faculty member allowed the group to piggyback on bandwidth to a Japanese satellite he had obtained for a research project he was working on. Throughout this process Purbo wrote articles for both the ITB press and the national press thanking the donors publicly and highlighting the benefits of the network for education and national development.

Up until 1996, the Bandung network grew but the bottleneck in connecting to the international gateway in Jakarta remained. In that year CNRG competed with a number of other Indonesian institutions, including BPPT and some others, to become the Indonesian partner in a pan-Asian network, AI3, being developed by Japan. The premise of AI3 was that partner institutions throughout Asia would participate in an exchange of research results in return for access to a high-speed connection to the Internet via a Japanese satellite. CNRG, which had put forward a proposal that recapitulated their work building Paguyuban Net and saying they would continue to expand the network to include universities across the archipelago, was a surprise winner of the competition. After some lobbying, ITB's administration agreed to provide funds for equipment needed on campus, and an alumnus arranged for his company, a ground station supplier owned by one of President Suharto's sons, to donate a ground station. Given that the link would conflict with government regulations on international transmissions, further lobbying, also involving alumni, was required to obtain a government license. After a great deal of wrangling and red tape, the connection was finally established and CNRG became the main Indonesian conduit for university access to the global Internet.

CNRG's success in this early stage of Internet development was disruptive and surprising in part because it represented an upstart strategy for technological innovation. The New Order regime embraced capital-intensive, foreign debt–intensive, top-down strategies of innovation in an effort to "leapfrog" toward national development. Habibie was the main proponent of such a vision, and many of his projects became infamous for their high costs, their heavy reliance on foreign experts, and their failures to achieve their lofty ambitions (Amir 2004). Purbo shared the stated objective of national development, but his approach to innovation was one that was low cost and debt free and relied on domestic know-how and mutual self-help. Such an approach was well-known in participatory development circles, in which Onno's father, Hasan Poerbo, was active, but it was only now being applied in a high-tech field. The approach allowed CNRG to establish a track record of successful innovation and network development, which gave its proposal to AI3 credibility that competing proposals may not have had.

In the years following CNRG's successful bid to AI3, the landscape of Internet development would change rapidly as commercial and state-owned service providers became more active. The first of these was started in 1995, and by 1996 fourteen others were operational. Hill and Sen (1997, 73) note that there appeared to be a policy of offering licenses to only a limited number of companies and that these were outside the "well-connected companies" that dominated the domestic economy, including private telecommunications. These ISPs initially focused on corporate clients and starting in 1996 helped to provide the backbone for Internet cafe" (warnet) that were sprouting up in major cities, including Bandung (members of CNRG were the first to set up a warnet in Bandung). From 1999 the number of warnets increased dramatically, and the Internet truly became a mass medium. Throughout these changes CNRG remained active and continued to expand its campus network to universities and other educational institutions all over the country.[7] Eventually Purbo left ITB so he could dedicate himself full time to giving talks, giving advice, and writing, both in Indonesia and around the world. As a fellow at Canada's International Development Research Centre, he gained international recognition for his work on ICTs and development. Within Indonesia, he became active in several high-level associations, governmental and nongovernmental, which sought to shape government policy toward telecommunications and the Internet. Through frequent appearances in the press, at conferences, on television, and on social media, he became a minor celebrity, and as a broader domestic market for computer products began to take hold, a greater proportion of his "road shows"—as his talking tours came to be known—were underwritten by the publishers of his books, by Hewlett Packard, and by others with an interest in the growing Indonesian market for hardware, software, and related books. He became, in essence, a freelance version of what some corporations might call a "technology evangelist."

Purbo's "Naïve Philosophy" and the Making of a Public

What did Purbo preach? An examination of his writings over the past two decades reveals that he has written a lot more than just technical manuals. He has spent a great deal of time reflecting upon and describing his and his cohort's experiences building the intercampus network discussed above, as well as his subsequent activities shaping developments at the national level. He has also written dozens of articles that discuss what the Internet is, how it can be used, what it will mean for society, and what the Indonesian government ought to do, and what individuals ought to do, to create an "information society."

The language he has used to discuss these things has changed over time. Writing in 1990, he explained the need for the Internet in terms of national development; in the early to mid-1990s, he talked about building the Internet as a process of "bottom-up" activism and community development; in the mid- to late 1990s he talked about building a national infrastructure that would underpin a second national awakening and would lead to the dawn of an information society; and then starting in the late 1990s he began to describe the Internet in terms of democratization, rebellion, struggle, jihad, and the quest for freedom.

These shifts were partly a reflection of broader discursive shifts that were going on in Indonesia at the time: from a discourse of development to one focusing on political reform. But they also drew on a range of oppositional sociotechnical imaginaries. As noted above, his belief in the power of "ordinary people" or common folk (*rakyat biasa*) to help themselves through "bottom-up" strategies can be traced back most immediately to his father, a professor at ITB who was quite well-known in NGO circles for his work helping to organize impoverished scavengers in Bandung and provide them with technology to give them sustainable livelihoods (see Moon's chapter in this volume). Onno shared his father's "community development" ideology and saw his efforts to expand the Indonesian Internet as part of a broader ambition to empower ordinary Indonesians in a sustainable fashion. Within ITB, his loosely structured group functioned—as one ITB professor described it—in a manner analogous to an NGO: it sought donations, provided training sessions, made use of community labor, and tried to become self-sustainable and self-replicating. In the context of the New Order, such an approach was generally seen as oppositional to the government's top-down, state-led approach to development.[8] However, in international development settings such approaches were highly valued, and Purbo soon found that his activities spoke to the burgeoning interest in the "digital divide" and using ICTs for community development. But within these settings he was also somewhat anomalous, since his approach was not to provide access to new technologies but to teach people how to build networks for themselves from the ground up (using cheap cable, retooled woks, self-help, etc.).

Purbo's politics of freedom also echoed a much older nationalist discourse that linked radio amateurism to revolutionary politics. An early example of this discourse is discussed by Mrázek (2002), who describes a prominent nationalist journal in the 1930s that promoted the idea of providing courses on radio technology for children of all ages as a means to help Indonesians unite, improve their fate, and come out into the world. "[L]et us become radio mechanics," the article exhorts its readers (quoted in 2002, 190). Revolutionaries took up this notion during the struggle for independence, when guerilla fighters–cum–radio mechanics built radio transmitters and carried them from village to village, broadcasting messages about the progress of the struggle. For these revolutionaries, and for subsequent generations of radio amateurs like Purbo, building radios and transmitting and receiving messages were strongly associated with being in the vanguard of political change and the struggle for freedom. In the context of the reform movement that enveloped Indonesia in 1997 and eventually led to the ouster of President Suharto, applying this revolutionary discourse to the Internet— the preferred medium of communication for anti-Suharto activists—gave it renewed significance as a tool of resistance.

Purbo's discourse about Internet freedom also grew out of global discourses associated with hackers, the free and open source movement, and the copyleft movement. He was exposed to these ideas while a graduate student in Canada in the late 1980s and early 1990s, and he has remained consistent in his view that information is meant to be free.[9]

Purbo's approach to all these discourses was very much within the spirit of *ngoprek*: take them apart, put them back together, make them work, and share them. At the same time, all his writings reinforce his basic outlook, which combines a faith in the power of ordinary people to deliver a better technology with a faith in the power of technology to deliver a better society. This is particularly evident in his "naive philosophy"—as he has called it— about how the Internet will transform society. According to Purbo, the difference between the present society and the future Internet society can be understood through the analogy of "platforms," which he (2001c) defines as "the place that we stand, labour, and interact":

> The platform that we stand on in Indonesia is constrained by fences, walls, stones, concrete, and asphalt highways. Things, people, paper, newspapers, magazines, information, and knowledge are usually sent from place to place on vehicles like cars, motorcycles, and bikes with speeds of 50–100 km per hour. The dimension of time limits the capacity to reach across districts and space. The slow process through bureaucracy recognizes stamps and physical signatures on papers that pass from one desk to another, which can be sped up slightly with some cash lubricant [bribes]. (Purbo 2000b)

The current platform is one that consists of transport and communication systems, and bureaucracies.[10] The new platform, in contrast, is speedy and allows for the transcendence of these material constraints:

> Developments in information technology, internet, and electronics are in fact able to transport writing, information, and knowledge in speeds of milliseconds from one place to another on the face of this earth. The platform where we stand has changed its shape significantly, as walls, tables, chairs, bureaucracy, authority, and power have become irrelevant in the new, non-physical platform. This platform is built using [. . .] (code), servers, ports, and resource locators: things difficult to imagine in the physical world as we have known it for thousands of years now. (Purbo 2001c).

The platform here means something that is sociotechnical: it involves what some (e.g., Rip and Kemp 1998; van de Poel 1998) might refer to as a sociotechnical "regime," a certain configuration of sociotechnical relations, including power relations, that achieves stability; a kind of sociotechnical analog of a scientific paradigm. For Purbo, the promise of the Internet is that it will establish a new platform that involves very different kinds of sociotechnical relations, and a different basis for social differentiation. As he puts it, "the need for physical appearances, suit jackets, ties, expensive cars, expensive homes, and prestigious offices will be thrown into question. The value accorded to someone will depend [instead] on the knowledge that they possess, along with their capacity to communicate information and knowledge to their community (*ummat*)" (Purbo 2001c).

Through such open communication, Purbo (2000b) envisions a flat society with an affective intimacy where the "feeling of being part of a close family is very strong, much like it was among breakers [radio amateurs] in the 1970s and 1980s."[11] In this society, he explains, everyone starts off equally and status is achieved on the basis of one's contributions to the group. Such contributions may take the form of writings, email conversations, or the creation and circulation of free software. The community will evaluate whose contributions are best and decide who is worthy of higher status.

This society, according to Purbo, will function based mainly on consensus and custom rather than on written law. Bureaucratic position will not matter; what will matter is your "capacity to engage in discussion, to make good arguments, [and] to represent concepts and perspectives to the public in writing" (Purbo 2001d). Bureaucrats, he explains (2001a), ought to be afraid of this future: "If a high official/institution makes even the smallest mistake, criticism is going to come immediately, one after another, via the open medium of the Internet. [. . .] If a high official is not very capable and is inattentive, it will be very easy for him or her to be audited by the people by way of bringing the problem to the public's attention" (Purbo 2000a). Purbo already saw evidence of such changes underway in electronic mailing lists, where at least one government minister who could not face all the criticism he received about his policies eventually "resigned" (*mengundurkan diri*) from the list.

While diminishing the power of bureaucrats, the new platform will also involve a kind of direct democracy. People will represent themselves rather than be represented. Whereas the current physical platform, with its spatial limitations, is unable to bring all Indonesians into a common discussion space, and so relies on "middlemen" for political representation, the same will not be true on the new platform, where the only "vertical" power will be the *ummat*'s relationship to God (Purbo 2000b).

Based on his experiences building the intercampus network, Purbo (2004) argued that the way Indonesians will arrive at this brave new world is through "bottom-up" strategies that encourage people to become self-taught in the new technologies and to build their own networks and make their own software. Computers will have to be put in schools, and people will need to write Indonesian-language technical manuals and share their knowledge so others can learn. Activists will have to fight to overcome the current regulatory regime for telecommunications, which obstructs bottom-up developments by allocating so much power to the monopoly providers. Bandwidth in the radio spectrum will have to be "liberated" for use by bottom-up initiatives like Internet cafés, self-organized and operated neighborhood networks, and community radio. Activists will also have to fight efforts by the police to shut down and extort money from these initiatives.

Purbo has spent much of the last decade spreading his "naive philosophy" through his road shows; through his appearances on TV, radio, social media; and in the press. As he criss-crosses the country, his presentations always draw a crowd, and they usually leave people dazzled by his enthusiasm, his virtuoso computer skills, and his strong antiauthoritarian streak. I have attended a number of these performances, and in Indonesia, the vast majority of those who attend are young men. During the more technical how-to talks, there appeared to be a minority of people who were highly competent and who followed each step of the way. A larger number of people seemed to be left behind but learned a little something and clearly liked the idea of being present among this insider group of technological savants-cum-activists.

According to Purbo, his presentations on hackers drew the largest crowds. I attended a talk on this topic at Indonesia's biggest mosque, the Istiqlal Mosque on Jakarta's main square, that was attended by a few hundred young people. Everyone sat on the floor in the basement underneath the mosque, and Purbo, along with other presenters, sat cross-legged on a slightly elevated stage, with computers on the table in front of him and a large screen behind him. Purbo's talk included step-by-step training in a couple of straightforward techniques for using one computer to break into another and a discussion of the hacker ethos. His understanding of the values underlying the hacker community is drawn from books (e.g., Steven Levy's *Heroes of the Computer Revolution* [1984]), but in these ideals Purbo clearly discovered something he could identify with.

They are very crazy about computers, their capacity to access computers, and any kind of knowledge about how computers work; they do all of this without limits and with totality. They do not like to hide information away, as all information has to be free, open and transparent—they are followers of the stream of copyleft rather than copyright. They do not believe in authority, bureaucracy, and those in power—power has to be decentralized. A person is evaluated based on his or her capabilities, not some made-up criteria like degree attained, bureaucratic role, public position, or ethnic group. They make art and beauty on the computer and they believe that computers will bring us all to a better condition. (Purbo 2001b)

While many people imagine the so-called information society as one in which new media will form the platform of our social and working lives, Purbo has always imagined it as a society in which everyone is a radio amateur or a hacker. This social imaginary entails a heightened reflexivity about both the political economy and the materialities of communication that form the basis for public life. It is what Chris Kelty (2008) refers to as a "geek" or "recursive" public, namely, a public organized around the deliberate construction, manipulation, and maintenance of the very platform that makes the public possible in the first place. For Onno, this self-organized public should ideally be one whose platform is accessible to all, either because it is low cost or because it is a public resource. It should provide participants with a form of sociality that is characterized by a set of shared technical practices, heightened reciprocity, less hierarchy, more speed, greater immediacy, and greater intimacy. Onno's road shows, seminars, books, and articles have all been part of his effort to mobilize people to feel a sense of belonging to, and desire for, this geek public.

Conclusion

Purbo's road shows, his talks, and his writings, along with those of his collaborators, were instrumental in establishing a sociotechnical imaginary for the Indonesian Internet that linked the budding network to a politics of antiauthoritarianism and freedom and a view of the network as a commons. This imaginary grew out of the encounter between older sociotechnical imaginaries in Indonesia and those of the emerging global Internet. It took shape and acquired its importance in relation to the broader landscape in which the network was rolled out. This landscape was one in which the major telecommunications infrastructure was controlled by monopolies and subordinated to the interests of the New Order's political and economic elite. As hobbyists and researchers built their networks, they were thus constantly coming up against economic, political, and regulatory obstructions to their work. Purbo and his group overcame and bypassed these obstructions by cobbling together an actor network that spanned many universities and institutes within Indonesia and abroad. In essence, the group sought to recreate for the Internet the kind of free space that had long existed in most countries for amateur radio: a domain of network provision that is designated as free and public and stands outside the infrastructure owned and controlled by the large providers. Purbo's political strategy for achieving this objective was to continually expand the Internet actor network while using electronic mailing lists, books, and talks to create a "counterpublic" (Warner 2002) that shared his sociotechnical imaginary. It was a strategy that Latour (2005, 5) might describe as a kind of "thing-politics" gathering an assembly of relevant parties around an object in order to provide an occasion for a public to discuss and debate issues in common.

The early imaginary for the Internet was also shaped by broader political changes underway at the time. The late 1990s, when Purbo was building out the nationwide campus network, saw the rise of the reform movement that eventually ousted the dictator Suharto. The intercampus network was one of the main means through which activists bypassed government censors to communicate about demonstrations and marches taking place in various parts of the country. That Purbo and his group had sought to build a network free of government regulation may well have facilitated its use by activists. At the same time, the association of the early Internet with the successful political campaign to end the New Order regime had the reciprocal effect of helping to reinforce the Internet imaginary that Purbo and his group were promoting.

The growing counterpublic included Indonesian proponents of copyleft, free and open source software, and (want-to-be) hackers. In time, and against a backdrop of liberalization in the telecommunications sector and democratization of the political system, this counterpublic would grow to the point at which it had enough clout to affect regulatory policy. This was particularly evident in the success of Purbo's struggle to "liberate" the 2.4-Mhz portion of the radio spectrum for free public use, which relied on the mobilization of the counterpublic he had helped to create. More generally, the counterpublic has helped to retain an association between the Internet and a politics of freedom. This association has underpinned numerous social media campaigns to expose government graft and dishonesty and to fight nascent forms of online censorship.

There is nothing natural about the association between the Internet and a politics of freedom. But it is an imaginary that still has a remarkable force.[12] An examination of this imaginary in the Indonesian context reveals a complex interweaving of local and global genealogies. It also shows how the spread of this imaginary, and its substance, was shaped by the struggle of building out network of this scale within a media ecology that was so dominated by an authoritarian state. In Indonesia, the association between the Internet and a politics of freedom is a product of these genealogies and this struggle.

Notes

Research for this paper was conducted with support from the Social Sciences and Humanities Research Council of Canada. I would also like to acknowledge and thank a number of people who commented on earlier drafts of this chapter: Sang Hyun Kim, Sheila Jasanoff, William Mazzarella and Grant Otsuki.

1. All translations from Indonesian sources are my own.
2. On Hasan Poerbo, see Suzanne Moon's chapter in this volume. The different spelling of their surnames is due to a shift in official Indonesian spelling that took place in the early 1970s.
3. This history is described in more technical detail in Barker et al. (2001), Lim (2005), and on a wiki created by Purbo, Lim, and myself, available at http://wikihost.org /wikis/indonesiainternet/wiki/start.
4. Habibie was Indonesia's Minister of Research and Technology (1978–98), who would go on to become vice president (1998) and president (1998–99).
5. On the history of the use of packet radio in early Internet development in the United States, see Abbate (1999).
6. See http://ngoprek.org/. Accessed September 30, 2008.
7. Purbo even experimented with creating a network for *pesantren* (Islamic boarding schools) in rural areas around Bandung, so that students in these Islamic schools could join in the fun.
8. The contrast between this approach and Habibie's top-down approach to development was to some degree a reformulation of a much earlier debate about technology and development that took place in the late-colonial Dutch East Indies (see Moon 2007).
9. He practices what he preaches: in my first "Interview" with him he said an interview really was not necessary; I should just bring a jump drive and he would copy his entire hard disk for me to use in my research.
10. A century earlier, the elements of this platform were considered to represent a new age of speed and friction-free movement (Mrázek 2002), but from Purbo's twenty-first-century perspective, they just seem to slow everything down.
11. For other examples of new communications technologies being associated with the erasure of traditional hierarchies in Indonesia, see Barker (2002, 2008) and Siegel (1997).
12. As a counterpoint, see Fred Turner's (2006) critical account of how the imaginative investment in the utopian and libertarian potential of cyberspace and information technology by countercultural icons like Stewart Brand "legitimized a metamorphosis within—and a widespread diffusion of—the core cultural styles of the military-industrial-academic technocracy that their generation had sought to undermine" (p. 238).

References

Abbate, Janet. 1999. *Inventing the Internet.* Cambridge, MA: MIT Press.

Amir, Sulfikar. 2004. "The Regime and the Airplane." *Bulletin of Science, Technology and Society* 24(2): 107–14.

Barker, Joshua. 2002. "Telephony at the Limits of State Control: 'Discourse Networks' in Indonesia." Pp. 158–83 in *Local Cultures and the 'New Asia.' The State, Culture and Capi-*

talism in Southeast Asia, edited by Wan-ling C. J. Wee. Singapore: Institute of Southeast Asian Studies.

———. 2005. "Engineers and Political Dreams: Indonesia in the Satellite Age." *Current Anthropology* 46(5): 703–27.

———. 2008. "Playing with Publics: Technology, Talk and Sociability in Indonesia." *Language and Communication* 28: 127–42.

Barker, Joshua, Merlyna Lim, Arie Rip, Teti Argo, and Sonny Yuliar. 2001. "Social Construction of Technology in the Indonesian Context nr.95-CS-03." Final Research Report submitted to KNAW. Ms. Universiteit Twente, Enschede.

Deibert, Ronald, John Palfrey, Rafal Rohozinski, and Jonathan Zittrain. 2012. "Access Contested: Toward the Fourth Phase of Cyberspace Controls." Pp. 327–99 in *Access Contested: Security, Identity, and Resistance in Asian Cyberspace,* edited by Ronald Deibert, John Palfrey, Rafal Rohozinski, and Jonathan Zittrain. Cambridge, MA: MIT Press.

Geertz, Clifford. 1960. *The Religion of Java.* Chicago: University of Chicago Press.

Hill, David T., and Krishna Sen. 1997. "Wiring the Warung to Global Gateways: the Internet in Indonesia." *Indonesia* 63 (April): 67–89.

Jasanoff, Sheila, and Sang-Hyun Kim. 2009. "Containing the Atom: Sociotechnical Imaginaries and Nuclear Power in the United States and South Korea." *Minerva* 47: 119–46.

Kelty, Christopher M. 2008. *Two Bits: The Cultural Significance of Free Software.* Durham, NC: Duke University Press.

Latour, Bruno. 2005. "From Realpolitik to Dingpolitik or How to Make Things Public." Pp. 4–31 in *Making Things Public: Atmospheres of Democracy,* edited by Bruno Latour and Peter Weibel. Cambridge, MA: MIT Press.

Lim, Merlyna. 2005. "Archipelago Online. The Internet and Political Activism in Indonesia." PhD diss., Universiteit Twente, Enschede.

Moon, Suzanne. 2007. *Technology and Ethical Idealism. A History of Development in the Netherlands East Indies.* Leiden: CNWS Publications.

Mrázek, Rudolf. 2002. *Engineers of Happy Land. Technology and Nationalism in a Colony.* Princeton, NJ: Princeton University Press.

Purbo, Onno. 1990. Jaringankomputerbiayamurahmenggunakan radio. http://www.bogor .net/idkf/idkf/network/jaringan-komputer-biaya-murah-menggunakan-radio-11 -1990.rtf. Accessed October 6, 2008.

———. 2000a. Demokrasi&KembalikeFitrahManusia: Paska Era Informasi. http://www .bogor.net/idkf/idkfwireless/aplikasi/hukum/demokrasi-kembali-ke-fitrah-manusia -paska-erainformasi-07–20.rtf. Accessed October 6, 2008.

———. 2000b. Mengapa DPR kalaubisajadiDewan Rakyat. http://www.bogor.net/idkf /idkf/aplikasi/mengapa-dpr-kalau-bisa-jadi-dewan-rakyat-mar-2000.rtf. Accessed October 6, 2008.

———. 2001a. Badai AFTA, Free Trade, Globalisasi—SiapaTakut. http://bogor.net/idkf /idkf-wireless/aplikasi/badai-afta-freetrade-globalisasi-siapa-takut-09–2001.rtf. Accessed October 6, 2008.

———. 2001b. DuniaBawah Tanah di Internet. http://bebas.vlsm.org/v09/onno-ind-1 /network/network-security/dunia-bawah-tanah-di-internet-08–2001.rtf. Accessed October 6, 2008.

———. 2001c. FilosofiNaifAspekSosial, Budaya&HukumDunia Maya. http://bebas.vlsm .org/v09/onno-ind-1/application/konsekuensi-aspek-sosial-budaya-di-dunia-maya -04–2001.rtf. Accessed October 6, 2008.

———. 2001d. MembuatTeknologiLebihMemihakPada Rakyat &Pencita-nya. http://www

.bogor.net/idkf/idkf-wireless/aplikasi/hukum/membuat-teknologi-lebih-memihak
-pada-rakyat-05–2001.rtf. Accessed October 6, 2008.

———. 2003a. CercahHarapan Dari World Summit on Information Society 2003. http://
yb1zdx.arc.itb.ac.id/data/OWP/library-onno-ind/onno-ind-3/application/policy
/cercah-harapan-wsis-12–2003.rtf. Accessed October 6, 2008.

Purbo, Onno W. 2003b. MembangunBangsaMerambahManca Negara. http://125.160
.17.21/speedyorari/view.php?file=library/library-onno-ind/onno-ind-3/application
/policy/membangun-bangsa-merambah-manca-10-2003.doc. Accessed October 6,
2008.

———. 2004. Motivating Community Based ICT Infrastructure Development. http://www
.choike.org/documentos/community_ict_infrastructure.pdf. Accessed October 6, 2008.

Rip, Arie, and René Kemp. 1998. "Technological Change." Pp. 327–99 in *Human Choice
and Climate Change, Volume 2: Resources and Technology*, edited by Steve Rayner and
Elizabeth L. Malone. Columbus, OH: Battelle Press.

Siegel, James T. 1997. *Fetish, Recognition, Revolution*. Princeton, NJ: Princeton University
Press.

Turner, Fred. 2006. *From Counterculture to Cyberculture. Stewart Brand, the Whole Earth Network, and the Rise of Digital Utopianism*. Chicago: University of Chicago Press.

van de Poel, Ibo. 1998. "Changing Technologies. A Comparative Study of Eight Processes of
Transformation of Technological Regimes." PhD diss., Universiteit Twente, Enschede.

Warner, Michael. 2002. "Publics and Counterpublics." *Public Culture* 14(1): 49–90.

TEN

Consuming Biotechnology:
Genetically Modified Rice in China

NANCY N. CHEN

Several weeks into the lunar new year of the dragon, I visited old friends and colleagues in Beijing during spring 2012. In marked contrast to visits during the previous decade, which took place in bustling restaurants at tables laden with food, the meals are at home, reminding me of earlier times when I first lived and worked in China in the mid-1980s. The dishes are elaborate, and my hosts remark how they also appreciate the opportunity to entertain at home. I am grateful for the more intimate settings and chance to have longer convivial conversations. The renewed intimacy reflects ongoing concerns for food safety in recent years, so that most urban residents avoid food items obtained from smaller stores and markets or from unregulated street vendors. Their kitchens display a vast array of cooking oil, spices, condiments, and other food supplies. I am shown official green labels and told how to avoid counterfeit or scam products sold on the streets. During another memorable meal, one friend reminisced about how much he learned about rural life as a sent-down city youth. "Farmers are incredibly talented. They've managed to cultivate rice even when there is little water." Although he and his family have prospered over the past few decades in the city, he continues to express his appreciation for the work of farmers, especially those who produce rice.

Rice is a critical marker of well-being and prosperity in much of the world. As a staple food source and commodity it has come to define everyday life, especially in China, where the term for meal is synonymous with rice. As scholars of China have shown, rice is a significant window onto state formation that spans the centuries (Anderson 1988; Bray 1994, 2008; Lee 2011). Rice was used in the production of wine for state rituals as well as paper to commemorate such events. The landholding and taxation system of imperial times based on rice productivity determined the wealth of families, villages, and the bureaucracy. Even the physical landscape reflects the coextensive relations between humans and rice cultivation. Iconic rural images of terraced paddies etched into hillsides or next to soaring mountains date back to late imperial China. Rice continues in the present as a key site of assemblages of knowledge, technology, and nation making.

This chapter examines rice cultivation in the context of the sociotechnical imaginaries that define twenty-first-century China and some of its formations of governance. Genetically modified (GM) or transgenic rice, in particular, has become the new locus of forging sovereignty in the face of disparity and projected futures of scarcity. Through the generative lens of imagining the collective good, however, I consider how recent concerns for food safety may interact with, and possibly shift, older imaginaries of food security and potentially set new agendas for evaluating GM rice in China.

Sheila Jasanoff and Sang-Hyun Kim engage the concept of sociotechnical imaginaries as a means of rendering visible the intimate connections between political institutions, national aspirations, and science and technology. In earlier work on nuclear power, they point to national policies as "active exercises of state power, such as the selection of development priorities, the allocation of funds, the investment in material infrastructures, and the acceptance or suppression of political dissent" (2008, 123), thereby highlighting the constitutive elements of national sociotechnical projects, as well as potential cross-national differences across technoscientific policies. In this volume, they further frame sociotechnical imaginaries as key instruments of coproduction by which "technological systems, policy styles, organizational behaviors, and political cultures" are shaped along with people's "hopes and desires for the future" (Jasanoff, Introduction, this volume).

Rice in China, historically as well as now, is a sociotechnical assemblage that reflects coproduced formations of nature, state policy, and national identity. While biotechnology continues to reframe categories of nature and culture with phenomenal techniques that produce new, new things, it is easy to consider these transformative technologies as emblematic of China's advancement, surpassing and marking a rupture with its past agricultural achievements. Yet, as Francesca Bray notes, there is a long history of Chinese statecraft based on rice technology. In her analysis on the role of agricultural techniques mediating between knowledge and matter, she points out that China "was from its inception an agrarian state in the strong sense of the term. The production and circulation of agronomic knowledge by the state was a key technique of government" (Bray 2008, 327).

Building on these frameworks, I argue that sociotechnical imaginaries concerning rice are part of the very DNA, or core, of contemporary Chinese state policies as seen in its five-year plans, focus on development of the biosciences, and, most recently, emerging regulatory frameworks for genetically modified organisms (GMOs). I specifically engage the concept of sociotechnical imaginaries as a means to illuminate how Chinese biotechnology serves as a platform for contemporary formations of governance and future making in the People's Republic. Ensuring the collective good in terms of adequate food sources and other material resources has facilitated assemblages of biotechnology as distinctively "made in China." In this moment of deepening environmental concerns and financial market volatility, the dramatic extremes of impoverishment coexisting with cosmopolitan transformation in China also give rise to new programmatic contexts for national science and technology projects, coproducing technological goals of governance with attempts to meet individual concerns for livelihoods and well-being.

Securing Rice and Nation

The story of rice in China is narrated through frameworks of urgency and necessity. Rice production is often cited in relation to China's ever-present demographic situation. As in dynastic periods, contemporary leaders face ongoing concerns for feeding the world's largest population that is also the biggest consumer of rice. The size of the nation's populace is often invoked to indicate the Malthusian dilemma of food security.[1] For instance, a recent article in *Nature* remarked, "China's population is set to top 1.45 billion by 2020, and the country needs to increase grain production by about 25%" (Qiu 2008, 850).[2] Such figures are amplified by social memories of hunger and starvation as well as concerns for inadequate food supplies that live on even in the present period of abundance. Although China has faced ongoing concerns for feeding its people throughout past centuries, considering the vast population as a perennially exceptional situation enables biotechnology and market formations to be utilized in effect as a double helix in the formation of new China and its emerging biosovereignty. Making GM rice facilitates a China-first position in the production of material objects as well as claiming modern trajectories of progress. GM rice authenticates the necessity of state governance in responding to disastrous scenarios of a burgeoning population.

In previous work, I traced the role of state funding that facilitated the initial biotech bloom in China during the first decade of this century (Chen 2010). Leading institutions such as the Ministry of Science and Technology, Ministry of Health, and Ministry of Education together with the National Reform Development Commission (formerly the State Planning Commission) were key entities that shaped early biotechnology research and growth in this sector. Five-year plans have been a key instrument of socialist agenda setting and of defining comprehensive goals for China's future. These documents and practices demonstrate what Jasanoff considers in her conclusion to this volume as the "integration of imagination with rulership," facilitating the joint structuring of life worlds and governance. The twelfth five-year plan (2011–2015) distinguishes itself from its predecessors by highlighting the theme of "scientific development" as a guiding principle. The biological industry (including biomedicine, biopharmaceuticals, biotechnology, and bioagriculture) is specifically noted as a "strategic new industry" in part III, chapter 10, section 2, of the plan.[3] Moreover, the revised plan includes increased focus on research and development (R&D), to be 2.2 percent of gross domestic product, as well as expanded goals of innovation, with the target of 3.3 patents for every 10,000 people.

While the twelfth five-year plan offers prescriptive targets that are future oriented, it is instructive to examine the existing biotechnology industry and the complex overlapping arenas of policy, knowledge, and practices. These convergences form what Xiaobai Shen, in her summary on the extensive evolution of rice technology, refers to "as a socio-technical ensemble—a specific assemblage of intertwined material and social elements (including plants, physical geographies, systems of irrigation, land system, agricultural knowledge and practices, government roles, in modern society, and in addition to these, the relation between farmers and R&D activities)" (Shen 2010, 1041). An ensemble still needs generative linkages, however, to knowledge making and to ways of animating social and material entities. In this regard, Jasanoff (Introduction, this volume) argues that "imagination, a crucial reservoir of power and action, lodges in the hearts and minds of human agents and institutions" and requires "skilled implementation." In the postgenomic world of rice technology, new platforms of knowledge making are being facilitated by key players, such as farmers and scientists, so that keeping focus on their transformed relations to rice technologies is crucial.

This chapter specifically highlights the increasingly prominent role of scientists engaged in the development of emergent rice technologies and their cultivation. While agricultural knowledge, farming practices, and farmers have previously been "a domain of collective complementary knowledge production that spanned the political spectrum rather than being restricted to specialist imperial bureau or to members of the educated elite" (Bray 2008, 143), it is unlikely that GM rice technologies will continue this rich tradition. As developers of new rice strains, scientists navigate between the worlds of bench science and the field, as well as the bureaucratic corridors that fund such research. With transgenic rice, scientists are significant interlocutors of governance who enable sociotechnical imaginaries to materialize in forms of state-owned knowledge, practices, and objects. In what follows, I first address the realm of hybrid rice technology that distinguishes contemporary Chinese agricultural biotechnology from its counterparts elsewhere in Asia. I situate the development of GM rice in a continuum from these earlier rice technologies.

Rice Technology in Postgenomic China: From Hybrid to GM

In 2007, China produced a breathtaking 186 million tons of rice, of which 65 percent was primarily hybrid rice (Normile 2008, 333).[4] Rice researcher and veteran scientist Yuan Longping is credited with the discovery of hybrid rice in Hunan province during the late 1960s and for advocating its cultivation during the 1970s. While the timing overlaps with the Green Revolution, which entailed technology transfer to third world nations, the story of Chinese rice technology is folded into a different narrative of independent knowledge formation in the midst of land reformation. Hybrid rice strains were sought for higher yield and drought resistance after an intensive period of widespread famine despite extensive socialist restructuring of the land tenure system and focus on agricultural production. After finding a wild rice strain in the field, Yuan developed a "super" rice that resulted in higher yields. Heterosis, the phenomenon that crossbred strains have better traits than parental strains, became a key component of subsequent rice research and cultivation. Despite such hybrid vigor, rice varieties from a narrow genetic strain are still subject to pests such as stem borers, requiring the extensive use of industrial pesticides and fertilizers to aid in production. Most significant, however, was the shift from farmers owning the means of production, by saving seeds to grow the next season's crop, to their reliance on seed suppliers of hybridized rice. The transition to hybrid rice production during the 1970s was greatly facilitated by the state becoming a regular provider of seeds and industrial agricultural chemicals (Shen 2010). Chinese farmers have adopted the new varieties at such a rate (every four to five years) that the turnover in their "technology portfolios" was closer to that of American farmers (every three to four years) in certain states than to other developing nations (Rozelle et al. 2005). Rice technology in this context of knowledge production thus entails an intimate but hierarchical relation among rice cultivators, scientists, and the state.

China's involvement in rice genome sequencing became a global story at the onset of the twenty-first century when the *indica* strain sequence was identified in 2001 and subsequently published in *Science* (Yu et al. 2002). In earlier work, I discuss the rice genome race and the significance of this sequencing for a China-first position that can take on multinational corporations such as Monsanto or Sygenta (Chen 2010). In her insightful analysis of emerging realms of "hybrid property regimes," Elta Smith (2012 and this volume) addresses key reframings of private property and the public domain introduced by multiple rice genome projects. In particular, she notes that combinations of state and private entrepreneurial funding are utilized to generate knowledge that is classified as private property or public good. The lines between public and private have blurred to the extent that hybrid is the most appropriate category to identify these new formations of funding, knowledge, and property regimes. The participation of Beijing Genomics Institute (BGI), a private company established with and supported by state funding, illustrates this complex dynamic of twenty-first-century genomics.

BGI was central to the selection of the *indica* over the *japonica* strain that was being sequenced by an international consortium of scientists from ten different countries. Since BGI's initial involvement with human and rice genomes, the company itself has taken on a heterogenous identity with state affliation and multiple funding structures from municipal loans to outsourced sequencing. After sequencing the SARS (severe acute respiratory syndrome) virus in 2003, BGI was incorporated as part of the prestigious Chinese Academy of Sciences. Three years later, the company headquarters moved to Shenzhen with the infusion of ten billion *renminbi* that purchased state-of-the-art second-generation sequencers as critical infrastructure for even more ambitious projects. Collaboration with international researchers and clients has also provided a flow of capital enabling the industrialization of sequencing such that BGI has been dubbed the "Sequencing Factory" (Cyranoski 2010), with faster completion times for more complex species. The list of sequencing projects at BGI reflects national interests such as the panda and silkworm genomes, as well as ongoing projects for one thousand plants and animals as well as 10,000 microbial genomes.[5] The company is notable also for its youthfulness, with an average employee age of only twenty-six years (Frank 2011).

BGI is a dramatic actor engaged in ambitious sequencing projects that drew worldwide attention to China's participation in rice genomic research. Everyday rice research, however, is carried out primarily by scientists affiliated in an intricate web of state laboratories, research institutes, and universities across the mainland. Since the establishment of the People's Republic of China in the mid-twentieth century, the concentration of R&D for rice technology has been defined by this network of scientific specialists. While rice sequencing has received more sensational press, the ongoing research efforts in developing rice strains has quietly been at the center of developing new sociotechnical assemblages.

Since the completion of rice sequencing in the previous decade, current scientific research on rice genetics has engaged in bioprospecting for functional aspects of genomics—in effect creating specific categories of information to mine for possible commercialization. Such activities include examining components of the rice genome for particular traits such as insect or drought resistance, nutritional efficiency, root development, flowering rate, and cold tolerance, among other elements relevant to rice cultivation. There is extensive specialization in genomic research depending upon the level and function of the genetic material being studied. "Mutant libraries," a compendium of sites where transferred DNA (tDNA) is inserted into various parts of the rice gene, have been utilized for examining such different traits (Li et al. 2011, 303). In addition, full-length cDNA libraries have been developed to analyze function at transcriptional and translational levels. Rice proteomics has emerged as another site of differentiated knowledge to examine the phenotypic qualities of plants based on the influence of proteins. Core collections of rice germplasm have also been developed in the ongoing study of the biodiversity of the variations in this plant. The extensive levels of specialized knowledge and the creation of new categories reflect the transitional role of rice scientists from breeders to curators of genomic data.

Although shot gun sequencing and next-generation sequencers can identify genomic transcripts faster than ever before, the process of R&D in GM rice is currently still lengthy, not unlike drug to market R&D timelines in the pharmaceutical industry. Safety concerns have added to the length of the timeline of rice to market, despite the technological assemblages that facilitate rapid sequencing. In 2005, the State Environmental Protection Agency announced that China had ratified the 2003 Cartegena Protocol on Biosafety, a regulatory document that governs the international movement of GMOs (Xinhua 2005).[6] Biosafety is broadly defined in this document as "the need to protect human health and the environment from the possible adverse effects of the products of modern biotechnology."[7] Prior to adopting this international regulatory framework, earlier forms of biosafety in China in 1993 focused on genetic engineering of human genetic material with four grades or levels of risk (Huang et al. 2008). Nearly a decade later, in 2001, four grades of biosafety were also adopted for five different stages of development of agricultural GMOs. In addition, the government formulated implementation guidelines concerning trials, processing, import/export, and monitoring (Pray 2006). The Biosafety Committee, consisting of fifty-two scientists under the Ministry of Agriculture, became the primary body of experts engaged in oversight of biosafety. As a result of prior biosafety regulations and the new protocol, there are several stages in the development of GM rice before a strain receives certification for release to market or commercialization. "China's biosafety procedures require transgenic crops to pass through three phases of trials before being commercialized: field, environmental release, and preproduction trials" (Wang et al. 2005, 613).

Despite this potential "bottleneck" (Lu and Snow 2005, 671), there are many different GM varieties at different stages of trial tests (Huang et al. 2008). Each year, more villages across different provinces have participated in trial plantings. The lower use of pesticides for GM rice compared with hybrid or conventional rice, with less labor devoted to multiple applications at different stages of cultivation, is appealing to farmers. "As of 2005, >100 GM rice varieties have been in field testing. Most of them are insect-resistant lines. Four GM rice varieties—two Bt rice, one rice transgenic for the Xa21 gene and one herbicide-resistant rice—are now close to approval for commercialization" (Wang and Johnston 2007, 717). What is crucial to note, however, is that the adoption of the biosafety standards has strengthened the hierarchical relationship between rice scientists and their counterparts in the field. In their assessment of GM rice field productivity, researchers noted that "preproduction trials in the experimental stations are being cultivated by *farmers-cum-technicians working under the direction of the scientists*" (emphasis added; Huang et al. 2007, 248). The regulatory regime of GM rice depends, in effect, upon the shifting role of farmers from skilled breeders to field technicians and the ascendance of scientists as custodians of increasingly specialized knowledge.

The majority of initial GM crops have focused on first-generation genetic engineering to reduce pests and weeds. Yet, there are many different potential risks associated with GM crops that may impact ecological and human health. Risks from GM rice are framed in terms of uncontrollable flow of genetic material from various parts of the plant to eventual incorporation in the food chain. The main environmental concern is "transgene escape," where genes from a transgenic plant are transferred to a non-GM or wild plant (Lu and Yang 2009, 1085). Gene flow can be mediated via pollen, seed, crop field, or weeds. The most common traits that can spread from GM rice to wild rice varieties include herbicide resistance and other fitness traits (Lu and Snow 2005). Scientists conducting field tests try to alleviate this problem by creating buffer zones such that genetic material from trial crops theoretically should not be able to reach other crops. Despite buffer zone planting, it is still possible for gene flow to occur.

Moreover, in a letter to the editors of *Nature Biotechnology* in 2005, the issue of agricultural management problems such as illegal planting of GM crops was raised as posing as much of a central concern as gene flow (Zi 2005). The issue of GM rice entering the food chain came to the fore a year later when *Nature* news online used the headline for a widely circulated story "Escaped GM Rice Reaches Europe" (2006). The article reported that Greenpeace had conducted tests on imported rice noodles from shelves in France, Germany, and the United Kingdom that were found to contain genes from Bacillius thuringiensis (Bt) bacteria from GM rice. The impact of GMOs on human health is a topic of intense dispute by the biotech industry and food activists. With regard to GM rice, a specific concern is allergenicity, when a new protein may cause the immune system to have an adverse reaction. In anticipation of this concern there is a GM rice strain developed in Japan with supposedly reduced allergencity as a primary feature.

While Chinese rice biotechnology has mostly involved farmers-breeders and scientists as the primary actors in rice cultivation, the role of consumers in a market-driven society is increasingly taking on more significance in reframing concepts of risk. Increased concerns about food safety are reframing earlier priorities for food security. In what follows, I address the recent challenges to the evolving regulatory framework concerning GM rice resulting in coproduction of new categories of risk, safety, and security.

Forging the Collective Good: Food Safety in a Context of Insecurity ·

Grain is a special commodity and material of strategic importance crucial to national economy and people's livelihood. The Draft for Opinions makes it very clear that safeguarding national grain security is the fundamental purpose of the legislation of the grain law.

—*Grain Law (Draft for Opinions)* (Lagos and Jiang 2012)

This passage is the preamble to a draft law issued by the State Council of the People's Republic of China on February 21, 2012. Comments on the draft online and in writing were widely solicited up to a month after its public release. The draft law is the most recent approach by the Chinese bureaucracy to address concerns for biosafety from GM foods. Rather than focusing solely on rice, language used in the document concerns the production, processing, distribution, and quality of all grains, edible oils, and seeds. Inviting concerned citizens to participate in providing feedback marks a remarkable shift in the regulatory process.

To understand the significance of this shift in the Chinese regulatory context, it is important to trace the growing role of consumers in voicing collective concerns about GM foods. Despite extensive research on and capital support for rice genomic research, little information about GM rice was circulated in the popular realm by state officials at the program's outset. Although GM labeling laws were introduced in 2003, compliance with this regulation was difficult for most Chinese crop and food producers, and the provision was widely unimplemented (Jia 2003). Early surveys conducted at the onset of the twenty-first century indicated that the majority of polled consumers in China had nearly no knowledge of biotechnology. Market research conducted as recently as 2005 concluded that Chinese consumers knew little about GM foods. Moreover, if any respondents had knowledge of biotechnology, their views were generally favorable (Ho et al. 2006). The shift to second-generation GM foods with functional or enhanced nutritional features rather than pest resistance has been a key component of ongoing GM rice research. A recent study, utilizing experimental auction methods for folate-enriched GM rice, indicated that consumers were willing to consume and even pay higher prices if the GM food item is considered to promote health with enhanced nutritional features through biofortification (De Steur et al. 2011).

Not all forms of GM rice, however, are the same, and consumer awareness of this fact has been rising through varied channels. Greenpeace commenced a targeted education campaign about GM foods in China in 2004, initially in Yunnan province and eventually countrywide. After a seven-year campaign, the organization announced on January 31, 2012, that China had suspended the commercialization of genetically engineered rice.[8] The anti-GMO campaign in China was further amplified by allegations raised in August 2012 regarding the testing of Golden Rice among six- to eight-year-old children in Hunan Province during trials in 2008. In this volume, Elta Smith writes about the role of corporate nonstate entities in shaping certain "imaginaries of science, technology, economics and ethics." Golden Rice in particular, she notes, has a specific history of knowledge formation in the corporate sector such that there is a "highly unregulated and disorderly set of power relations between corporations and recipients." Similar worries emerged in China. Broad ethical concerns with preventing widespread hunger were offset by more immediate concerns about intellectual property and informed consent. Greenpeace raised concerns that informed consent was not properly secured and parents were not informed prior to their child's participation that the Golden Rice being tested in trials was GM to produce beta-carotene to enhance the bioavailability of vitamin A. After a three-month investigation, the China Center for Disease Control (CCDC) formally announced that three Chinese principal investigators (two of whom were CCDC officials) were relieved of their posts on December 6, 2012. A Central China Television documentary concerning the controversy was broadcast nationally two days later. The matter was investigated by the National Institutes of Health and Tufts University to address Institutional Review Board concerns for informed consent on the part of United States–based scientists (Qiu 2012).

The recent sea change in awareness about GM rice is part of a broader spectrum of deep-seated concerns for contaminated and fake foods manufactured in China. Since 2008 there has been worldwide concern for the safety of Chinese medicines, toothpaste, pet food, infant formula, toys, and, most recently, food products in the global commodity chain. Both official Chinese news media as well as netizens have reported on the extensive web of unsupervised manufacturers or renegade producers who have gone through elaborate means to produce ingenious but dangerous fake foods, such as fake bottled water (using untreated tap water instead of mineral water), faux eggs (made with chalk shells and chemical yolks), mock walnuts (shells with cement rocks inside), tainted noodles (made of wax and ink), and even fake rice (made out of potatoes and plastic resin). During the spring of 2011, urban residents were careful to avoid street food vendors and ate out less at restaurants for fear that "gutter oil" (reclaimed oil from kitchen drains) was used in cooking their dishes. Eventually, the public security ministry conducted a sweep across fourteen provinces to locate illegal producers and sellers of reprocessed oils. Despite extensive efforts to prepare foods at home with elaborate attention to sourcing, the gutter oil scandal seemed to be a tipping point in demands for broad-scale measures to ensure food safety. In response to the outrage of parents and consumers, state agencies have been engaged in an ongoing campaign to allay fears and respond to widespread food scandals. In April 2012, the official news agency Xinhua reported that the State Food and Drug Administration (SFDA) issued a warning for eleven "bogus" food products. In addition to urging local officials to "strengthen inspection" and "severely punish the producers," consumers were directed to the SFDA website and urged to report illegal or fake products to local administrators (Xinhua 2012). Increased transparency in the inspection process as well as offering venues of communication to consumers have introduced new possibilities for sharing governance over food safety.

How will the fake rice and ongoing food safety scandals affect the trajectory of GM rice cultivation? In the short term it seems that fake and counterfeit foods have raised awareness about the food commodity chain and the vulnerability it creates for ordinary consumers. Notions of the collective good in this context are focused on the need for enforcing regulations for safe food production and tracing the provenance of foods. It is unclear, however, whether food safety concerns will have an impact on the long-term issue of food security for the world's largest national population. GM rice technology will continue to be developed in the foreseeable future mainly because it is framed primarily in terms of technological modernization, knowledge production, and intellectual property rather than in a continuum of fake or tainted foods.

Nonetheless, sociotechnical imaginaries can shape opportunities for future directions of governance and citizenship by way of "producing diverse visions of the collective good" (Jasanoff, Introduction, this volume). The process of nation building in twenty-first-century China is evolving to address not just material capabilities but also the cultivation of its citizens. Rather than making the world's largest dam, expanding space exploration, or hosting another successfully stage-managed Olympic event, careful attention to the cultivation of ordinary consumers and rice farmers as primary knowledge producers in an age of intensive agricultural technologies may be one of the breakthrough imaginaries needed to shape outcomes focused on collective well-being, and not just survival.

Notes

I am deeply grateful to Sheila Jasanoff and Sang-Hyun Kim for their encouragement and patience as coeditors. Sheila, in particular, offered incisive and helpful remarks as discussant for this paper in the workshop as well as for the volume chapter.

1. Susan Greenhalgh's research (2009) offers extensive contextualization and analysis of China's population policy.
2. This figure is consistent with the Population Reference Bureau projections; see http://www.prb.org/.
3. "China's Twelfth Five Year Plan (2011–2015): The Full English Version." http://cbi.typepad.com/china_direct/2011/05/chinas-twelfth-five-new-plan-the-full-english-version.html. Accessed on June 29, 2012. For a useful comparison of the eleventh and twelfth five-year plans, see Casey and Koleski (2011).
4. The totals for rice output can differ dramatically each year. See also the National Bureau of Statistics, http://www.stats.gov.cn/english/.

5. For additional information on these projects, see the BGI website, http://en.genomics .cn/. Accessed on August 2, 2012.

6. See also the Biosafety Clearing House online, http://english.biosafety.gov.cn/. Accessed on August 2, 2012.

7. Cartegena Protocol on Biosafety to the Convention on Biological Diversity. See http:// www.cbd.int/doc/legal/cartagena-protocol-en.pdf. Accessed on August 2, 2012.

8. See http://www.greenpeace.org/international/en/news/features/China-says-no-to -genetically-engineered-rice/.

References

Anderson, Eugene N. 1988. *The Food of China*. New Haven, CT: Yale University Press.

Bray, Francesca. 1994. *The Rice Economies: Technology and Development in Asian Societies*. Berkeley, CA: University of California Press.

———. 2008. "Science, Technique, Technology: Passages between Matter and Knowledge in Imperial Chinese Agriculture." *British Journal for History of Science* 41(3): 319–44.

Casey, Joseph, and Katherine Koleski. 2011. *Backgrounder: China's 12th Five-Year Plan*. Washington, DC: US China Economic and Security Review Commission, June 24.

Chen, Nancy. 2010. "Feeding the Nation: Chinese Biotechnology and Genetically Modified Foods." Pp. 81–93 in *Asian Biotech: Ethics and Communities of Fate*, edited by Aihwa Ong and Nancy N. Chen. Durham, NC: Duke University Press.

Cyranoski, David. 2010. "Chinese Bioscience: The Sequence Factory." *Nature* 464: 22–24.

De Steur, Hans, Xavier Gellynck, Shuyi Feng, Pieter Rutsaert, and Wim Verbeke. 2012. "Determinants of Willingness-to-Pay for GM Rice with Health Benefits in a High-Risk Region: Evidence from Experimental Auctions for Folate Biofortified Rice in China." *Food Quality and Preference* 25: 87–94.

Frank, Lone. 2011 "High Quality DNA." *Newsweek*, April 24.

Greenhalgh, Susan. 2009. "The Chinese Biopolitical: Facing the Twenty-First Century." *New Genetics and Society* 28 (3): 205–22.

Ho, Peter, Eduard B. Vermeer, and Jennifer H. Zhao. 2006. "Biotechnology and Food Safety in China: Consumers' Acceptance or Resistance?" *Development and Change* 37(1): 227–53.

Huang, Jikun, Ruifa Hu, Scott Rozelle, and Carl Pray. 2008. "Genetically Modified Rice, Yields, and Pesticides: Assessing Farm-Level Productivity Effects in China." *Economic Development and Cultural Change* 56 (2): 241–63.

Jasanoff, Sheila and Sang-Hyun Kim. 2009. "Containing the Atom: Sociotechnical Imaginaries and Nuclear Power in the United States and South Korea." *Minerva* 47(2): 119–146.

Jia, Hepeng. 2003. "GM Labeling in China Beset by Problems." *Nature Biotechnology* 21 (8): 835–6.

Lagos, Joshua Immanuel, and Jiang Junyang. 2012. *China—People's Republic of: Grain Law (Draft for Public Comments)*, GAIN Report No CH12023. USDA Foreign Agricultural Service Global Agricultural Information Network.

Lee, Seung-Joon. 2011. *Gourmets in the Land of Famine: The Culture and Politics of Rice in Modern Canton*. Stanford, CA: Stanford University Press.

Li, Jinjie, Hongliang Zhang, Deping Wang, Bo Tang, Chao Chen, Dongling Zhang, et al. 2011. "Rice Omics and Biotechnology in China." *Plant Omics Journal* 4(6): 302–17.

Lu, Baorong, and Alison Snow. 2005. "Gene Flow from Genetically Modified Rice and Its Environmental Consequences." *BioScience* 55 (8): 669–78.

Lu, Baorong, and Chao Yang. 2009. "Gene Flow from Genetically Modified Rice to Its Wild Relatives: Assessing Potential Ecological Consequences." *Biotechnology Advances* 27(6): 1083–91.

Normile, Dennis. 2008. "Reinventing Rice to Feed the World." *Science* 321 (5887): 330–33.

Pray, Carl. 2006. "Costs and Enforcement of Biosafety Regulations in India and China." *International Journal of Technology and Globalisation* 2 (1/2): 137–57.

Qiu, Jane. 2008. "Is China Ready for GM Rice?" *Nature* 455: 850–52.

———. 2012. "China Sacks Officials over Golden Rice Controversy." *Nature News and Comments* (published December 10). http://www.nature.com/news/china-sacks-officials -over-golden-rice-controversy-1.11998.

Rozelle, Scott, Jikun Huang, and Keijiro Otsuka. 2005. "The Engines of a Viable Agriculture: Advances in Biotechnology, Market Accessibility and Land Rentals in Rural China." *China Journal* 53: 81–111.

Shen, Xiaobai. 2010. "Understanding the Evolution of Rice Technology in China—From Traditional Agriculture to GM Rice Today." *Journal of Development Studies* 46(6): 1026–46.

———. 2012. "Rice Genomes: Making Hybrid Properties." Pp. 184–210 in *Lively Capital: Biotechnologies, Ethics, and Governance in Global Markets*, edited by Kaushik Sunder Rajan. Durham, NC: Duke University Press.

Wang, Yanqing, and Sam Johnston. 2007. "The Status of GM Rice R&D in China." *Nature Biotechnology* 25 (7): 717–8.

Wang, Yonghong, Yongbiao Xue, and Jiayang Li. 2005. "Towards Molecular Breeding and Improvement of Rice in China." *Trends in Plant Science* 10 (12): 610–14.

Xinhua. 2005. "China Ratifies Protocol on Biosafety" (May 20). http://www.china.org.cn /english/2005/May/129371.htm. Accessed on August 2, 2012.

———. 2012. "11 Fake Foods Uncovered in China" (April 11). http://news.xinhuanet.com /english/china/2012-04/11/c_131520732.htm. Accessed on February 17, 2013.

Yu, Jun, et al. 2002. "A Draft Sequence of the Rice Genome (oryza sativa L. ssp. Indica)." *Science* 296 (5565): 79–92.

Zi, Xun. 2005. "GM Rice Forges Ahead in China amid Concerns for Illegal Planting." *Nature Biotechnology* 23 (6): 637.

ELEVEN

Imaginaries of Science and Society: Framing Nanotechnology Governance in Germany and the United States

REGULA VALÉRIE BURRI

Introduction

Emerging technologies have been framed in varied ways in different political cultures; biotechnologies, for example, have been developed, responded to, and regulated in different manners in Europe and the United States (Bauer 1995; Jasanoff 2005; Parthasarathy 2007). The politics of pharmaceuticals (Daemmrich 2004) and chemicals (Brickman et al. 1985) likewise were shaped by varying regulatory cultures. These examples point to major differences in the ways modern societies produce, implement, and regulate technological innovation. Sociotechnical imaginaries play an important role in these processes, in forming the ways societies assess and govern emerging technologies.

The imaginaries of individuals and of social groups can be seen as cultural tools to understand our present and past and to anticipate futures. Sociotechnical imaginaries as discussed in this volume open up spaces of possibility. They create knowledge and expectations about future social and technical orders, and they are crucial for practices involving science and technology. Imaginaries, for example, play an important role in the development, assessment, and regulation of cutting-edge technologies.[1]

This chapter explores sociotechnical imaginaries related to nanotechnologies in the political cultures of Germany and the United States. It examines how the role of science and technology is imagined in each country. It looks at how the meanings and functions of science and its relation to society are shaped in policy documents aimed at deliberating future developments of nanotechnologies and defining national science policies and research strategies. By probing the tacit assumptions underlying crucial policy documents, the chapter asks if and in what ways nanotechnology assessment and governance are shaped by national sociotechnical imaginaries.

Nanotechnologies were considered as key technologies of the twenty-first century by many stakeholders, including national governments. At the same time, their development and societal impact reflected a heterogeneous set of expectations (Kearnes and MacNaghten 2006). Controversial views were embedded within broader cultural assumptions about science and its role in society and about the future in general, and they drew on the self-understanding and identity of individuals, groups, and nations.[2]

Processes of identity construction and visions of technological innovation are always intertwined. Sheila Jasanoff (2004) has called attention to the coproduction of scientific knowledge and social identities, when "the ways of knowing the world" are inseparably linked to the ways in which members of a society seek to organize and control it. Drawing on the example of biotechnology and political power, Jasanoff (2005) displays technology as a site of both world making and nation building. Similarly, when analyzing two crucial documents in current nanotechnology policies in Germany and the United States,—Germany's "Action Plan" and the US "NNI Strategic Plan"—this chapter explores some very concrete ways of knowing the world and managing it.

Although both documents delineated research and governance strategies of nation-states, the institutional settings in which they were formulated were very different. Nevertheless, both documents were produced in a similar time period and represented similar intentions to outline a strategy for dealing with innovations in nanotechnology. The documents can be understood as political artifacts produced in particular historical and cultural contexts. They are the results of negotiations among heterogeneous people, departments, and organizations—each with different interests and aims. In the resulting constructions, competing imaginaries became black boxed; it is no longer visible that the words and sentences in each document were once the objects of long discussions. The documents express compromises of antagonistic views and can be understood as stabilized agreements on national strategies. Nevertheless, like any other policy documents for science and technology, the documents analyzed in this chapter can be read as expressions of particular sociotechnical imaginaries. The imaginations in question can be discerned in both the documents' explicit goals and their underlying assumptions.

This chapter compares key aspects of sociotechnical imaginaries between Germany and the United States. It analyzes the dominant visions, images, and ideals associated with nanotechnologies; the expectations of gains, profits, and risks projected onto nanotechnologies at the current early stage of research and development (R&D); the governance of nanotechnologies as imagined in policy documents; and the role society is imagined to be playing in nanotechnology assessment and development. In alignment with the program of real-time technology assessment formulated by David Guston and Daniel Sarewitz (2002), this chapter contributes to understanding technological innovation by focusing on the ways in which sociotechnical imaginaries shape societal responses to emerging technologies.

Deliberating and Ranking

The sociotechnical imaginaries for nanotechnology in Germany—as one prominent example in Europe—and the United States can be compared along three dimensions: first, the ways in which the role of science and the benefits and risks of nanotechnologies are imagined; second, the envisioned governance of these technologies; and, third, the perceived role of citizens and state-society relations.

The Role of Science

With a commitment of about 440 million euros in public funds, Germany led Europe as the country with the highest investment in the nanotechnology sector at the beginning of the millenium.[3] The German government established a so-called nano initiative as part of its general high-technology strategy. Seven federal ministerial departments were involved in discussing and formulating what eventually became the "Action Plan 2010" (BMBF 2007).[4] Directed by the Federal Ministry of Education and Research (Bundesministerium für Bildung und Forschung), the Action Plan provided a unified framework across all ministerial departments and can be seen as a result of a negotiation process among the concerned policy makers. Each of the seven departments nominated a nanotechnology representative who was responsible for taking part in defining the future of national nanotech governance. In their replies to the report's leading question, "Why is there a need for a Government Nano-Initiative?" (BMBF 2007, 7), the varied perspectives of the representatives become evident. Each participant underlined the importance of his or her domain. For example, the representative of the Federal Ministry of Food, Agriculture and Consumer Protection emphasized the importance of protecting the user, while his colleague from the Federal Ministry of Health perceived an increasing need for innovative technologies in medicine. Such divergent goals had to be negotiated into a common strategy.

Despite their differences, the negotiators shared some common visions and assumptions. Most important, they all agreed that science should contribute to national economic growth through innovation. Annette Schavan, then Federal Minister of Education and Research, articulated this view. In her introduction to the report, she maintained that the "availability of nanotechnology [. . .] determines the performance and international competitiveness of the German economy." Germany, in her opinion, "has no choice but to concentrate on a strategy of permanent innovation." Nanotechnologies, she claimed, were part of this strategy because they are "one of the most promising interdisciplinary fields of technology in the world." Schavan conceived them as "'tickets' to the future" since they "are particularly significant for the creation of jobs" and will "open up entirely new markets" (BMBF 2007, 3). The approximately six hundred companies that, according to the report, were involved in the development, application, and sales of nanotechnological products in Germany provided around 50,000 jobs (BMBF 2007, 13).

Competitiveness in the labor and product markets thus constituted a driving force in Germany's nanotechnology strategy. "Germany must face up to increasingly demanding technological and economical challenges in the future," the report stated, and it criticized Germany's innovation in nanotechnology within an international context: "In comparison with the USA and South East Asia, Germany takes more time to turn the results of R&D into products" (BMBF 2007, 13). The vision of economic development through scientific and technological innovation was thus shaped by a global perspective. The world was imagined as a place in which nations and regions compete for jobs, products, and markets. Science was viewed as a means and resource to advance the country in this global competition with the aim of achieving economic growth and prosperity.

The ministerial representatives also expressed their hope that nanotechnologies would contribute to improving the environment. The representative from the Federal Ministry for the Environment, Nature Conservation and Nuclear Safety plainly supported this view. Nanotechnologies, he argued, could contribute to increased resource efficiency and improved environmental protection (BMBF 2007, 7). More unexpectedly, his colleague from the Federal Ministry of Economics and Technology shared this view, arguing for the implementation of nanotechnological products because they may contribute to the promotion of "environmentally friendly economic circuits" (BMBF 2007, 9). The potential of nanotechnology to solve environmental problems was mentioned prominently throughout the report. A number of specific applications were suggested, such as novel filter systems for wastewater treatment, the replacement of toxic substances with nanomaterials, and saving on raw materials by using miniaturized technical components (BMBF 2007, 19).

The dominant German imagination of the role of science in society was thus twofold: nanotechnologies should contribute to both economic welfare and environmental improvement. They were perceived as innovations with a huge potential and able to integrate what for many decades had been declared by economists as conflicting goals: to advance the economy while improving the environment. Nanotechnologies were imagined as having the power to advance industrial capabilities and economic progress while at the same time diminishing what Ulrich Beck (1992 [1986]) called the unintended side effects of modernity.

In the United States, a different perspective emerged on the role of science and technology in society. In 2001, the National Nanotechnology Initiative (NNI) was established as a program to coordinate US federal nanotechnology R&D. By 2014, around twenty federal departments and independent agencies were involved in this initiative. Funding for the NNI was exceptionally high; the federal budgets in this period provided about 1.5 billion US dollars each year.[5] The NNI included a Strategic Plan prepared by the Subcommittee on Nanoscale Science, Engineering, and Technology, a part of the National Science and Technology Council. A first version was released in December 2004, and an updated one in December 2007. The updated report listed four goals for nanotechnologies: to advance a world-class nanotechnology R&D program; to foster the transfer of new technologies into products for commercial and public benefit; to develop and sustain educational resources, a skilled workforce, and the supporting infrastructure to advance nanotechnology; and to support responsible development of nanotechnology (NSET 2007, 3).

At first glance, the Strategic Plan seems similar to Germany's Action Plan. As in the German report, the role of nanotechnology for advancing the economy was taken to be central. "Nanotechnology contributes to United States competitiveness by improving existing products and processes and by creating new ones," the report stated, and it claimed that the NNI "implements strategies that maximize the economic benefits of its investments in nanotechnology" (NSET 2007, 3). The NNI Strategic Plan also pointed to the importance of "responsible development." It highlighted a program of research, education, and communication dedicated to exploring issues in both "Environment, Health, and Safety" and "Education and Societal Dimensions." Furthermore, a separate strategy report was published on nanotechnology-related environmental, health, and safety research (NSET 2008).

Despite some similarities, the US NNI Strategic Plan differs from the German report in focus and terminology. The most explicit purpose of sponsoring emerging technologies was to ensure the country's number one position in nanotechnology innovation. "The United States has been and is now the recognized leader in nanotechnology research and development [. . .], but this lead cannot be assumed to be permanent," the document warned, and it presented the NNI "as important as ever to ensuring U.S. leadership in nanotechnology R&D" (NSET 2007, 5). The report indicated that the "NNI ensures United States leadership in nanotechnology research and development by stimulating discovery and innovation" (NSET 2007, 3). Such leadership was deemed to be "in the national interest" (NSET 2007, 11).

While the German plan stressed competitiveness and solutions to environmental problems, the US document thus put its primary focus on leadership and the defense of national interests and on the maximization of economic benefits. The late John H. Marburger, presidential science adviser and former Director of the Office of Science and Technology Policy, wrote in his letter introducing the document: "With the implementation of this [NNI Strategic] plan, the United States will remain at the forefront of nanoscale science and engineering and a leader in achieving the economic benefits offered by the emerging technology" (NSET 2007, introductory letter). Science was understood as a resource to achieve and consolidate the global leadership of the United States—explicitly in research and in the economy, but implicitly also in geopolitical terms. While the German strategy assigned similar importance to economic and environmental issues, the U.S. strategy was more asymmetric in that it prioritized economic profits and placed leadership in R&D above environmental protection.

Benefits and Risks

Further differences emerge when looking at the expected benefits and risks associated with nanotechnologies in the two national contexts. In Germany, expectations of benefits and risks were equally weighted. German authorities stressed the potential of nanotechnologies in a variety of sectors. The Action Plan explicitly listed six fields in which applications and new products were expected: medicine, optics, energy technology, environmental technology, consumer products, and information and communication technology (BMBF 2007, 12). In their statements regarding the need for a governmental nano-initiative, however, the delegates from the various governmental departments did not highlight all these applications to the same extent. Improvements in energy and environmental technology and their ecological advantages

were the most frequently addressed. Medical innovations were also emphasized. Annette Schavan, for example, hoped for "better diagnostics and improved therapies" (BMBF 2007, 3). She also mentioned new applications in
the construction industry, the textile industry, and sports and leisure, and
the use of nanotechnology to help manufacturers make faster electronic
components and more efficient lighting elements (BMBF 2007, 3). Ministry
delegates also cited potential improvements in the agricultural and food
industries, the general improvement of existing products, and the enhanced
performance of future military systems, although these expectations were
only mentioned by individual delegates.

Along with their hopes and expectations, German authorities expressed their awareness of the risks associated with nanoparticles. In the
Action Plan, hopes and risks—or positive and negative expectations—were
equally weighted. Most often, the positive expectations were immediately
discounted by mentioning potential risks associated with nanotechnologies.
For example, the representative of the Federal Ministry of Food, Agriculture
and Consumer Protection addressed both positive and negative impacts
in the same sentence, saying that his department "supports the exploitation of nanotechnology for the agricultural and food industries and for end
consumers and emphasises that consumers should be protected from any
risks that might arise in relation to the use of nanotechnology." He pleaded
for an "early identification of the potential risks of using nanomaterials in
foodstuffs, packaging for foodstuffs, cosmetics, and other necessary commodities" (BMBF 2007, 8). In a similar move, Annette Schavan, after listing
potential improvements, added that "[n]aturally, any possible side effects
must be considered and researched first" (BMBF 2007, 3).

US authorities, in contrast, mostly addressed the potential benefits of
nanotechnology. Although environmental, health, and societal concerns
were mentioned in the Strategic Plan, and research on these impacts were
included as part of the Initiative, they were not given the same weight as potentially useful applications. "The power of nanotechnology is rooted in its
potential to transform and revolutionize multiple technology and industry
sectors," the Strategic Plan stated (NSET 2007, 23). "[H]igh-impact application opportunities and critical research needs" (NSET 2007, 23) were seen
in the following areas: aerospace, agriculture and food, national defense and
homeland security, energy, environmental applications, information technologies, medicine and health, and transportation and civil infrastructure
(NSET 2007, 24). While the German plan mentioned medicine first and
energy and environmental technology soon after (BMBF 2007, 12), the US
plan put the potentially innovative sectors in an alphabetical order, thus im-

plicitly not ranking them. Expected applications included, for example, the production of smarter computers, the early detection of life-threatening diseases, and more energy-efficient transportation and energy security (NSET 2007, 25–34). In contrast to the German plan, in which only the representative from the defense ministry addressed the potential of nanotechnologies to transform military systems, nanoscience for defense was mentioned more prominently in the US document. While the German delegate noted that the research activities of the Federal Ministry of Defense are primarily based on civilian activities (BMBF 2007, 8), according to the US Strategic Plan, several research laboratories of the air force, the army, and the navy "have developed unique and complementary approaches for fostering nanoscience research directed toward understanding and exploiting the unique properties that some engineered nanoscale materials exhibit" (NSET 2007, 18). The NNI expected that a wide range of nano applications would be important to future defense technologies.

The ultimate goal of all such developments, as John H. Marburger put it, was to improve "the Nation's economy and the quality of life for all citizens" (NSET 2007, introductory letter). By interpreting potential military applications as beneficial for the nation and by focusing on a wide range of improvements in civil life, the US plan was less reserved and more optimistic than its German counterpart.

To sum up, in the German plan, the potential positive and negative impacts of nanotechnologies were equally weighted and carefully deliberated in envisioning future nanotechnology R&D. Hopes regarding future applications in environmental technology, medicine, and industrial products were discounted by underlining possible safety risks. In contrast, expectations were more asymmetric in the United States. Environmental, health, and societal concerns were addressed, and research on these impacts was included in the NNI Strategic Plan, but they were not given the same weight as expected applications across a wide range of sectors. Benefits, in other words, were more highly ranked than risks.

Avoiding and Managing

The governance of nanotechnology's risks was imagined differently in the strategic plans of Germany and the United States. On both sides of the Atlantic, there was agreement that research should be done on the environmental and health effects of nanotechnologies. However, attitudes toward the potential risks associated with these emerging technologies were not the same.

"Behaving in a responsible manner" was one of the (precautionary) principles on which German authorities drew in their framings of the governance of nanotechnology. This included investigating the effects of nanoparticles on health and the environment and evaluating these effects as part of a comprehensive strategy (BMBF 2007, 25). Delegates of the governmental departments unanimously agreed on the quest for more information and sound science. The representative from the Federal Ministry of Food, Agriculture and Consumer Protection, for example, was convinced that risk assessment "is based on sound knowledge" and thus pleaded for the development of adequate testing methods to get the necessary data on how intentional and unintentional exposures to nanoproducts affect consumers (BMBF 2007, 8). Having more scientific information at one's disposal, the German Action Plan assumed, will allow for the identification and assessment of potential risks. This should happen as early as possible, that is, at an early stage of R&D, the report suggested.

The Action Plan addressed risks, as we have seen, alongside potential benefits. Risks were perceived as an issue of high priority when assessing future nanotechnology applications. Both the usefulness and the harmfulness of the emerging technologies had to be carefully deliberated, in the German view, and products may only be released to the market after a thorough risk evaluation. The delegate from the Federal Ministry of Health, for instance, observed that "pharmaceuticals and medical products must only be brought into use following a comprehensive evaluation of risks and benefits" (BMBF 2007, 9).

Once identified, risks should be minimized or, better yet, avoided. For example, the Federal Ministry of Labor and Social Affairs was charged with identifying nanoparticles that are highly prevalent and formulating a code of practice for working with synthetic nanoparticles "in order to limit risks" (BMBF 2007, 25). This is consistent with a precautionary approach to risk governance.

In the United States, as noted above, governmental authorities involved in formulating the NNI Strategic Plan envisioned a "responsible development" of nanotechnology (NSET 2007, 3). Similar to the German plan, the US strategy asked for investments in risk research in order to increase the knowledge base on which risk assessments of nanomaterials and nanoproducts would be assessed. Responsible development in the view of US policy makers "recognizes the value of supporting basic research to develop nanotechnology as well as research to address environmental, health, and safety concerns related to the use of the technology" (NSET 2007, 19). It was thus seen as the government's job to foster responsible development.

US authorities also supported the investigation of risk issues at an early stage of innovation. However, unlike the German Action Plan, the U.S. NNI saw risks as a problem to be managed. Responsible development, according to the Strategic Plan, had "to maximize the benefits of nanotechnology and at the same time to develop an understanding of potential risks and to develop the means to manage them" (NSET 2007, 3). Risks were mainly imagined as something to be handled and controlled, not avoided.

German and US imaginations thus differed in their ways of conceptualizing the governance of risks. In accordance with the predominant European view, Germany's policy was directed toward a minimization or avoidance of the potentially negative impacts of nanotechnologies. A "responsible" or "integrated" way to govern risks, in the German view, had to be based on sound science, and safety concerns should be identified as early as possible. The US Strategic Plan also sought to foster research on potential risks and to examine potential negative implications at an early stage of nanotechnology R&D, but the NNI imagined risk management rather than risk avoidance. The imaginations of risk governance in Germany and the United States thus diverged along a temporal dimension. In alignment with European precautionary policies, the German strategy was oriented toward ex ante action, since it focused on intervention before possible harm emerges; the US strategy, by contrast, was oriented toward the management and control of risks, thus focusing primarily on ex post interventions.

Citoyens and Consumers

Strategies for how to govern nanotechnology in Germany and the United States included assumptions about the ideal form of the science-society relationship. Public interests and citizens were imagined as playing a specific role by getting engaged in the development of the emerging technologies.

The idea of dialogue was key to how the science-society relationship was envisioned in the German sociotechnical imaginary. An explicit goal the German Action Plan aimed to achieve was to "[e]nable an intensive dialogue with the public about the opportunities offered by nanotechnology but also taking possible risks into account" (BMBF 2007, 14). In this framing, dialogue did not refer simply to what authorities call "branch-level industrial dialogues," which served to clarify the opportunities offered by nanotechnology and to explain possible applications in a specific sector (BMBF 2007, 15). Rather, dialogue was imagined as inclusive, involving all relevant stakeholders. Informing the public and enabling a "social dialogue" were viewed as indispensable for the success and social acceptance of nanotechnology— thus, for the construction of "socially robust knowledge" (Nowotny et al.

2001). Dialogue was imagined as a negotiation process in which benefits and risks are carefully deliberated. The government declared that it "wants to discuss both potential benefits and risks with the public" (BMBF 2007, 26). Citizens were thus perceived as having the skills to deliberate both the advantages and the possible disadvantages of nanotechnologies. They were imagined as well informed, responsible, and engaged, in other words, as interested and active people who are willing to learn and acquire new knowledge in order to participate in the dialogue. They were perceived as *citoyens* whose civic duty is to participate responsibly in democratic decisions and public life—including the assessment and governance of technology.[6]

To foster exchange among stakeholders, the German government envisioned several initiatives. One was the so-called Nano-Dialogue 2006–2008, which consisted of two ministry-led working groups that considered issues relating to the "safety and responsibilities of research" and to the "promotion of innovation and opportunities for environmental protection" (BMBF 2007, 25). The working groups were open to representatives from industry, science, government, and nongovernmental organizations (NGOs). Another initiative planned to create a "nanotechnology future forum" with the aim "to discuss the relationship between commerce, science, technology, and the public." The forum was to provide an opportunity to discuss the benefits and risks of nanotechnology in an "interdisciplinary dialogue between natural scientists, humanists, politicians, managers, and journalists." The forum was also supposed to formulate recommendations for future nanotechnology research (BMBF 2007, 24).

While these initiatives addressed professionals, other initiatives focused more on the lay public. Citizens were seen as laypeople who stood outside science and industry. The Action Plan recommended initiatives to raise laypeoples' general interest in nanotechnology, leading to "the creation of a knowledge base for further social debate on the topic" (BMBF 2007, 26). These initiatives included the organization of discussion events, the production of information leaflets to "inform the public about the complex interactions of the world of nanotechnology in a way that is easy to understand," and the creation of an Internet "nanoportal" with access to all governmental initiatives on nanotechnology in Germany (BMBF 2007, 26–27). In addition, a mobile information campaign called nanoTruck was planned and began traveling to schools and public places all over the country. Former minister Annette Schavan's concern to "conduct a rational, scientifically ethical debate" imagined the participation of both professionals and laypeople (BMBF, 3). Benefits and risks were to be discussed and negotiated among the members of German society. The government's imagined role was that of a

244 / Regula Valérie Burri

mediator who would foster these social negotiations by enabling exchanges among citizens, science, and the industry. Such dialogue, it was believed, would enhance the social and democratic legitimacy of policy decisions on emerging technologies. In Germany, nanotechnology imaginaries thus were intrinsically political.

The framing of the science-society relationship differed in the United States. At the core of the relationship was the idea of communication, not dialogue. Responsible development of nanotechnology, according to the Strategic Plan, "entails establishing channels of communication with relevant stakeholders." Communication was understood in terms of "both providing information and seeking input" (NSET 2007, 19). While dialogue in the German view was an open and multidimensional concept, simultaneously including all stakeholders, communication in the US Strategic Plan was conceptualized as a more linear exchange between two major parties— government and civil society.

Societal dimensions were considered important in the development of nanotechnology in the United States. The NNI included a program on "Education and Societal Dimensions," which aimed at identifying the implications of nanotechnology for society. Explicitly, the program embraced activities of "public communication, including outreach and engagement" (NSET 2007, 7). This included the funding of two research centers studying the societal dimensions of nanotechnology "to encourage the distribution and exchange of insights from leading experts" and "to develop avenues for societal input into nanotechnology development" (NSET 2007, 20).[7] Other means to promote public outreach were a website, a web-based dialogue, and media roundtables (NSET 2007, 20). Governmental authorities were convinced that communication "allows the public and the NNI agencies to make well-informed decisions and builds trust among all stakeholders" (NSET 2007, 19). In this framework, the government was imagined as a director who organizes or enables communication activities and seeks input for its regulatory decisions. Outreach activities aimed to foster citizens' understanding and acceptance of nanotechnology.

Citizens and the public were seen as future consumers who will finally be the purchasers of nanotechnology products. Their understanding and acceptance of nanotechnology and its applications were considered "necessary components of successful commercialization" (NSET 2007, 19). Citizens' perceptions of the emerging technologies were viewed as crucial for their role as consumers. Communication thus became instrumental for the successful commercialization of nanotechnology research. The purpose of informing and involving the public was ultimately commercial.

In sum, a basic difference between the German and US sociotechnical imaginaries had to do with implicitly framing the issues in terms of either political or market objectives. The German approach can be seen as that of democratic enhancement, which pursues the political aim of increasing dialogue between state and society. The US imaginary focused rather on optimizing the economic exploitation of nanotechnology, with the implicit aim of turning citizens into informed and willing consumers of nanotechnology's products.

Attitudes, Experiences, Identities

The imaginaries expressed in the policy documents analyzed in this chapter correspond to general attitudes of the German and US publics toward emerging technologies. Earlier studies have shown that perceptions toward emerging technologies differ between Europe and the United States. Social responses to agricultural biotechnology, for instance, have been more reluctant in Europe than in the United States (Durant et al. 1998; Gaskell and Bauer 2001; Bauer and Gaskell 2002; Jasanoff 2005). A more skeptical, although balanced, European concern regarding emerging technologies has also been identified with regard to nanotechnologies (e.g., Royal Society 2004; Gaskell et al. 2004, 2005; Burri 2007; Burri and Bellucci 2008). In contrast, US public attitudes toward nanotechnologies displayed a high level of enthusiasm for their potential benefits and little or no concern about possible risks at the beginning of the millennium (Bainbridge 2002; Cobb and Macoubrie 2004; Macoubrie 2006).

These public attitudes toward (potentially) risky technologies are reflected in the sociotechnical imaginaries of the policy makers in the respective political cultures. Both public attitudes and collective imaginaries are coproduced along with the framing and marketing of emerging technologies (Jasanoff 2004). While German attitudes and imaginaries revealed a balance between risks and benefits, the US public's approval of nanotechnologies was less symmetrical, valuing benefits higher than risks and attaching greater weight to economic than to ecological goals.

Experiences

Besides the differences in general public attitudes, the policy makers' experiences have diverged on both sides of the Atlantic. In Germany, as all across Europe, policy makers were still deeply influenced by social controversies on issues such as green biotechnology. The involvement of NGOs and a considerable number of citizens in organizing political resistance to genetically modified organisms (GMOs) made state authorities more cautious in approaching the assessment and governance of emerging technologies. Policy makers aimed to avoid a repetition of the heated discussions and political consequences that had accompanied the diffusion of biotechnology. These experiences contributed to shaping a democratic enhancement approach in which the public is involved early on in deliberations on nanotechnology R&D in order to increase legitimacy and reduce resistance. These concerns reinforced an ex ante approach that entails the goal of minimizing or avoiding the possible negative impacts of nanotechnology from the outset.

In the United States, policy makers had no comparable experiences with domestic politics. The public uptake of biotechnology had been more positive than in Europe, and no major political impacts had resulted from controversies on GMOs. US policy makers were thus less negatively affected by earlier technology experiences. Nevertheless, they followed the European controversies and were keenly aware that they should not let the GMO case infect nanotechnology in order to avoid a Europeanization of the nano debate. Imaginaries, however, remained similar across the two cases. Biotechnology was mostly perceived as a social good that provides new products and applications, thus advancing the industry. A similarly economistic approach was also imagined for nanotechnology. New products were expected to benefit the national economy, which would also help to maintain US global leadership. As in the case of biotechnology products, an ex post managerial approach was imagined as sufficient to handle and control potential risks.

Identities

Finally, national identities play a major role in shaping sociotechnical imaginaries. In Europe, imaginaries of the place of science in society are intertwined with political history. Social movements have played an important role in shaping the discourses and framings of technological innovation since the 1970s. In Germany, social movements were particularly influential. Originating in the "new social movements" such as the political resistance against nuclear power plants, the ecological movement has been rising since the 1980s. In 1983, the newly formed political party called the Greens (*Die Grünen*) first succeeded in electing representatives to the *Bundestag*, the German parliament. Environmental issues thereafter became more important in policy discourses and have been taken up by other parties since the 1990s. Care for the environment became an issue that policy makers could no longer neglect. Environmental aims had to be deliberated along with other policy goals such as the advancement of the economy. The balanced, precautionary approach described above results in part from this history. In addition, ethics have been a central value of Germany's national self-understanding and self-representation (Sperling 2004, 147). Bioethics, for example, became a field in which the question of German identity worked itself out (Jasanoff 2005, 183). Ethical considerations toward nature resulted in according high priority to environmental concerns in policy discourses and strategies, including science and technology policy.

Institutional traditions play an important role in the national self-understandings and imaginaries as well. In Germany, a social market economy was established after World War II. Its core idea was to combine the liberal tradition of a market economy with the principle of a social equilibrium. The notion of creating a social balance was also reflected in economic politics. By German law, any board of directors of a company has to include representatives from both employers and employees, the latter consisting mostly of union members. The principle of including the socially less powerful, rather than excluding them, goes along with the heightened appraisal of democratic legitimacy in postwar Germany. The democratic enhancement approach imagined for nanotechnology policy is deeply interwoven with the national purpose to maintain a social equilibrium. The principle of fostering the dialogue among all stakeholders also corresponds to what Sheila Jasanoff has identified as a consensus-seeking "civic epistemology," that is, "the institutionalized practices by which members of a given society test and deploy knowledge claims used as a basis for making collective choices" (Jasanoff 2005, 255).

National identities are further shaped by histories of success and failure in technological innovation. In Europe, experiences of failure have been more formative and lasting than in the United States. Policy makers, as mentioned above, are still sensitized by the public reaction to green biotechnology. Memories of the GMO debate are present as well in the larger society. From such experiences, lessons were drawn for the future governance of technology, such as the institutionalization of upstream public dialogue. German experiences with nuclear technologies were similarly ambivalent. The 1986 Chernobyl and the 2011 Fukushima disasters are well remembered, and the unsolved problem of radioactive waste deriving from nuclear power plants—as manifested in the "Castor trains" that transport the waste—makes many citizens skeptical toward new technological risks. The public gives preference to strategies that focus on minimizing or avoiding risks associated with emerging technologies. This ex ante approach is consistent with the precautionary principle that was chosen to deal with perceived technological risks in the past.

In the United States, success stories of research innovation and technological applications are built into the national identity, and they are associated with US power and culture by other nations. The 1969 moon landing is an iconic example. Advances in genetic research that are expected to improve medical treatments are considered another success story. Scientific innovations and national identities are coproduced, as Jasanoff (2004) has shown. The perceived advantages of emerging technologies have often been given a higher evaluation in the United States, as is also evident in the asymmetric approach to nanotechnology development. In certain cases, specific cultural work has been done to repress potential risks. As Joseph Masco (2006) argues in his study of the Manhattan Project, US cultural practices were aimed at making the danger of a nuclear war "unthinkable"—a theme also underscored in Jasanoff and Kim's (2009) account of nuclear containment. The perceived mostly positive experiences with technology in the United States imply that potential risks are considered but are evaluated as manageable. The US imaginary of nanotechnology can be understood as a further expression of this perception.

US imaginaries of the role of science are rooted in the vision described by Vannevar Bush, Director of the Office of Scientific Research and Development, after World War II. In his famous and influential report to President Franklin D. Roosevelt, *Science—The Endless Frontier*, Bush (1945) requested institutionalized public funding of basic scientific research. It was government's responsibility to provide public resources to the scientific community. While science, according to Bush's plan, would profit from society, its duty was envisioned as a contribution to the advancement of society by improving the nation's economic and social welfare. In his accompanying letter of transmittal to President Roosevelt, Bush wrote, "Scientific progress is one essential key to our security as a nation, to our better health, to more jobs, to a higher standard of living, and to our cultural progress" (Bush 1945, 1). R&D were seen as vital to the nation's economic welfare and social progress.

As soon as science fails in its duty to advance society, the "social contract for science" (Guston 2000), which was established following Bush's report, is transformed. A decaying productivity and the declining economic performance, together with cases of scientific misconduct, contributed to a growing public mistrust in science and a changed science-society relationship in the 1980s. "Boundary organizations" were established in order to bridge the gap and mend the relationship (Guston 2000). Economic progress and social welfare remain the prevailing goals of US science and technology policy. An important task imagined for nanotechnology was to advance the economy. Giving priority to an economistic approach in the national nanotechnology strategy was a result of this imagination.

A further important imaginary for science imagined in the United States is, as we have seen, to work in the national interest. National science policy imaginations center on the nation's perceived leadership in R&D. Implicitly, this imagination incorporates the notion of geopolitical leadership along with leadership in science. The vision of science and technology as means to achieve and stabilize such political leadership embraces a long tradition in US society and is integral to national identity. The Manhattan Project, for example, contributed to changing the world order to one in which the United States would play the leading role of international power broker. The National Aeronautics and Space Administration (NASA) space program, as another example, was established in a climate of competition over global leadership between the United States and the Soviet Union. Science and technology, in the US government's imagination, can help stabilize the country's global position in economic, social, and geopolitical terms. The notions of progress and leadership have long been part of the US national self-understanding and are deeply embedded in the nation's outlook and attitude toward science and technology. These elements of identity are reaffirmed in the US imaginary of nanotechnology.

Sociotechnical Imaginaries and Civic Epistemologies

Sociotechnical imaginaries are crucial resources for the assessment and governance of emerging technologies. They shape the ways a political culture evaluates and regulates new and emerging technologies. Although sociotechnical imaginaries are different depending on the cultural context in which they are embedded, they inform what Sheila Jasanoff (2005) has called the civic epistemologies of their respective political cultures. The institutionalized practices for testing and deploying knowledge claims in a society correspond to the ways that members of a given society imagine science, technology, and their relation to social order, in other words, to national sociotechnical imaginaries. The assessment and governance of nanotechnology, according to this analysis, are strongly shaped by prevalent sociotechnical imaginaries in a given society.

This chapter described three approaches to the assessment and governance of nanotechnologies in Germany and the United States. A comparison of two crucial policy documents defining each nation's nanotechnology strategy revealed contrasting imaginaries of the place of science in society, of the governance of nanotechnologies, and of citizens in the social order. The German approach was balanced in its consideration of benefits and risks, attributing equal importance to economic and ecological claims and discounting expected gains with potential harmful consequences. This approach envisioned upstream political intervention in order to guard against the negative implications of emerging nanotechnologies. Further, the social and democratic legitimation of policy decisions were seen as important values. The German imaginary thus contains a political core. In contrast, the US imaginary appeared more asymmetric because commercialization and potential benefits of nanotechnologies were given more weight than environmental impacts and other potential risks. Risks were imagined to be manageable, and interventions were imagined to take place more downstream, after products are developed. This analysis showed that the US imaginary framed nanotechnology development in terms of creating a working market in which citizens will behave as responsible consumers of new products.

These revealed imaginaries should be seen as ideal types in the Weberian sense. They do not mark rigid characteristics of the respective political cultures but rather help us discern some distinctive elements of how these political cultures envision relations between science and society when considering emerging nanotechnologies. Nevertheless, the different approaches can be related to heterogeneous cultural perceptions, experiences, and identities. Sociotechnical imaginaries are thus deeply embedded in the political cultures. During processes of technology assessment and governance, sociotechnical imaginaries are enacted and (re)constructed at the same time, thus shaping the forms of political action.

Both political cultures and sociotechnical imaginaries are formed by local and national histories, experiences, and identities, but they also have to be seen in a global context. National imaginaries are continually being shaped and possibly transformed by supranational institutions and transnational developments. It will be interesting to follow how the national sociotechnical imaginaries described in this chapter will influence future (nano)technologies and policies and how they may be transformed by future technological, political, and social developments, in both national and transnational contexts.

Notes

My warmest thanks go to Sheila Jasanoff and Sang-Hyun Kim for inviting me to participate in the Sociotechnical Imaginaries workshop in fall 2008 and to the participants of the workshop. I thank the *Harvard Weatherhead Center for International Affairs*, especially Adelaide Shalhope, and the Harvard Kennedy School's Program on Science, Technology and Society, especially Lauren Schiff, for the support provided on the occasion of the workshop. I also thank Arie Rip for helpful comments when I presented the paper at the European Science Foundation Science and Values Conference in Bielefeld in 2009. Research for this chapter was made possible by the support of Collegium Helveticum of the Swiss Federal Institute of Technology and the University of Zurich, the University of Basel, and the Holcim Foundation.

1. For a definition of the term "sociotechnical imaginaries," see Jasanoff and Kim (2009 and the introduction to this volume). Similar terms have been used by Anderson (1983), Appadurai (2003 [1996]), Marcus (1995), and Fischer (1999).
2. Arie Rip (2006) used the term "folk theories" to describe how culture informs the ways societal actors deal with nanotechnologies.
3. www.bmbf.de/de/nanotechnologie.php (accessed March 30, 2014).
4. This chapter analyzes the plans published in 2007 (BMBF 2007 and NSET 2007). In recent years, two further plans have been published: the German "Action Plan Nanotechnology 2015," published in 2011, and the new US NNI Strategic Plan, published in 2014. Both recent plans display mainly the same respective sociotechnical imaginaries as reconstructed in this chapter. The participating departments involved in discussing the "Action Plan 2010" were the Federal Ministries for Labor and Social Affairs; Environment, Nature Conservation and Nuclear Safety; Food, Agriculture and Consumer Protection; Defense; Health; Economics and Technology; and Education and Research.
5. www.nano.gov (accessed March 30, 2014).
6. The French term *citoyen* is used to underline the idea of a citizen in the tradition of the French Revolution, which is still prevalent in European social thought today. Sharing the values of *égalité*, *fraternité*, and *liberté*, the *citoyen* is actively exercizing his (her) civil rights.
7. These centers have been established at the University of Arizona and the University of Santa Barbara, California.

References

Anderson, Benedict. 1983. *Imagined Communities*. London: Verso.

Appadurai, Arjun. 2003 [1996]. *Modernity at Large: Cultural Dimensions of Globalization*. Minneapolis: University of Minnesota Press.

Bainbridge, William Sims. 2002. "Public Attitudes toward Nanotechnology." *Journal of Nanoparticle Research* 4: 561–570.

Bauer, Martin, ed. 1995. *Resistance to New Technology: Nuclear Power, Information Technology and Biotechnology*. Cambridge: Cambridge University Press.

Bauer, Martin, and George Gaskell, eds. 2002. *Biotechnology: The Making of a Global Controversy*. Cambridge: Cambridge University Press.

Beck, Ulrich. 1992 [1986]. *Risk Society: Towards a New Modernity*. London: Sage.

Bundesministerium für Bildung und Forschung (Federal Ministry of Education and Research). 2007 [in German 2006]. *Nano-Initiative—Action Plan 2010*. Bonn and Berlin: Bundesministerium für Bildung und Forschung (BMBF).

Brickman, Ronald, Sheila Jasanoff, and Thomas Ilgen, eds. 1985. *Controlling Chemicals: The Politics of Regulation in Europe and the United States*. Ithaca, NY: Cornell University Press.

Burri, Regula Valérie. 2007. "Deliberating Risks under Uncertainty: Experience, Trust, and Attitudes in a Swiss Nanotechnology Stakeholder Group." *NanoEthics* 1: 143–54.

Burri, Regula Valérie, and Sergio Bellucci. 2008. "Public Perception of Nanotechnology." *Journal of Nanoparticle Research* 10: 387–91.

Bush, Vannevar. 1945. *Science—The Endless Frontier*. A Report to the President by Vannevar Bush, Director of the Office of Scientific Research and Development, July 1945. Washington, DC: United States Government Printing Office.

Cobb, Michael D., and Jane Macoubrie. 2004. "Public Perceptions about Nanotechnology: Risks, Benefits, and Trust." *Journal of Nanoparticle Research* 6: 395–405.

Daemmrich, Arthur. 2004. *Pharmacopolitics: Drug Regulation in the United States and Germany*. Chapel Hill: University of North Carolina Press.

Durant, John, Martin Bauer, and George Gaskell, eds. 1998. *Biotechnology in the Public Sphere: An European Sourcebook*. London: Science Museum.

Fischer, Michael M.J. 1999. "Emergent Forms of Life: Anthropologies of Late or Post Modernities." Pp. 37–58 in *Emergent Forms of Life and the Anthropological Voice*. Durham: Duke University Press.

Gaskell, George, and Martin Bauer, eds. 2001. *Biotechnology 1996–2000: The Years of Controversy*. London: Science Museum Publications.

Gaskell, George, Toby Ten Eyck, Jonathan Jackson, and Giuseppe Veltri. 2004. "Public Attitudes to Nanotechnology in Europe and the United States." *Nature Materials* 3: 496.

———. 2005. "Imagining Nanotechnology: Cultural Support for Technological Innovation in Europe and the United States." *Public Understanding of Science* 14: 81–90.

Guston, David H. 2000. *Between Politics and Science: Assuring the Integrity and Productivity of Research*. Cambridge: Cambridge University Press.

Guston, David H., and Daniel Sarewitz. 2002. "Real-Time Technology Assessment." *Technology in Society* 24: 93–109.

Jasanoff, Sheila, ed. 2004. *States of Knowledge: The Co-Production of Science and the Social Order*. New York: Routledge.

———. 2005. *Designs on Nature: Science and Democracy in Europe and the United States*. Princeton, NJ: Princeton University Press.

Jasanoff, Sheila, and Sang-Hyun Kim. 2009. "Containing the Atom: Sociotechnical Imaginaries and Nuclear Power in the United States and South Korea." *Minerva* 47: 119–46.

Kearnes, Matthew, and Phil MacNaghten. 2006. "Introduction: (Re)Imagining Nanotechnology." *Science as Culture* 15: 279–90.

Macoubrie, Jane. 2006. "Nanotechnology: Public Concerns, Reasoning and Trust in Government." *Public Understanding of Science* 15: 221–41.

Marcus, George E., ed. 1995. *Technoscientific Imaginaries: Conversations, Profiles, and Memoirs*. Chicago: University of Chicago Press.

Masco, Joseph. 2006. *The Nuclear Borderlands: The Manhattan Project in Post-Cold War New Mexico*. Princeton: Princeton University Press.

Nowotny, Helga, Peter Scott, and Michael Gibbons. 2001. *Re-Thinking Science: Knowledge and the Public in an Age of Uncertainty*. London: Polity Press.

Parthasarathy, Shobita. 2007. *Building Genetic Medicine: Breast Cancer, Technology, and the Comparative Politics of Health Care*. Cambridge, MA: MIT Press.

Rip, Arie. 2006. "Folk Theories of Nanotechnologists." *Science as Culture* 15: 349–65.

Royal Society. 2004. *Nanoscience and Nanotechnologies: Opportunities and Uncertainties*. RS policy document. London: Royal Society and Royal Academy of Engineering.

Sperling, Stefan. 2004. "Managing Potential Selves: Stem Cells, Immigrants, and German Identity." *Science and Public Policy* 31: 139–49.

Subcommittee on Nanoscale Science, Engineering, and Technology. 2007. *The National Nanotechnology Initiative—Strategic Plan.* Prepared by the Subcommittee on Nanoscale Science, Engineering, and Technology/Committee on Technology of the National Science and Technology Council. Washington, DC, December 2007 (NSET).

———. 2008. *The National Nanotechnology Initiative—Strategy for Nanotechnology-Related Environmental, Health, and Safety Research.* Prepared by the Subcommittee on Nanoscale Science, Engineering, and Technology/Committee on Technology of the National Science and Technology Council. Washington, DC, December 2007 (NSET).

TWELVE

Corporate Imaginaries of Biotechnology and Global Governance: Syngenta, Golden Rice, and Corporate Social Responsibility

ELTA SMITH

Genetically modified organisms (GMOs) are one of the most polarizing scientific developments of the past thirty years (Doh and Guay 2006; Jasanoff 2006; Bernauer 2003). Strong opposition, notably in Europe, has had resounding impacts on global agricultural production, trade, and even disaster relief efforts from Africa to the Americas (e.g., Newell 2007; Becker 2003; Manda 2003; BBC News 2002). Concern has predominantly centered on risks to human health and the environment, largely sidelining issues of food security, trade equity, and the balance of benefits from GMOs to farmers versus consumers. One way to understand these debates is by tracking the historical evolution of modern biotechnology alongside changes in intellectual property law at national and international levels, the rise of multinational life sciences corporations, and the development of social movements.

In this chapter, corporate social responsibility (CSR) programs are a lens through which to view corporate governance and its underlying sociotechnical imaginaries (Jasanoff, this volume). Corporations are combining their for-profit objectives with humanitarian aspirations. The effort to integrate CSR projects with standard business practice creates new possibilities but also new risks that ultimately lead to the clear division of consumer-oriented practices from charitable ones. This reflects larger social, political, economic, and technological ideas and sensibilities, including international development objectives and market-based ideologies, and the belief that science and technology can solve social problems. The corporate imaginary highlighted in this chapter is future oriented but also constrained by present and historically produced conditions, whether technical or political (Marcus 1995). Like other chapters in this volume, this story demonstrates the role that biotechnology has played in establishing a narrative of "public good" (Chen, Hurlbut, Kim, this volume) but shows how the risks involved in the project led those involved in its development to restrict the potential of that good to only certain groups of people and only in particular, circumscribed ways. Corporate humanitarian efforts have established a clear boundary that separates the recipients of donations (the developing world) from profitable populations (the developed world).

Corporate imaginaries have radically changed over the last three decades. This is partly due to developments in biotechnology, for example, through the creation of transgenic organisms, and also due to changes in property rights law, especially the landmark 1980 US patent case *Diamond v Chakrabarty* (447 U.S. 303, 1980). The global dimensions of these developments are becoming increasingly salient as states tie their policies to international frameworks and treaties. Corporations play a powerful role outside and alongside the role of states, particularly at a global level where formal government is nascent, largely inchoate, and difficult to monitor and enforce. CSR programs are one of the means by which companies engage in these global policy arenas (on globalism, see also Miller, this volume).

This chapter looks at how corporate imaginaries have interacted with those of the nation-states in which companies plan to sell or donate their technologies, focusing on a novel agricultural technology, Golden Rice, and its development in the context of larger shifts in agricultural biotechnology from the early 1980s to the present. Golden Rice is a genetically engineered plant designed to produce higher levels of beta-carotene, which the body converts into vitamin A. It is intended for consumption in countries where certain forms of malnutrition caused by vitamin deficiency are considered to be endemic. Golden Rice is the first crop plant whose genetic modification is intended to improve human health and is therefore primarily aimed at benefiting the consumer, as opposed to improving yields, which primarily benefit the agricultural industry and wealthy farmers. Golden Rice can be attributed to the ideas, investments, and pressures of many individuals and institutions, and the role of each is highlighted in this chapter. But the multinational company, Syngenta, is the focus of this essay as the primary developer of Golden Rice.

I begin with a history of Golden Rice as it emerged within the field of biotechnology in the 1980s and 1990s. Second, I take an institutional look at the evolution of social responsibility programs within corporations—a movement that coincided with the rise of agricultural biotechnology and the breakup of agriculture and pharmaceuticals into separate industries. Third, I discuss Syngenta's CSR projects for international agriculture and the particular ways in which "value" is conceived through its CSR agenda. Finally, through the case of Golden Rice, I explain how Syngenta's CSR objectives play out in the development of this new technology and its associated intellectual property contracts.

In conclusion, I argue that despite the apparently seamless transition to corporate governance through social responsibility programs, and the integration of economics and ethical ideas of international development, the policy landscape consists of a highly unregulated and disorderly set of power relations between corporations and recipients of their CSR projects. As such, the corporate imaginaries emerging through Syngenta's CSR programs create a kind of "humanitarian contract" that presumes a static relationship between donor and recipient, raising questions about the democratic viability of these programs.

A History of Golden Rice

Vitamin-enriched rice was first proposed at a meeting in the early 1980s convened by the Rockefeller Foundation at the International Rice Research Institute (IRRI) in the Philippines. As biotechnology began to achieve international recognition and economic importance (Bud 1993), the Foundation initiated a research program to invest in this new set of techniques, particularly for food crops. Molecular biology and plant breeding were only tentatively coming together, and there was much speculation about what biotechnology could accomplish. Gary Toeniessen (GT), the Rockefeller Foundation's Director of Food Security who had been involved in their grant-related agricultural research efforts since the mid-1970s and who participated in the IRRI meeting in the early 1980s, observed, "I think it's safe to say that plant breeders everywhere were pretty skeptical of this new technology. They'd seen . . . new technologies come down the line, but to an old, hard-core plant breeder, [these were] just a new way of generating some genetic diversity . . . [A] lot of the breeders that attended this meeting were justifiably pretty skeptical when some of the plant molecular biologists, who really knew very little about breeding . . . made presentations about how this technology was going to revolutionize plant breeding" (GT, interview, 2005). During an informal evening gathering of these skeptical breeders, an IRRI breeder said to the others: "These molecular biologists tell us they can stick any gene into rice that you want stuck into rice—what would you want?" (GT, interview, 2005). Another rice breeder, who had worked at IRRI since the 1960s, suggested that he would most like to see rice with higher levels of beta-carotene, since vitamin A deficiency was such a serious threat to health throughout Asia.

In an initial priority-setting exercise, however, nutrient enrichment did not quite fit in the new molecular biology program for rice. The priority list of traits was based entirely on yield foregone in the absence of a given trait, and beta-carotene enrichment did not affect yield (GT, interview, 2005). During the 1980s, the foundation moved forward with rice as the focus of its crop improvement efforts, but early genetic manipulations focused on yield-related traits, such as herbicide tolerance and disease and pest resistance. This was in part because of biological and scientific limitations, but it was also due to the perception that enhanced yield was the most important goal of plant improvement in a geopolitical world where the US government focused on food security as an essential component of anti-Soviet policy to prevent revolution by reducing hunger.

Biotechnology was also seen to have potential in the agricultural industries, particularly for agrichemical producers. In the United States, for example, private sector research focused on six major commercial crops—corn, cotton, potato, rapeseed, soybean, and tomato—and on only a handful of traits that held the most commercial value added, mostly for herbicide resistance (Huttner et al. 1995). By the late 1980s, Monsanto had begun field testing its first genetically modified crops in the United States (Monsanto Company 2008). Herbicide-resistant soybeans and pest-resistant cotton were introduced in 1996. Industry dominated biotechnology research in this period, accounting for some 87 percent of field trial permits granted by the US Department of Agriculture from the late 1980s to the mid-1990s (Huttner et al. 1995).

In the 1990s, interest in health arose alongside yield and disease and pest resistance as an important goal in the agricultural life sciences. The Rockefeller Foundation began funding research on vitamin A–enhanced rice and convened several meetings to discuss the potential for developing genetically engineered carotenoid-rich rice. In 1993, the foundation held a meeting that brought together university scientists and industry representatives who had worked on the manipulation of micronutrient content in crop plants, including those from DuPont, Kirin Brewing Company, and Amoco (Rockefeller Foundation 1993). Following the meeting, the foundation began funding two projects to test vitamin A enrichment for rice.

Six years later, in early 2000, one of the projects led by Ingo Potrykus, a plant biologist at the Institute of Plant Sciences in Zurich's Swiss Federal Institute of Technology, and Peter Beyer, a geneticist at the University of

Freiburg, Germany, published its results (Ye et al. 2000). Using a gene from a bacterium and two daffodil genes, scientists had created a rice plant with increased beta-carotene levels. This was an important achievement for technical and social reasons: scientifically, a three-gene insertion had never been done before; and this was considered to be the first genetically engineered crop produced expressly and solely with humanitarian objectives. Potrykus and Beyer set up a humanitarian advisory board (HumBo),[1] which included the two scientists, as well as representatives from the Rockefeller Foundation, IRRI, and the recently formed multinational corporation, Syngenta. The HumBo was tasked with thinking about how to get the new vitamin A–enriched rice seeds from the laboratory to the poor and malnourished consumers who were thought to be the beneficiaries of these initiatives.

But Beyer and Potrykus soon ran into trouble in their humanitarian ambitions. It turned out that their project potentially infringed on more than seventy patents worldwide. In the process of producing this novel biotechnology, they had signed material transfer agreements, which stipulated not only that commercialization was disallowed but that products could not be shared with third parties (GT, interview, 2005). Intellectual property rights (IPRs) made it tricky for Beyer and Potrykus even to test the technology through research efforts in other countries. Golden Rice faltered as a humanitarian project with so many intellectual property claims on it.

At that time, the multinational corporation Zeneca was part of a European Union–funded consortium working on enhanced carotenoid expression in plants (Adrian Dubock [AD], Head of Public-Private Partnerships, Syngenta, interview, 2006). The EU Agriculture and Agro-Industry including Fisheries Programme of Research and Technological Development (FAIR) that supported the carotenoid project sought to enhance scientific and technological research throughout Europe, including strengthening links to industry (FAIR Programme 2006). While Golden Rice was not part of this project, one of its inventors, Peter Beyer, was, and through this link Zeneca came to know about the Golden Rice work (AD, interview, 2006).

Agricultural biotechnology companies consider health an important focus of their strategic initiatives, and for this reason Golden Rice was of considerable market interest. Nutritional enhancements of food crops, particularly of rice, had been a strategic investment for large multinational corporations like Zeneca, which eventually became Syngenta, since the mid-1990s (AD, interview, 2006). Some of the infringed patents in Golden Rice belonged to Zeneca, as well as to agricultural life science giants Novartis and Monsanto. Beyer and Potrykus entered into negotiations with Zeneca and Novartis over the problematic patents, and by 2000, Zeneca had arranged the first licenses for Golden Rice (AD Interview 2006). This was perceived as a good deal for both parties: Zeneca would get the IPRs from Beyer and Potrykus for carotenoid-enhanced rice; and Beyer and Potrykus would receive the rights to most of the patents they needed to share, and potentially commercialize, the new technology (GT, interview, 2005). Monsanto followed shortly thereafter, arranging a humanitarian license with the two inventors, which allowed for the use of the technology free of charge under certain conditions.

Agricultural Life Sciences and CSR

The first genetically engineered products won regulatory approval in the United States in the mid-1990s. Companies housing both pharmaceuticals and agrichemicals were created through mergers in an attempt to capitalize on the synergies from biotechnology that were expected to lead to product development across both sectors (Copping 2003). AstraZeneca and Novartis are notable examples of this phenomenon.[2]

By the end of the 1990s, however, the boom period for life sciences companies had ended. Companies that had expanded their interests in both pharmaceuticals and agriculture only a few years earlier began to decouple those aspects of the business, spinning them off and merging them with other pharmaceutical and agricultural businesses, respectively (Morrow 2000; Niiler 2000). This was a result of the cyclic nature of agribusiness: in the 1990s it was going through an unprofitable period and was seen as a financial liability to the more profitable pharmaceutical components of these companies (Dutfield 2003). Agricultural research was ripe for being spun off from pharmaceuticals. In addition, the anticipated synergies between pharmaceuticals and agriculture, with their predicted efficiencies and greater profits, never materialized (Morrow 2000). Syngenta was formed in 2000 out of the merger between the agricultural divisions of Novartis and AstraZeneca.

Meanwhile, multinational firms began undergoing massive internal reorganization and consolidation; many companies narrowed their strategic horizons to increase profitability. Gerard Barry (GB), former head of rice genomics for Monsanto, noted that the company was nearly US$7 billion in debt at this time. Monsanto decided that the cost of managing all of their crop biotechnology projects was too high and that adding new "high-profile" crops such as rice would only put an additional financial burden on the company (GB, interview, 2005; also Dutfield 2003). Companies reduced the range of products in their pipeline, cutting their crop focus to only a handful. Syngenta dropped rice from its strategic crops and began focusing on only five main field crops: maize, soybean, sugar beet, and sunflower and rapeseed oils. In addition to massive workforce layoffs, Monsanto also dropped its research on rice.

Throughout the 1990s, industry also faced intense pressure from an altogether different quarter: activist groups urging them to reform their business practices. Monsanto, most notoriously, received harsh international criticism for developing what was dubbed "terminator technology"—genetically engineering seeds to render them sterile in the second generation. This was part of a broader phenomenon of activist pressure against multinational businesses. Nike was pushed to engage in better labor practices for workers in its Asian shoe manufacturing facilities. Philip Morris underwent massive national and international lawsuits by smokers who wanted to recover health care payouts. Monsanto, in short, was not an isolated target.

These activist-led movements pushed companies to develop CSR programs to rebuild public trust both at home and abroad. Ethics became a central feature of good business practice. For companies like Monsanto, CSR was a method of staving off these activist pressures. Golden Rice came to prominence during this period and has thus been criticized as a mere public relations effort by the industry to improve its image. By the time the Golden Rice research was publicized in the late 1990s, corporations had become prominent participants, primarily through licensing agreements as a result of IPR infringements through Potrykus and Beyer's research. But the project soon became a high-profile example of CSR efforts in the agricultural life sciences, particularly for its major sponsor, Syngenta.

By 2006, Syngenta's CSR projects included environmental efforts geared toward better management of soil, water, and biodiversity through farmer education programs; funding higher education in the life sciences; and monitoring corporate conduct and human rights violations at its many international sites (Syngenta CSR Report 2007). In the life sciences, CSR can, and often does, take the form of technology donations for crop plants, where the word donation specifies a dimension of CSR in which the donor and recipient are engaged in a charity relationship. Biotechnology donations represent only one dimension of CSR, but it is a crucial part because of the potential impact on human health, food production and consumption, food security, international trade, and intellectual property regimes. Both Syngenta and Monsanto donated rice genome data to public sector genome sequencing initiatives (Smith 2012). Syngenta has donated biotechnologies for Bt[3] potatoes in South Africa to a Michigan State University project and for delayed ripening of papayas in South Asia through the University of Nottingham, as well as data on the banana genome to an international banana genome collaboration (AD, interview, 2005).

While both Monsanto and Syngenta have extensive CSR programs and spend a great deal of time and money working on humanitarian projects, there is a conundrum inherent in the effort to use CSR to deflect activist criticism: "We're still very interested in doing the right thing and sharing [technologies], but we're becoming increasingly cautious because of the attitudes of society to GMOs in particular . . . especially with biological materials, even if you just hand them over, the biological material originally came from you . . . so we have to try and protect ourselves while trying to do the right thing (AD, interview, 2006). The underlying question for Dubock is not whether companies have a social impact, but rather, for what a corporation is responsible (Vogel 1978). Put differently, there are tensions between industrial development and humanitarian efforts and between CSR and the responsibilities of government. The question of responsibility is vital: the contradictions between simultaneously seeking economic growth and an ethical business model are important sites for the emergence of new forms of global governance.

Syngenta Circle of Value Creation

The corporate vision of benefit in which Golden Rice has evolved is closely tied to the emergence of the CSR concept. CSR arose out of the civil rights and antiwar protests of the 1960s and 1970s (Vogel 1978)[4] and became commonplace in the 1970s (Carroll 1999). The direct application of public pressure on companies like Nike, Philip Morris, and Monsanto resulted from prevailing opinion that the modern corporation acts as a private government to some extent, immune both from market constraints and those of the state (Vogel 1978; also Laird 1989).

Corporate philanthropy among businesses has been prevalent in the United States since the 1960s, with many firms donating as much as 5 percent of their pre-tax earnings to social causes. But the case for social responsibility as a practice that could lead to greater profitability dates from the 1990s. Today many companies focus on some degree of CSR, particularly in industries in which activists and other stakeholders have been vocal, such as in the life sciences.

Two of the most cited arguments in favor of corporate engagement with social issues are improved corporate-civil relationships (i.e., better public relations) and reduced regulatory burdens, for example, by preempting future regulation. Other reasons include "enlightened" corporate self-interest, whereby the benefits of good deeds make the overall social environment in which a corporation operates more productive; obligation (corporations should invest in society because it is the right thing to do); and, most recently, profitability. In the words of global consulting firm McKinsey & Co., the new business of business is "business in society" (Davis 2007). The idea is that social issues are not side issues to the generation of market value but rather central to good business practice.

The word *value* is used by Syngenta to indicate the multiple ways in which it conceives of its business practices. In its CSR literature, the company has used a wheel diagram, which it calls the "Syngenta circle of value creation," to illustrate the role of its social responsibility efforts. The top half is labeled "business performance," and the bottom "social and environmental." In the middle of the wheel lies another circle with "Syngenta value creation" forming the core that joins these two halves together (Syngenta CSR Reports for 2003, 2004, 2005).[5] Such value generation, the company claims, benefits a set of stakeholders that includes customers and business partners, state authorities, society/communities, employees, and the public more generally.

The "circle of value" is not simply schematic: it also projects an imaginary of ethical economic development. In two of Syngenta's reports, the circle is represented as the large back tire of a tractor. The very first CSR report in 2003 shows the image embedded in a sketch of a tractor, immediately following a cover image of that same tractor with an archetypal agricultural family standing next to it (fig. 12.1). To a mind inflected by American culture, this image can only be read as a wife, husband, son, and family dog. All but the dog are clad in knee-high agricultural boots; the husband, wearing a New York Yankees baseball hat and holding a coffee mug, leans casually against the tractor. Behind them, agricultural fields extend to the horizon. This image clearly represents one ideal of stakeholder value for Syngenta.

In the 2004 report, the tractor wheel remains, but it follows a cover picture of a young, blond girl lying on the back of a cow, her arms clasped around the cow's neck, with the words "cultivate trust" and an arrow pointing toward both girl and cow (fig. 12.2). By 2005, the wheel image has gone, but the "circle of value" diagram remains. In a nod to its focus on its operations in China, the cover image now depicts an Asian woman at an outdoor market, her hands full of fruit, with the words "growing responsibly" written across the stand (fig. 12.3). In each report, value is shown as integrating economics and ethics. For example, the introductory letter cosigned by the chairman of Syngenta's Board of Directors and its Chief Executive Officer in

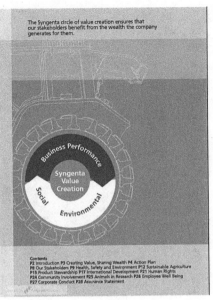

12.1. Syngenta CSR Report 2003, cover (*left*) and table of contents (*right*). Used with permission of the Syngenta Corp.

12.2. Syngenta CSR Report 2004, cover (*left*) and table of contents (*right*). Used with permission of the Syngenta Corp.

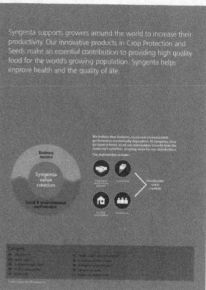

12.3. Syngenta CSR Report 2005, cover (*left*) and table of contents (*right*). Used with permission of the Syngenta Corp.

2004 observes that by "integrating social and environmental demands with business performance, [Syngenta] create[s] an expanding circle that benefits our customers, our suppliers, our shareholders, our employees, and society at large" (Syngenta CSR Report 2004).

In contrast to the synthetic, integrated images of value presented in Syngenta's reports, CSR projects for agricultural biotechnologies parse the concept of value along clearly demarcated lines, separating philanthropy from profit making, public from private knowledge, and ethics from economics. In the process, many salient questions go unasked or are simply assumed away. For example, contradicting activist and academic critiques of IPRs as an impediment to innovation and the flow of knowledge and technologies, Adrian Dubock, a Syngenta corporate executive tried to explain to me why IPRs aid development: "IP [intellectual property] issues are not at all a problem in agriculture—for development projects for use in developing countries. [It's] different if it's a development project exported to a developed and industrialized country" (AD, interview, 2006). In this formulation, technology donations are possible only where IPRs systems are weak and profitability is expected to be low. Dubock explained that this has to do with the fact that particular technologies are economically viable for the company in some countries but not others. Furthermore, companies are only able to enforce their patents reliably in developed countries, while enforcement is much harder in most developing countries.

Through this interpretation of IPRs, Syngenta has drawn boundaries that implicitly separate its profitable activities from its socially responsible ones. In places where new biotechnologies are sources of profit (i.e., in developed countries), Syngenta will enforce its intellectual property claims. In places where they are not, Syngenta may donate technologies and property rights as one of many types of CSR projects. There are no agreed-upon rules for CSR donations among companies; Syngenta can define donation in whatever way it chooses. Even if there were standards for particular forms of CSR, there are no enforcement mechanisms.

This kind of boundary work (Gieryn 1999) is a mode of decision making because it frames broad sets of issues in advance and thus forecloses debate on some aspects of those issues from the outset; they are bounded out, not within the frame. In this case, CSR projects in which IPRs play a role are only viable in "developing" countries, not "developed" ones. Profit making happens in the latter, not in the former. The continuous "circle of value" articulated in Syngenta's CSR materials gets divided in selling versus donating new technologies, settling without debate some ethical value judgments about ownership, economic viability, and the presumed beneficiaries of CSR. IPRs are fundamental to this process of foregrounding some ethical obligations, while eliding others.

Golden Rice, a Reprise: Licenses, IPRs, and the Politics of CSR

A close look at the IPRs and related licensing agreements for Golden Rice in the context of Syngenta's policies points up three major themes corresponding to modes of corporate governance effected through CSR projects: the relative value of different property claims, the obstacles created through patents, and the contradictions of simultaneously pursuing ethical commitments and economic growth. These themes, in turn, characterize what might be called a tacit "humanitarian contract" between the corporation as a donator of technology and the recipients of those technologies. These points emerge if we pick up the Golden Rice story where we left it earlier—with the publication of Potrykus and Beyer's vitamin A–enriched rice experiments and the realization that large numbers of patent claims might stand in the way of the researchers' humanitarian aims.

All Property Claims Are Not Created Equal

Worried about the large numbers of patents at stake in the Golden Rice project, the Rockefeller Foundation commissioned a report from three consultants to investigate the potential intellectual property constraints associated with Golden Rice. They found seventy patents that might require licensing and fifteen cases of technical property[6] that further complicated the picture (Kryder et al. 2000). Approximately forty patents apply in the United States and Europe. Another twenty-one are held in Japan, eleven in China, ten in Brazil, ten in Cote d'Ivoire, and nine in Vietnam.[7] A majority of these property rights across all countries belong to biotechnology and life sciences companies.[8]

In 1999, the two scientists who had engineered the first version of Golden Rice, Beyer and Potrykus, and the Rockefeller Foundation's representative, Gary Toenniessen, approached Zeneca (later Syngenta) to negotiate licenses for some of these property rights (GT, interview, 2005). The company agreed to provide a "humanitarian" license, whereby Golden Rice could be transferred to third parties for research and testing. The sublicense outlines rules for research, production, profit making, and trade (Golden Rice Webpage 2007).

The negotiations with Zeneca also highlight that not all property claims are equally powerful. Some patent holders have the institutional capability and the economic incentives to pursue violations of property claims; others do not. In the former category are multinationals like Zeneca, in the latter, academic institutions. Gary Toenniessen describes the situation as follows:

> [We] were having the meetings at Zeneca . . . [and] we had the list of thirty-six-something patents in Europe that were of concern. [T]he Zeneca lawyers said, "This is only about six patents that you've really got to worry about." And we said, "Well look at all these other patents." They started looking through the list [and] started laughing: "Hebrew University?" They were talking back and forth between one another: "I remember that patent—that patent would never hold up." Well, they knew that—we didn't. Or, "That [university's patent] might [hold up], but they'd never win." What they meant was, "We can out-fund them." They may have a good patent, but their pockets aren't deep enough to fight us. So it came down to five companies. (GT, interview, 2005).

Here Zeneca's imaginary of technology development involves a relatively more powerful position than that of the universities doing similar or related work. Companies act as a kind of "gatekeeper" to Potrykus, Beyer, and the Rockefeller Foundation's attempts to deploy Golden Rice for philanthropic purposes.

Second, property regimes are not equally viable in each country. The United States, Japan, and the European Union, for example, all have strong intellectual property regimes, with regulatory enforcement mechanisms for noncompliance. Many parts of Asia and Africa do not. For this reason, companies are more likely to donate technologies and license their use without claiming royalties in Africa and Asia but not in the United States and Europe. In the case of Golden Rice, the licensing agreement between Syngenta and Beyer and Potrykus provides a humanitarian license to Beyer and Potrykus with the right to sublicense to public research institutions and low-income farmers in developing countries but stipulates that all commercial rights to Golden Rice are retained by Syngenta.

The viability of national legal regimes also crucially influences Syngenta's patent application policy according to Adrian Dubock. At Syngenta, he observed, "We have a policy of not applying for patents in least developed countries and we seldom apply for them in developing countries because it's just not worth it. It's costly to apply for a patent [especially] if there's no real business opportunity in that country, and there's no real ability to prosecute the patent—to enforce it" (AD, interview, 2006). For obvious reasons, companies eschew practices that are costly and likely to provide no returns.

Companies then have good reasons for protecting their patent rights in countries with strong intellectual property regimes and ignoring or donating rights in countries without. For companies, patents in many cases *are* their product: it takes so much money to develop, test, and gain regulatory approval for new products that companies stake their future earnings on the anticipated profits from a strong patent portfolio (Fowler 1994). Conversely, copying or reproducing a material product, for example, by saving genetically modified seeds from one year to the next, may be very inexpensive for users. It also costs a lot of money to enforce those property rights when a product finally does make it to market. If an IPR regime in a country is not strong enough to provide enforcement for infringement claims, then a company can only sit back and watch its investments evaporate as an informal market in genetically engineered seeds emerges (CS Srinivasan personal correspondence 2005).[9] If a firm donates technologies to a country with a weak property rights regime, a company can at the very least hope for some positive public relations from the venture.

Third, and directly related, it may not be economically efficient to pursue some patents. The agricultural industry has been consolidating for more than thirty years and continues to do so; companies are downsizing, rationalizing resources, and trying to increase efficiency—"part of that is not spending money on patents that you're never going to be able to use" (AD, interview, 2006). But there is still good reason to hold on to rights. For example, in the agreement, Syngenta retains all commercial rights to Golden Rice, and the technology is only "free" to farmers who make less than US$10,000 per year in profits from its cultivation. Beyond that level, the company retains the right to seek royalties. While Golden Rice is not a strategic crop for Syngenta and is unprofitable as a "humanitarian" project, that status may change some day. Syngenta can always capitalize on its investment at a later point in time.

In donating technologies then, multinational companies must navigate the ethical and economic complexities that arise at the intersection between technology donations for CSR and IPRs, which are essential for maintaining corporate profits. In an interview, Gerard Barry, a former Monsanto executive who helped negotiate the donation of rice genome data for the company in 2000, summed up his company's imagination of CSR programs for biotechnologies: "There was technology that the company had. It was applicable to a crop the company would never make any money on, or would never be interested in. There was nobody who could license it who could make money out of it, so there was no commercial interest in any respect. And the company really liked the reaction to having done their initial sharing [of rice genome data] . . . You'll look at other companies—Syngenta and Pioneer—who also do a lot of this" (GB, interview, 2005). Of course, future profits are decidedly *not* foreclosed in the sublicense for Golden Rice; in this case, the technology always has potential profitability.

Unsurprisingly, companies are more likely to pursue property claims related to crops of economic importance to them, such as maize, sorghum, and wheat, and more willing to ignore or donate rights to less economically or strategically important crops, such as rice or cassava. The high cost of enforcing patents means that donating the technology for noneconomically important crops garners good public relations in the face of low profit expectations, at least in the present. Companies may donate the technology now but still retain control over the intellectual property in the event that it becomes potentially profitable later, thereby ensuring a window of opportunity should a crop gain economic value. The patent system provides for this kind of exploitation of broad claims to a new technology as I detail below.

The Ambivalence of Property Logic and Licensing Agreements

The logic of patents does not always mesh with the logic of donating technologies and rights. The logic of the patent system is to increase innovation through release of data and processes to the public, while generating profits for new inventions through a financial incentive to "create." Dubock explained Syngenta's reasoning behind its property rights policies as follows:

> [U]sually, the international patent application methodology is used, and that means you make one international patent application. Until the process gets to . . . the national phase, there's no benefit in limiting the countries, or indeed the claims. All companies do this . . . you apply for wide claims because if you don't get the patent published but you've put the information into the public domain, then your competitors see what you're at and so you try and disguise it . . . The name of the game, at least in commercial patenting circles is to put it in wide and scope it down, because the patent office never says, "Why didn't you go in for broader claims?" (AD, interview, 2006)

An IPR system that encourages broad claims in order to protect investments creates upstream control over future products. This allows a company like Syngenta to negotiate humanitarian aid in the present, for example, for rice, but does not foreclose the opportunity to obtain economic growth through royalty claims later on through another crop.

Additionally, although patents provide access to the information underlying a new technology, licensing that technology can be challenging—both in determining *whom* to approach for a license and *what* the licenses actually mean. For example, when companies are acquired by or merge with other companies (e.g., when Zeneca became Syngenta), intellectual property ownership may or may not be retained by the original company. It can then be difficult to determine who has the right to provide a license since the terrain is shifting constantly as companies re-structure, sell or assign patents, or grant licenses with or without the right to sub-license (Kryder et al. 2000). Gary Toenniessen puts it this way: companies like to make-believe that IP is not a constraint and say things like, 'There's no IP in the countries that you want to market this [product] in,' or 'We're willing to license.' All of that is true, but it is still a huge constraint within the public sector system, because the public sector system doesn't . . . have the capacity to deal with it (GT, interview, 2005).

At PhilRice, the Philippine national rice research center, I met in 2005 with the lawyer, Ronaldo Bernillo, who is responsible for the licensing agreements between Syngenta and the research center. PhilRice's goal is to transfer the Golden Rice traits into local Philippine varieties. PhilRice is funded by the government and other entities, including Syngenta, but it is a profit-making institution. Seeds produced at PhilRice are sold to farmers, and the profits are cycled back into research and development of new rice varieties. It takes US$1–2 million to conduct trials such as the one being done on Golden Rice, but Syngenta has agreed to provide only US$15,000 toward research costs. For PhilRice to stage an *initial* release of Golden Rice to farmers, costs would top the US$10,000 profit threshold set by the Syngenta license according to Bernillo (personal correspondence, 2005). Syngenta has not responded to questions about *who* is responsible for paying royalties if profits from Golden Rice exceed the US$10,000 profit limit (i.e., PhilRice or farmers or both). Moreover, the Syngenta sublicense contractually limits "humanitarian research," excluding commercial gain of any kind (Schedule 12: Sub-Sub-License 2002).[10] Since an intellectual property regime is still nascent in the Philippines, there is little capacity—or required knowledge—to interpret and negotiate these licenses, making people like Bernillo nervous about testing their limits.

I discussed this situation with Gary Toenniessen at the Rockefeller Foundation. He was adamant that it was only the farmer who must pay royalties for profits greater than US$10,000, not the seed company, and went on to say,

I don't think anybody ever thought that [Golden Rice] was going to be such an economically valuable trait that it was going to be a problem, but Syngenta . . . said "Well, let's just protect ourselves. If some big company . . . starts shipping this back into the U.S., we have a right to have a royalty on it, particularly if it's competing with something they were expecting to produce in the United States. I wish the Syngenta license was a much simpler license, because as it is now it causes a lot more concern on the part of national programs than it really warrants . . . I don't think *I* know today what all the terms mean . . . because it never really gets settled until the courts have made a final decision. (GT, interview, 2005)

This last sentence is particularly striking. Syngenta has indicated that the US$10,000 limit is only applicable to farmers and that the logic behind the limit is to prevent cooptation by competitors. But the contract also stipulates that "humanitarian research" cannot lead to commercial gain "either directly or indirectly," and yet PhilRice, itself a nonprofit, operates from revenues generated through the sale of the seeds it develops (Ronaldo Bernillo, personal correspondence, 2005; "Schedule 12: Sub-Sub-License 2002). And in the last analysis, the contract's enforceability is never fully determined absent a court decision.

While the logic of patents is to facilitate innovation and the logic of licenses is to facilitate efforts to experiment on, and ultimately produce, Golden Rice, the donation agreement meets neither of these objectives. The ambiguity of patent claims and licenses complicates efforts to conduct CSR projects through complex and unclear sets of power relations between those who desire to give aid (companies) and those who would ostensibly receive it (e.g., countries, research institutions, farmers). Ultimately, the mixture of profit motives with ethical concerns that I described earlier as a key feature of contemporary CSR leads to multitiered restrictions on and ambiguities in the use and dissemination of Golden Rice.

The humanitarian contract as conceived and enacted through Golden Rice creates different classes of consumers across nation-states, publics, and plants—a kind of spectrum of market access as envisioned through a corporate lens. The Golden Rice sublicense outlines the parameters of "humanitarian research and use" of the technology including the US$10,000/year profit limit on farming and commercial revenues for research organizations. Golden Rice must be consumed in "developing countries"; it must be developed in "public" germplasm (particularly germplasm used by farmers in "vitamin A deficiency prone areas"; and there can be no additional charge for the crop beyond what nongenetically engineered rice varieties cost (Golden Rice Webpage 2007). The Syngenta license also stipulates that no export of the product is allowed beyond licensing to others for research purposes. For example, even if Golden Rice became a potentially profitable export for India or the Philippines, it could not be so used. Separating domestic consumption from international trade imposes another economic constraint on a philanthropic objective.

The (Mis)alignment between Ethics and Economic Growth

There is an ethical trade-off in the determination that a humanitarian project involves donating enhanced vitamin content for a plant but then forecloses economic growth through international trade, farmer profits, or commercial gain for research organizations. These kinds of rules governing the production and consumption of Golden Rice, including the stipulations requiring research using particular kinds of germplasm (i.e., only those in the public domain) and consumption in clearly defined country contexts (i.e., "developing countries" as designated by the United Nations) fix the recipients in a relationship of dependence.

Along the same lines, there is no accounting for what happens when a farmer crosses over from "subsistence" to "profitable." The licensing terms for Golden Rice do not allow for a smooth transition from one state to another—the farmer is caught in a predicament: produce Golden Rice for its assumed nutritional value but foreclose future economic viability.[11] A similarly unhappy choice confronts a country that has allowed the production of Golden Rice for humanitarian purposes but is then faced with a surplus that it cannot trade in international markets, consistent with the terms of its licensing agreement.

In Syngenta's dilemma with regard to sharing information and technologies, "doing right" conflicted with the potential harm of negative publicity. Ironically, however, the very aspects of some countries' IPR systems that make them appealing locations for humanitarian efforts (e.g., strong vs. weak; enforceable vs. nonenforceable) also make companies the most vulnerable targets in the event that their donation "fails." Without state-sponsored biosafety regimes, recourse to IPRs, and a host of other *enforceable* mechanisms, companies are left exposed to public antagonism for their humanitarian pursuits at the same time that they are using CSR to relieve that antagonism. Public relations efforts backfiring are not the only thing to be feared. Syngenta's orderly CSR reports obfuscate the complicated, disorderly, and unregulated sets of power relations exemplified by a "wheel of value" that seamlessly incorporates ethics and economic growth. Ultimately, for Syngenta, economics precedes ethics and ethics becomes simply what is left over after economic efforts are exhausted.

Conclusion

A close look at intellectual property agreements shows three major points at which power relations are unsettled and unevenly produced through the Golden Rice licensing agreements—the unequal status of different kinds of property rights, the ambiguous terms of patenting decisions and licenses, and the misalignment between ethics and economic growth. These points, in turn, illustrate the contours of a tacit "humanitarian contract," delineating not just how Golden Rice can be developed and used but also the very meaning of what constitutes a "humanitarian" project.

The production of new biotechnologies requires decisions and negotiation over ownership and control. For Golden Rice, this means delineating the boundary between "developed" and "developing" countries, economic and ethical practices, and public versus private goods. As I have shown throughout this chapter, shifts in the terms of debate, including the questions that get asked about what constitutes an "ethical" versus an "economic" objective and who gets to decide, are present throughout the CSR process.

"Feeding people" requires different kinds of social work. But defining who gets fed and how requires the parsing of different stakeholder categories, which, while presented as a coherent circle of value in Syngenta's CSR literature, fall out in very particular kinds of ways in practice. In sum, ethics and economics are not necessarily jointly attainable through technology donations in the name of CSR: different property rights have unequal status for countries, publics, and plants; patenting decisions and licensing agreements have ambiguous terms so that the company's economic interests are protected but ethical objectives are impeded; and there is a misalignment between ethics and economics. Ultimately, CSR itself is redefined, with economic outcomes taking precedence over ethics.

Multinational corporate strategies are undergoing important transformations under the rubric of CSR and associated corporate activities. Understanding these transformations is a significant challenge for policy analysis as corporate imaginaries of value, and the social and market demarcations enacted through contracts, can serve to validate or exclude different positions. The case of Golden Rice and the issues it raises for CSR in the agricultural biotechnology industries force us to ask the question of responsibility: not simply what companies are responsible for but who gets to decide those responsibilities.

Notes

1. A discussion of this process can be found on the Golden Rice Website: http://www.goldenrice.org.
2. See Copping (2003) for a history of the evolution of these companies.
3. Bt cotton is a genetically modified cotton variety that produces an insecticide - the bacterium *Bacillus thuringiensis* (Bt).
4. These movements focused on firms' employment and investment practices (Vogel 1978).
5. The 2006 report does not include the diagram but explains the same logic in the introductory pages (3): "Syngenta is guided by the conviction that value creation depends on the successful integration of business, social and environmental performance." Reports can be found at http://www.syngenta.com/en/social_responsibility/publications.aspx.
6. Technical property involves the exchange of materials through material transfer agree-

ments (MTAs). MTAs stipulate the conditions under which materials can be shared with or transferred to a third party, but material transfers also may include intellectual property that is not contractually included in the MTA. Licenses from multiple parties may be required to use materials provided in an MTA. For a discussion of technical property and hypothetical examples of this situation, see Kryder et al. (2000).

7. In addition, six patents apply in Indonesia, five in India, five in South Africa, and one in the Philippines. All of these countries are major rice exporters, importers, or both, in the case of the Philippines and Brazil.

8. These include AMOCO, Calgene, Cetus, DuPont, Hoffman-La-Roche, and Monsanto, as well as the companies that through mergers now comprise Syngenta (ICI Ltd., Novartis, and Zeneca Corp.). Thirty-one patent-holding institutions are listed (Kryder et al. 2000).

9. C.S. Srinivasan is an agricultural and development economist at the University of Reading, United Kingdom. He related how such informal markets emerged in India for Bt cotton.

10. Dubock noted that this license is used with all sixteen licensees with few modifications (AD, email correspondence, 2006).

11. For an example of how multinational companies might pursue patents in a country with strong IPR, see Simon (2004) and *Monsanto Canada Inc. and Monsanto Company v. Percy Schmeiser and Schmeiser Enterprises Ltd* 2001 FCT 256. Could a farmer who planted Golden Rice under the humanitarian license be prosecuted for patent infringement if that farmer profits more than US$10,000? Would that farmer face removing the Golden Rice crop if this were a possibility?

References

BBC News. 2002. "Famine and the GM Food Debate," November 14. http://news.bbc.co .uk/2/hi/africa/2459903.stm. Accessed January 29, 2008.

Becker, Elizabeth. 2003. "Battle over Biotechnology Intensifies Trade War." *New York Times*, May 29.

Bernauer, Thomas. 2003. *Genes, Trade and Regulation: The Seeds of Conflict in Food Biotechnology* (Princeton, NJ: Princeton University Press).

Bud, Robert. 1993. *The Uses of Life: A History of Biotechnology* (Cambridge: Cambridge University Press).

Carroll, Archie. 1999. "Corporate Social Responsibility: Evolution of a Definitional Construct." *Business and Society* 38 (3): 268–95.

Copping, Lee. 2003. "History: The Evolution of Crop Protection Companies." *Pesticide Outlook* (December): 276–8.

Davis, Ian. 2007. "What *Is* the Business of Business?" *Business in Society: The McKinsey Quarterly* (Chicago: McKinsey & Company).

Doh, Jonathan, and Terrence Guay. 2006. "Corporate Social Responsibility, Public Policy, and NGO Activism in Europe and the United States: An Institutional-Stakeholder Perspective." *Journal of Management Studies* 43 (1): 47–73.

Dutfield, Graham. 2003. *Intellectual Property Rights and the Life Science Industries: A Twentieth Century History* (Burlington: Ashgate).

FAIR Programme. 2006. "FAIR: Agriculture and Agro-Industry including Fisheries Programme of Research and Technological Development." http://www.biomatnet.org /secure/Fair/S32.htm. Accessed March 17, 2008

Fowler, Cary. 1994. *Unnatural Selection: Technology, Politics and Plant Evolution* (Langhorne, PA: Taylor and Francis).

Gieryn, Thomas. 1999. "Boundaries of Science." In *Handbook of Science and Technology Studies*, edited by Sheila Jasanoff, Gerald Markle, James Petersen, and Trevor Pinch (Thousand Oaks: Sage Publications).

Golden Rice Webpage. 2007. "Intellectual Property-Related Issues." www.goldenrice.org /Content2-How/how9_IP.html. Accessed January 28, 2008.

Huttner, Susanne, Henry Miller, and Peggy Lemaux. 1995. "U.S. Agricultural Biotechnology: Status and Prospects." *Technological Forecasting and Social Change* 50:25–39.

Jasanoff, Sheila. 2006. "Technology as a Site and Object of Politics." In *The Oxford Handbook of Contextual Political Analysis*, edited by Robert Goodin and Charles Tilly (Oxford: Oxford University Press).

Kryder, R. David, Stanley Kowalski, and Anatole Krattiger. 2000. "The Intellectual and Technical Property Components of Pro-Vitamin A Rice (*Golden*Rice™): A Preliminary Freedom-to-Operate Review." International Service for the Acquisition of Agri-biotech Applications, No. 20, Ithaca.

Laird, Frank. 1989. "The Decline of Deference: The Political Context of Risk Communication." *Risk Analysis* 9: 543–50.

Manda, Olga. 2003. "Controversy Rages over 'GM' Food Aid." *Africa Recovery* 16 (4): 5.

Marcus, George. 1995. "Introduction." *Technoscientific Imaginaries: Conversations, Profiles, and Memoirs* (Chicago: University of Chicago Press).

Monsanto Company. 2008. "Company History." http://www.monsanto.com/who_we_are /history.asp. Accessed April 25, 2008.

Morrow, David. 2000. "Rise, and Fall, of 'Life Sciences'; Drugmakers Scramble to Unload Agricultural Units." *New York Times*, January 2.

Newell, Peter. 2007. "Introduction: Beyond CSR? Business, Poverty and Social Justice." *Third World Quarterly* 28(4): 689-661.

Niiler, Eric. 2000. "Demise of the Life Science Company Begins?*Nature Biotechnology* 18: 14.

Rockefeller Foundation. 1993. "International Program on Rice Biotechnology, Workshop Report: Potential for Carotenoid Biosynthesis in Rice Endosperm." June.

Schedule 12: Sub-Sub-License from an Inventor (or Humanitarian Organisation) to a Bona Fide Public Sector Research Organisation [and] License, with Resource Poor Farmer and Developing Country Defined with No Right to Grant Sub-Licenses." 2002. Provided by Adrian Dubock, January 18, 2006.

Simon, Bernard. 2004. "Monsanto Wins Patent Case on Plant Genes." *New York Times*, May 22.

Smith, Elta. 2012. "Rice Genomes: Making Hybrid Properties." In *Lively Capital: Biotechnologies, Ethics, and Governance in Global Markets*, edited by Kaushik Sunder Rajan (Durham, NC: Duke University Press).

Syngenta. 2003. *Corporate Social Responsibility Report*. http://www.syngenta.com/en/social _responsibility/publications.aspx. Accessed January 15, 2008.

———. 2004. *Corporate Social Responsibility Report*. http://www.syngenta.com/en/social _responsibility/publications.aspx. Accessed January 15, 2008.

———. 2005. *Corporate Social Responsibility Report*. http://www.syngenta.com/en/social _responsibility/publications.aspx. Accessed January 15, 2008.

———. 2007. *Corporate Social Responsibility Report.* http://www.syngenta.com/en/social _responsibility/publications.aspx. Accessed January 15, 2008.

Vogel, David. 1978. *Lobbying the Corporation: Citizen Challenges to Business Authority* (New York: Basic Books, Inc.).

Ye X., S. Al-Babili, A. Klöti, J. Zhang, P. Lucca, P. Beyer, and I. Potrykus. 2000. "Engineering the Provitamin A (Beta-Carotene) Biosynthetic Pathway into (Carotenoid-Free) Rice Endosperm." *Science* 287: 303–5.

THIRTEEN

Globalizing Security: Science and the Transformation of Contemporary Political Imagination

CLARK A. MILLER

Introduction

Edward Said, Benedict Anderson, and Charles Taylor have created an important starting point for understanding why an account of social and political imagination is crucial to the development of robust social theory (Taylor 2004; Anderson 1991; Said 1979). Together, they capture the historical outlines of modern social imaginaries—individual, national, and imperial—from the sixteenth through the early twentieth centuries. Their work falls short, however, in two crucial ways. First, and particularly important for this volume, it fails to account for the ways in which social and political imagination are coproduced with technoscientific ideas, organization, and materiality. Science and technology are central, as Sheila Jasanoff has observed most eloquently among present-day social theorists, to the construction of modern forms of social and political order (Jasanoff 2011, 2004b). To neglect the significance of science and technology in an account of the making of contemporary societies is therefore to leave out key building blocks in the sculpting of the modern social and political imagination (Jasanoff, introduction to this volume). Second, their work fails to account for a major shift in social imaginaries that has been underway since the middle of the twentieth century: the rise of a new globalism that expands, challenges, engages with, and transforms earlier imaginaries grounded in conceptions of the nation and nation state. Globalism imagines that human societies and economies, the systems they create, the environments within which they flourish, and the risks and threats to security they experience are increasingly global, capable of being understood and governed on scales no smaller than the planet (Miller 2004a). The new globalism is, in the full sense of the definition given by Jasanoff in the introduction, a sociotechnical imaginary.

I offer in this chapter an account of the postwar rise of globalism as a technoscientific ideal, incrementally embedded and put to work in institutions of international governance; its extension over time into a fundamental element of contemporary world making; and the frictions and resistances that now arise between globalism and older, still powerful forms of individual, national, and imperial imagination and order. At the heart of this account, I argue, is a significant transformation in the imagination and governance of security since World War II. Where security threats once stemmed exclusively from power-hungry leaders and their armies, today, security and insecurity are imagined to flow as much from tight couplings within global systems, processes, and networks. Conventional narratives trace the history of security through forms of conflict: war in Europe before World War II, the bipolar world of the Cold War, networked forms of asymmetric warfare in the age of terrorism. By focusing on security imaginaries, I propose to tell a very different story: one in which a science-driven imagination brought together new ideas of the scale of risks and manageable problems with a new political vision of how to govern the world.

Science and technology are integral to this account in several ways. Globalism, in the first instance, is a form of scientific imagination that naturalizes and objectifies a range of technical understandings of global ecological and social processes and systems (see, e.g., Jasanoff 2004a, 2001; Miller 2004b). As science has become capable of monitoring and modeling Earth systems in intricate detail, globalism has become the basis for locating insecurities as the product of complex, interacting social, ecological, political, economic, and technological systems operating at global scales. Globalism thus transforms the Earth from a place that people live to a set of global systems that they inhabit and shape and that, in turn, imposes limits to which people must increasingly adapt themselves and their actions.

Scientists and scientific institutions also played central roles in embedding this imaginary in the United Nations (UN) Specialized Agencies—the World Bank, International Monetary Fund (IMF), World Health Organization (WHO), International Atomic Energy Agency (IAEA), World Meteorological Organization (WMO), and numerous others—and through their work, in extending it around the world as a force for sociotechnical transformation. Formed after World War II, these organizations were built on a model of international political cooperation, in which states collaborated to solve problems. Over time, however, they have become expert institutions built around a new supranational model of global sociotechnical surveillance and response. These institutions identify, frame, and seek to govern security problems on a worldwide basis and thus seek to align political authority and organization with the underlying realities of global risk (Miller 2007). This transformation occurred first in the domains of environment and health, later spreading to the governance of nuclear weapons proliferation and financial instabilities, and, to a lesser extent, global terrorist networks.

At the same time, scientists shared their ideas with larger publics. The success of early works such as R. Buckminster Fuller's *Operating Manual for a Spaceship Earth*, published in 1969, *Scientific American's* September 1970 issue, *The Biosphere*, and Garrett Hardin's "Living on a Lifeboat," published in 1974, helped usher in an era of popular science writing that located new threats to humanity's survival in unstable global systems being pushed out of balance by human affairs (Fuller 1969; Hutchinson 1970; Hardin 1974). Scientists thus not only advised governments but also publicized novel metaphors of insecurity—ideas of systems, instability, complexity, vulnerability, resilience, and contagion—that continue to resonate today, both in ongoing debates about climate change and epidemics and across a range of domains of military and economic security. The underlying logic of both Steve Schneider's *Laboratory Earth* (1997) and Joseph Stiglitz's *Globalization and Its Discontents* (2002), while they are very different books, ultimately derives from a failure to bring the governance of human affairs into line with the logics of global (respectively, climate and financial) systems.

Finally, the rise of globalism has been coproduced with novel technological systems that are themselves the products of human engineering in the twentieth century, including technologies of observation, computation, visualization, communication, and transportation (Jasanoff 2004a; see also Edwards 2010; Miller and Edwards 2001). These technologies made possible the collection of data about and analysis, modeling, and visualization of global systems and processes; they carried armies, scientists, businessmen, and tourists around the globe in vastly greater numbers; and they brought the average person into relatively routine contact with the sounds (via radio), images (via television), and now tweets (via the Internet) of faraway places and events. These technologies have been critical to the fashioning of global imaginaries and to their socialization among diverse publics.

Reimagining the World

In early October 2008, the "dreamscapes of modernity" (Jasanoff, introduction) turned to nightmares. On October 1, the Dow Jones Industrial Average stood at 10,831. Ten days later, it stood at 8,451, a 22 percent drop. The proximate cause: a sharp tightening of credit markets, as banks reevaluated the safety of debt securities after the bankruptcy of Lehman Brothers and escalating subprime mortgage defaults. In response, governments and markets turned to globalist imaginaries to help understand and frame responses. This globalism was at once ontological (the kinds of security threats that exist in the world) and political (the ways in which those threats should be managed). Drawing on the metaphor of disease pandemics, business and policy leaders framed the world as containing novel risks of planetary scope. "'There is a growing recognition that not only has the credit crunch refused to be contained, it continues to spread,' said Ed Yardeni, an investment strategist. 'It's gone truly global.'" (Andrews and Grynbaum 2008). In the following days, political leaders responded by convening global institutions. In the evocative language of the *New York Times*, "[October 10] was a surreal day here in the capital of the free world, as the people who have been setting global economic policy—the Group of 7, the World Bank and the IMF—gathered to plot strategy in the middle of the scariest economic free fall the world has seen since 1929" (Cooper and Savage 2008)

Deepening the imagined links between global risks and global action, European leaders in particular argued for new measures to strengthen global regulation and restore order to global financial markets. President Nicolas Sarkozy of France and Prime Minister Gordon Brown of the United Kingdom each spoke of the need for a new global approach to banking regulation. They sought to assure attentive listeners that political leaders understood the new forms of economic insecurity racing through financial markets and to make the case for enhanced global governance as an appropriate solution. Sarkozy observed, "the economy is global; no country can protect itself alone." Brown agreed, "We now have global financial markets but what we do not have is anything other than national and regional regulation and supervision. . . . The IMF has got to be rebuilt as fit for purpose for the modern world. We need an early warning system for the world economy." Brown continued, calling for "globally accepted standards of supervision and regulation applied equally and consistently in all countries." Sarkozy, in turn, called for "a new capitalism." The Associate Press reported, "French President Nicholas Sarkozy said all European Union nations backed a radical restructuring of international institutions like the International Monetary Fund and the World Bank" (White 2008). Reviewing these proposals, the *Christian Science Monitor* editorial board argued that the United States should work with European nations to create a "new architecture" and "a new cop for global finance" (Monitor's Editorial Board 2008).

Standing days later at Camp David with Sarkozy and US president George Bush, announcing a global summit to be held in November to address the crisis, European Union president Jose Manuel Barroso summed up: "We need a new global financial order" (Stolberg 2008).

My task in this chapter is to explain the rise of globalism that makes it possible to imagine both global insecurity and global governance. As Anderson described eloquently in *Imagined Communities*, for the past two hundred years modern societies have imagined themselves and their economies as coextensive and bound up with the *nation* as a sovereign, geographically constrained territory (Anderson 1991). Explaining the rise of globalism, as a competitor and challenger to nationalism, thus requires understanding the dynamics of social imagination—how and why particular imaginaries arise, change over time, and decline, why particular dreamscapes of modernity lodge in the social and political imagination while others do not, and why differences arise in response to particular imaginings across comparative social settings and contexts. US leaders, for example, systematically resisted European leaders' calls for new, stronger global financial institutions.

Social imaginaries are emergent. They do not spring forth, whole cloth, in some final and unchanging form but rather come into being and take shape slowly. In *Modern Social Imaginaries*, Taylor (2004, 5) describes imaginaries as beginning small, in the form of theories held by small communities of elites, only taking on greater importance and "generat[ing] more and more far reaching claims on political life" over considerable amounts of time and through processes of social deliberation and conflict. Emergence, in this sense, is contingent, uncertain, and dynamic. Processes of redaction, to use Taylor's phrase, transform ideas as they are reflected upon, debated, taken up, adapted, and put into practice in wider communities. Ultimately, new imaginaries may evolve into what Taylor (2004, 23) describes as "the ways [ordinary] people imagine their social existence, how they fit together with others, how things go on between them and their fellows, the expectations that are normally met, and the deeper normative notions and images that underlie these expectations."

What differentiates the study of sociotechnical imaginaries from Taylor's more narrowly social variant, as Sheila Jasanoff emphasizes in her conclusion to this volume, is its attention to the centrality of science and technology—understood as human and social institutions—as both anchors of particular forms of social imagination and contributors to how and why those imaginations change over time. In their most widely read works, *Imagined Communities* and *Orientalism*, Anderson and Said get part of the way. Both are concerned, fundamentally, with how people imagine their identities and the identity of the communities to which they belong. For both, these imaginings were closely tied to language and the literary conventions and forms people encountered in their daily readings about the world. This consumption took place in the context of institutions that invented and disseminated texts to both elites and publics. Departments of Oriental studies, newspaper distribution chains, census bureaus, museum collections, and mapping enterprises all helped to transform, in Taylor's formulation, social theory to social imaginary, even as they redacted it in encounters with the particulars of history, geography, and politics.

Yet the ties of social imaginaries to particular forms of scientific and technological representation and rationality are, in truth, an almost marginal side note for Said and Anderson, relegated in the latter, for example, to a final chapter added only in the second edition. The larger lesson to be drawn for explaining how, when, and which social imaginaries take hold is that imaginaries are, in reality, sociotechnical, enmeshed in the coproduction of knowledge and the organizations that generate, disseminate, deliberate, and put that knowledge into practice (Jasanoff 2004b; Shapin and Schaffer 1985). Sociotechnical imaginaries, then, cannot be fully understood without inquiry into the relationships between science and technology and key forms of social organization, such as the state. And yet, compared, say, with Sheila Jasanoff's *Designs on Nature* or Yaron Ezrahi's *The Descent of Icarus*, or even with Michel Foucault's lecture on "Governmentality" or James Scott's *Seeing Like a State*, Taylor, Anderson, and Said give remarkably short shrift to the ways in which science and technology are implicated in the imagination and organization of society, the state, the exercise of power, and the possibility of governance (Jasanoff 2006; Ezrahi 1990; Foucault 1991; Scott 1998).

For Foucault, for example, the origins of modern imaginations of government and security lie in the ways that "population" and its "problems" come to be understood and managed by the state, via what he termed a "science of government": "the welfare of the population, the improvement of its condition, the increase of its longevity, health, etc." (Foucault 1991, 100). Populations came to be understood, through statistics, as a scientifically tractable quantity that states could manipulate and manage through social and economic policy. What emerged was the complex that Foucault terms "governmentality": "The ensemble formed by the institutions, procedures, analyses and reflections, the calculations and tactics that allow the exercise of this very specific albeit complex form of power, which has as its target population, as its principal form of knowledge, political economy, and as its essential technical means, apparatuses of security" (Foucault 1991, 102).

Building on this idea, historians and sociologists have identified a range of ways in which the concept of the welfare state, its particular knowledge engines—statistically represented populations and their problems—and its imaginations of security emerged together in a "double institutional transformation" (Wittrock and Wagner 1996): the coproduction of science and the state (Jasanoff 2004b). In this work, much of it in the traditions of science and technology studies, we can see in greater detail how and why the specific social imaginaries identified by Taylor, Anderson, and Said took the form that they did. Society, understood as more than just a collection of individuals, emerged as "objectified reality" (Taylor 2004, 69), as did the economy and the nation, but not merely as the ad hoc result of processes of social deliberation and conflict. Rather, they emerged as the specific production of tightly interlinked sociotechnical arrangements, themselves also emergent, that comprised the administrative, welfare state: the modern research university and, within it, the social sciences, offering conceptual foundations for viewing the nation and the economy as governable statistical entities; programs for training in these sciences the management class that administers state authority to ensure the welfare of the nation; and scientifically imagined and implemented government projects and agencies dedicated to social welfare and resource management (Hays 1959, Hacking 1990, Jasanoff 1990, Nowotny 1991, Skocpol 1992, Porter 1995, Rueschemeyer and Skocpol 1996, Hilgartner 2000).

The Globalization of Security

Globalism reflects this same triad of security, population, and government, reconfigured around new forms of insecurity, new objects of scientific analysis, and new institutions of governance. It is also a coproduction of epistemic and political authority and organization. The new sociotechnical imaginary of the globe—and especially of global systems—and the forms of technoscientific and political organization that anchored it took form in the half century following World War II. Over this period, ideas of security were adjusted continuously as diplomats and experts worked to make sense of emergent threats and find ways to tame their consequences for postwar societies. In the process, a globalist imagination slowly emerged and became embedded within a new suite of international organizations known as the UN Specialized Agencies. Birthed during a brief flourishing of international diplomacy from 1943 to 1947, these institutions struggled for authority and relevance in the resurgent tensions of the Cold War. These struggles gave rise to powerful epistemic resources through alliances with scientific and expert communities eager to reestablish global networks after the disruptions of wartime. Together, scientists and international organizations fashioned impressive new technoscientific capabilities for data collection and analysis on global scales. At the same time, they sought ways to put their analyses to use in improving international peace and prosperity, fashioning new understandings of global processes and systems, new insights into the risks posed to and by such processes and systems, and new authority to monitor and regulate these risks.

Security and the Founding of the UN

The founding of the UN in early 1945 reflects one of the earliest sites in the emergence of new imaginations linking science and technology to security. For US president Harry S. Truman, the project to build a postwar UN organization responded directly to new technological capacities that had transformed the nature of war. At the UN Conference on International Organization on April 25, 1945, three months before the Trinity test in New Mexico, Truman pitched this new perspective (Truman 1945). His audience was not yet aware of the atomic bomb or of the massive psychological impact it would unleash. Listeners were well aware, however, of the tremendous power and unfathomable destruction of modern armaments. Only a month earlier, the United States had launched a campaign of bombings of Tokyo that would later be judged among the most destructive military operations ever carried out, outdoing even the destruction of Dresden and the London Blitz. Offering the opening address to the UN Conference, Truman built his argument for the creation of a new "world organization" on this threat of annihilation through technological war: "[M]odern warfare, if unchecked, would ultimately crush all civilization." Truman viewed a new "world organization for the enforcement of peace" as essential in the face of the "international chaos" wrought by "the torture and tragedy of two world conflicts." Military technologies threatened "ever-increasing brutality and destruction" to everyone, everywhere on the globe.

In the face of this powerful threat to humanity, however, Truman's imagination of security was still built on a foundation of nations, existing in parallel, each confronting its own challenges and one another, and perhaps working together. World and international were the operative ideas for Truman. Nowhere did he use the word global. Truman's world was not yet composed of ontologically global objects, systems, or processes. Consistent with his period's talk of global war, Truman's vision coupled modern weapons and the will of individual leaders to the potential for all-out conflict between nations. People constituted the threat: "madmen . . . who in every age plan world domination" and "differences between men, and between nations, [that] will always remain."

Truman desired a "world organization . . . to provide sensible machinery for the settlement of disputes among nations" and build "for tomorrow a mighty combination of nations founded upon justice for peace." This was still the Westphalian order, international and not global, with human security depending on the cooperation, and interdependence, of nations. Truman observed that the participants at the meeting "represent the overwhelming majority of humankind." "Man has learned long ago that it is impossible to live unto himself. This same basic principle applies today to nations. We were not isolated during the war. We dare not become isolated in peace."

Truman's world of international cooperation was thus largely aspirational. The speech was studded with appeals for bringing that world into being. "You members of this Conference are to be the architects of the better world." "Let us labor to achieve peace." "We hold a powerful mandate from our people. They believe we will fulfill this obligation. We must prevent, if human mind, heart, and hope can prevent it, the repetition of the disaster from which the entire world will suffer for years to come." Note here the "we" of Anderson's imagined communities. Truman imagined the assembled representatives of nations as a single human community, capable of uniting as a collective of the whole. "We must build a new world—a far better world—one in which the eternal dignity of man is respected." But the building of that "better world" only made sense as Truman understood it, as primarily the work of men from all nations acting together to improve relations among nations: "For lasting security, men of good will must unite." Truman neither envisioned insecurity as lodged in global systems (as did, for example, his successor, Eisenhower, who laid the blame for the continued buildup of highly destructive weapons not in nations and their leaders but rather in a global military-industrial complex; see Eisenhower 1961), nor did he mean that the UN could build a better Earth (as, for example, do contemporary proponents of geoengineering who believe they can manage Earth systems to produce improved human and ecological outcomes in the face of global climate risks; see, e.g., Keith 2000).

Truman's language echoed other proponents of one-world discourse, such as the 1940 Republican presidential candidate Wendell Willkie, whose book One World, published in 1943, explicitly advocated that the world be reimagined as one. These advocates sought to colonize the minds of readers, to persuade them that people everywhere were the same and held the same aspirations to live in peace (Willkie 1943; cf. Takacs 1996, who describes similar efforts to colonize minds on behalf of global biological diversity). For the world in 1945 was not one. Far from it. In 1945, nations were still divided, perhaps irrevocably. Truman could imagine the birth of a new problem of modern technological warfare that had enveloped the world during World War II and would continue unabated with the deployment of the atomic bomb. Yet his social and geopolitical imagination remained resolutely international. Fashioning "one world" out of the world's many nations was the only unifying discourse at his disposal.

The Birth of the UN Specialized Agencies

Truman's vision of the UN as a space for international diplomacy to confront world conflict was successful, at best, only at the margins. Since their formation, neither the UN General Assembly nor the more exclusive UN Security Council has functioned as much more than a space for symbolic diplomatic maneuver. By contrast, since 1945, the UN Specialized Agencies have contributed deeply to the evolution of global security imaginaries. These organizations began within the logic of the UN—as places for nations to work collectively to solve problems—but almost immediately departed from this path to become scientific and expert institutions (see, e.g., Miller 2001). Today, these organizations are imagined as institutions for documenting and managing global security risks, as evidenced by the calls by Brown and Sarkozy to upgrade the regulatory authority of the IMF discussed above, even if their exact power and authority remain contested.

This evolution in security imaginaries arose out of efforts to resolve tensions between geopolitical and scientific imaginations that confronted the UN agencies from the outset. Scientists and their ideas stood in opposition, especially, to the idea that geopolitical considerations might limit the scope and membership of these organizations to only some of the nations and territories of the planet. In opening the key negotiation session for a World Meteorological Convention in 1947, for example, US Assistant Secretary of State Garrison Norton offered a similar logic to Truman's (Miller 2001, 184–8). As Norton portrayed it, in a speech cowritten by the Chief of the US Weather Bureau, F.W. Reichelderfer, the central challenge for nations in 1947 was the potential for renewed international conflict. Key to addressing that challenge, Norton argued, was to make international cooperation a reality, not just an aspiration. Yet weather scientists, Norton suggested, were not diplomats, and this made them uniquely situated to help in that task. Meteorologists, he argued, had several salient advantages over their more political contemporaries. First, weather science was already a worldwide enterprise in that its practitioners inhabited every country but collaborated together. Meteorologists offered in this respect an exemplar and a model for other forms of international cooperation, based on their ability as scientists to put aside political differences. They also had the benefit of a common language, science, which helped prevent misunderstanding and miscommunication. Finally, weather scientists had concrete solutions to help mitigate the kinds of challenges that could lead nations into war. Specifically, in 1947, Norton was worried about the continuing impact of harsh winters in Europe on the potential for renewed conflict. If international cooperation among meteorologists could generate better weather forecasts, and through these an improved ability to solve key international problems, then the WMO they hoped to establish could be instrumental, in Norton's eyes, in helping shape a new world peace.

Here we see an early articulation of a new sociotechnical imaginary that, like Truman's, linked science and technology to security, but in a wholly different configuration. Even as the proliferation of atomic bombs and rockets was further reinforcing Truman's vision of a world torn asunder by technologies of mass destruction—and the Soviets were rejecting the US Baruch Plan to establish an agency for international control of nuclear weapons at the UN Atomic Energy Commission—the US State Department and Weather Bureau were collaborating to fashion new ideas and institutions through which science and technology made the world more not less secure (the domestic US sources of this new imaginary are examined in Miller 2006). The meteorologists gathered in Washington for the meeting shared Norton's ideals but also had their own ideas. Specifically, they viewed the WMO's geographic coverage as critical. Debates over China's membership in the future WMO, for example, pitted geopolitical concerns (China had, by then, become communist) squarely against the desire of meteorologists to ensure access to weather data for the entire globe. Without such data, weather prediction in areas downwind from the gaps would be considerably more difficult. For meteorologists, then, the weather was already beginning to emerge as a global entity that made claims on political life, calling for old divisions to be set aside for the sake of new modes of cooperation. Over time, they built the WMO into a powerful instrument of worldwide weather data collection, synthesis, and dissemination. Meteorologists would continue to refine this idea over the next half century, making the Earth's atmosphere, ozone layer, and climate system into key elements in a new global imagination of risk and insecurity.

A similar tension between scientific and geopolitical imagination also appeared a year earlier in the 1946 negotiations over the constitution of the World Health Organization (WHO 1946). In debating the name of the organization, for example, the proceedings noted that the Iranian delegate favored the name "World Health Organization" over that of "United Nations Health Organization": "Dr. Hafezi (Iran) favoured the retention of the title 'World Health Organization,' because all of the nations of the world were not represented in the United Nations. The present conference was the first international conference to allow the admission of non-self-governing territories as associate members. The new organization would be working for the entire human race, whereas the United Nations comprised fifty-one nations; and the objective of the new organization, as stated in its charter, was the attainment by all peoples of the highest possible level of health" (WHO 1946, 47).

For Hafezi, *world* was to be preferred precisely because it represented "all the nations of the world." While doctors did not yet view health as a global entity in the same way that meteorologists were coming to view the atmosphere and the weather, nonetheless, it had some similar characteristics. Hafezi portrayed health as important "for the entire human race," and its "attainment by all peoples [at] the highest possible level" could only be achieved if the organization was universal in its membership.

These thoughts echoed those of many participants, including the Chinese delegate, who followed Truman in linking a new vision of the world to the growing power of technologies of war:

> Even at that early stage it had been recognized that, in the province of health, activities must be universal and cover a wider field than the United Nations organization itself. [Dr. Sze] believed that a similar trend of thought had been expressed in the titles of the International Civil Aviation Organization, the International Bank for Reconstruction and Development, and the International Fund for Stabilization of Currency . . . The realization, after the dropping of the atomic bomb at Hiroshima, that the world had entered a new age, the Atomic Age, had been reflected in further titles that began to appear on the international scene—such titles as "World Bank." (WHO 1946, 47)

The representative of Argentina offered yet another variant of the claim: "Dr. Zwanck (Argentina) supported retention of the name 'World Health Organization,' on the ground that the Conference had been called for the purpose of uniting all nations in the interest of health. Argentina believed in the interdependence of the human race and of the nations of the world, and that the Organization should not be closed to any State. It was stated in the Preamble that every human being, without distinction of race, religion, political belief, or economic or social condition, should have the fundamental right to good health and well-being" (WHO 1946, 47–48).

Much as the 1948 UN Universal Declaration of Human Rights would, the World Health Convention declared that every human being held certain fundamental rights—in this case to health. Rights, of course, are a central element in modern social imaginaries, as Taylor describes at length. For Taylor (2004, 21), however, rights are located precisely at the interface between the individual, the society that serves the individual, and the state, which acts as a guarantor of the individual's rights. In Zwanck's discussion of the emerging World Health Convention, we see a new formulation of this ontological relationship that begins to relocate those rights outside of and against the power of national societies and states, as part of the authority of the UN Specialized Agencies to act as supranational governing bodies of a global humanity. This idea has continued to expand over time, today underpinning the authority of the International Court of Justice to prosecute individuals for war crimes and of Amnesty International to criticize states for violations of human rights (for a comprehensive treatment of international human rights policy, see, e.g., Moyn 2010).

Just as importantly, the WHO articulated those rights not in political terms (e.g., as freedoms) but instrumentally, as an outcome to be performed via improvements in public health knowledge, technologies, and organizations across the globe. Negotiators established two important concepts for the new WHO. First, they defined health as a universal property: everyone desired and was entitled to health, and everyone experienced health problems. Second, they understood health risks, especially infectious diseases, and the differential technical capacities of states to control and manage them, as a common threat. These two ideas both envisioned membership in the WHO including all nations, even against geopolitical reasons to leave some countries out, both to ensure worldwide coverage and to enroll all nations in upgrading their public health infrastructures:

> Dr. Gines (Paraguay): neither disease nor health took account of national boundaries . . . unequal development in different countries in the promotion and control of disease, especially communicable disease, was a common danger. Why then should the Conference persist in excluding any State from the new organization? . . . It would not be in keeping with the objective of the Organization for political criteria to be in evidence in the Constitution . . . The policy of the Organization should be inspired by that of the Red Cross, which had brought relief and help to millions without enquiring the reason for the war or for their misery. (WHO 1946, 67)

In making this case, representatives focused especially on the sociotechnical foundations of disease, noting that it is precisely human activities and technologies together that created the new, common risk. The reduction in travel times to less than the incubation period of communicable diseases made it impossible to use the old policies of quarantine at the border to protect a nation's health. The traveler might not be showing symptoms when arriving at a border, a problem that would recur time and time again in the age of air transportation, most recently in the cases of Ebola, SARS, and avian influenza. New forms of international regulation were necessitated: "The greatly enhanced importance of international health controls, said Dr. Chisholm (Canada), had been recognized ever since the steamship had reduced the time of travel between continents below the time of incubation of many diseases. Air transport throughout the world would now largely nullify protective barriers against disease, and thus no country could depend on its own arrangements alone, but must be assured of satisfactory controls in other countries as well" (WHO 1946, 33).

To sum up, the imaginaries of the mid-twentieth century understood the problem of international security as a problem of and created by people and nations: Truman's "madmen" bent on dominion through war with ever more powerful technologies. Health risks were not yet linked to global processes and systems, but they were crossing borders and posed common dangers. Health officials began to see health as a universal problem and a universal right, and they sought to collaborate to help achieve worldwide protections. Nor did weather as yet pose global risks, yet its accurate prediction required worldwide collection of data. Humans had created global war and the "steamship" and "air transport" that made nations more dependent on one another. But the community of nations would only become one world if they fashioned institutions of collective action that enabled them to work together to fashion a secure future. Experts insisted on the creation of organizations that spanned the entire globe and encompassed all countries. That these ideas were consistent with the logic of Truman and others of promoting security through new international organizations gave scientists the power to pursue their ideas.

Ontological Globalism in the UN Specialized Agencies

Over time, however, these expert organizations would go well beyond Truman's vision and fashion a new sociotechnical imaginary in which threats to security stemmed from processes and systems that were ontologically global, that is, both defined and controllable only at the scale of Earth itself.

This process of imaginative change can be seen in some of the earliest work of the WMO, which began operations in 1951. During the 1950s, the WMO took a lead role in planning and implementing the 1957–58 International Geophysical Year (IGY). Through this effort, the WMO planned and persuaded countries to build numerous new weather stations in priority locations that would generate insights into the global circulation of the atmosphere rather than simply expand local or national weather prediction efforts. For example, North American and European nations worked with the WMO to build new weather stations along longitudinal lines in the middle of the Atlantic, while several countries set up stations in the Antarctic. During the IGY, the WMO also collated data from national observations into global data sets. As a consequence, scientists working with the WMO were first able to capture a real-time, global picture of atmospheric dynamics as a basis both for improved weather prediction and atmospheric research.

By the 1960s, data from the IGY—and from the World Weather Watch, the WMO program set up in 1960 to further expand its global weather data collection network—became the backbone for creating the first computer models of the global atmosphere, known as general circulation models (GCMs; Edwards 2010). These models, in turn, became the focal point for fashioning a new understanding of the weather and climate as ontologically global. As has been documented elsewhere in detail, scientists working with GCMs helped construct the very idea of the Earth's climate system in the 1970s and 1980s, in explicit contrast to older definitions that held climate to be a thirty-year aggregate of local weather patterns (and, hence, cities like Boston and Phoenix have different climates; Edwards 2001; Shackley and Wynne 1995, 1996).

Over the next two decades, the idea of the Earth's climate system helped give credence, meaning, and influence to the new sociotechnical imaginary of globalism by linking scientific visions of undesirable futures to social and political reconfiguration on global scales (Miller 2004a). Analyzing changes in global temperature and models of the effects of rising atmospheric concentrations of greenhouse gases, scientists argued that human activities now put the Earth's climate system at risk and that, in turn, changes in the climate system threatened the lives and livelihoods of people across the globe. Responding to these studies, global leaders established and empowered new international organizations, such as the Intergovernmental Panel on Climate Change and the Conference of Parties to the UN Framework Convention on Climate Change, who in turn negotiated new forms of knowledge and new international treaties. Subsequent developments, while never quite reaching the ambitious goals initially hoped for, have nonetheless transformed the global politics of carbon and energy, putting on the defensive some of the world's richest and most powerful nations and corporations, while focusing widespread public and policy attention on transforming the technological systems that humans rely on for energy.

The history of the WHO offers parallel examples of putting a global imagination to work in science and technology. In the early imagination of the WHO, health was a universal problem that transcended borders, but individual diseases were not (yet) seen as global phenomena. For example, in 1958, the Soviet representative to the WHO suggested that the organization tackle the challenge of smallpox. The resulting Smallpox Eradication Program focused on vaccination in areas of the world where smallpox was endemic. The very use of the word endemic suggests a still localized approach to disease. Nonetheless, given the universality of health and the worldwide distribution of endemic diseases, the WHO began to see itself as a global operation, as did others. Reflecting this perspective, also in 1958, a medical journalist, Albert Deutsch, titled his report reviewing the organization *The World Health Organization: Its Global Battle against Disease* (1958). In 1965, the WHO director-general proposed "a comprehensive plan for a global eradication programme." This strategy succeeded and, and eventually, in 1980, the WHO's Global Commission for the Certification of Smallpox Eradication declared smallpox dead (Fenner et al. 1988).

The success of the smallpox campaign lay not only in eradicating disease but also in building sociotechnical infrastructures on which to ground further shifts in imagination. During the 1970s and 1980s, WHO scientists and their colleagues in national disease control agencies established the epistemic and organizational foundations for reimagining some health threats as explicitly global. WHO scientists working with smallpox established an extremely effective surveillance network that coordinated national and international public health officials to rapidly detect and respond to new outbreaks. This network became the forerunner to later institutional initiatives, building up to the Global Outbreak and Alert Response Network, which WHO operates today to facilitate rapid response to all new epidemic outbreaks. The establishment of a committee of international scientific experts to coordinate the smallpox eradication effort also helped establish the WHO as the key convener of international scientific expertise to respond to global health threats (Fenner et al. 1988). These precedents came together around the AIDS epidemic. In 1985, the WHO, together with the US Centers for Disease Control and Prevention (CDC), organized the first International AIDS Conference to begin to develop worldwide disease statistics, epidemiology, and response strategies. It continued to lead major international responses to the disease thereafter. It is not an accident that AIDS quickly became described in public discourse as a global epidemic (e.g., Grose 1987; World Bank 1999). Indeed, AIDS subsequently emerged as the exemplar of a new sociotechnical imaginary of emergent global pandemic diseases—a category that quickly grew to include Ebola and other hemorhagic fevers, SARS, avian influenza, and drug-resistant tuberculosis—even as researchers came to understand that disease etiologies often varied from place to place across the globe (see Lakoff, this volume).

Throughout these epidemics and outbreaks, the WHO consolidated its authority to lead international responses in describing, tracking, and managing new infectious disease risks. New International Health Regulations (IHR) promulgated in 2005 significantly upgraded the WHO's power and political authority in line with its built-over-time epistemic authority. Consistent with a now globalized Foucauldian vision of political authority grounded in expert knowledge and management of diseased populations, these rules aim to enable the WHO "to prevent, protect against, control and provide a public health response to the international spread of disease" (WHO 2005, 2). The new IHR rules provide explicit, global standards for the capacity of public health agencies at the national level as well as external quality assessment procedures to ensure that states conform with these standards. They generalize the authority of the WHO, which had previously been limited to acting with regard to specifically authorized diseases, to act on any disease that presents international health risks. Most importantly, they grant the WHO the authority to declare public health emergencies for the globe and to make temporary recommendations for addressing those emergencies, such as imposing travel restrictions.

A review of the WHO by Theodore M. Brown, Marcos Cueto, and Elizabeth Fee (2006) summarizes this shift from an international to a global imagination of health. The WHO, these authors suggest, only slowly and haltingly adopted global health as its central mission. They note that, in the PubMed database, articles using the phrase "international health" predominated over those referencing "global health" by large majorities through the 1970s. In the 1950s and 1960s, references to "global health" occurred in only 5 percent of the number of references to "international health," rising to 10 percent in the 1970s, 30 percent in the 1980s, and 44 percent in the 2000s. More significantly, they note that the WHO only redefined its mission in terms of global health threats in the early 1990s in response to efforts by the World Bank, UNICEF, and other organizations to wrest control of key programs for infectious disease observation and response. Responding to these threats, the WHO selected, in 1998, Gro Harlem Brundtland to serve as the first director-general from outside the organization. The choice was well considered. Brundtland, as head of the World Commission on Environment and Development, had led the effort to define a new global perspective on environmental sustainability in the 1980s. Now she brought that same perspective to the WHO, helping, for example, to define smoking as a global health threat and launching a global campaign to regulate cigarettes.

The Uptake and Socialization of Global Security

Drip . . . drip . . . drip . . . dystopian fantasies of insecurity fall upon the surface of public imagination, rippling through civic consciousness, reconfiguring flows of planetary ontology and authority as communities reconstruct impressions of danger and defense. But fantasies do not fall on unprepared ground (Jasanoff 2001). Scientific instruments detect them, and expert explanations make sense of them, all the more persuasively for their presumed objectivity. Governments respond and reassure, deploying people and power across the face of the planet. TV and Internet websites carry images and voices from around the globe to our eyes and ears, orchestrated by new global centers of calculation. All of it telling us that we face global risks and must act appropriately, as a unified global community, to counter them. We inhabit a new sociotechnical imaginary of global security that performs for us routinely, linking the power of science and technology to describe risks and propose solutions to the power of social and political institutions to fashion order.

On February 12, 2014, scientists at the US Centers for Disease Control and Prevention, working with counterparts in the WHO, announced that avian influenza H7N9 had traveled internationally for the first time, from China to Malaysia. The next day, the Obama administration announced the Global Health Security Agenda, a partnership of twenty-nine countries, covering 4 billion people, to strengthen the global network of institutions aimed at detection of and response to emergent pandemic diseases. On the CDC's new Global Health Security website, stark images met visitors: scientists in protective suits investigating an infectious disease outbreak in Uganda; a newspaper headline, somewhere in Africa, "Ebola kills 14"; Chinese doctors working in a lab: "SARS cost the world $30 billion in just 4 months"; and the last, a diagram of the global aviation network: "Disease can spread nearly anywhere in 24 hours" (CDC 2014).

The Global Health Security Agenda is a deliberate response to problems with the WHO's 2005 IHR. Under the IHR, countries had agreed by June 2012 to upgrade their public health infrastructures to detect, analyze, and contain infectious diseases. Global recession had gotten in the way, however, and 80 percent of countries had fallen short and asked for an extension of the deadline. At the heart of the Global Health Security Agenda was a pledge by the United States to provide new financing to enhance government infectious disease laboratories and upgrade diagnostic equipment in poor countries in Asia and Africa. Already, by some accounts, the US Department of Defense had spent over $300 million on the task. The goal of the Global Health Security Agenda: a global biosurveillance network capable of responding within two hours to the outbreak of any new disease.

Imagining global risks and insecurities, world leaders are reauthorizing the administrative organs of the UN Specialized Agencies in new, more powerful forms, granting them added capacities and new powers of action to develop knowledge about and regulate global systems and processes. The 1988 Intergovernmental Panel on Climate Change, the 2005 IHR, and the 2008 proposals to expand the IMF discussed above are potent examples of this trend. So, too, is the expansion of the IAEA's authority after the 1991 Gulf War. When UN inspectors found previously undetected nuclear weapons projects in Iraq, the IAEA was granted wide-ranging new powers to conduct inspections in countries. In a few years, the IAEA went from an organization with the authority to ensure, at most, that declared materials did not disappear from declared facilities to a global investigator of the possibility that states were surreptitiously enriching uranium and plutonium for military use (Scheinman 1993). At the heart of this endeavor is a robust techno-scientific capacity, in the IAEA and its partner networks around the globe, for tracking nuclear materials and identifying their source of origin.

Paralleling these developments is a broader and deeper socialization of the global security imaginary. The global spread of infectious diseases has become a subject for popular novels and high-profile Hollywood blockbusters. Discussions of the climate system have become commonplace on the evening news. Global markets have become a staple of dinner table conversations and the focal point of widespread social protest movements. These movements target institutions—including the IMF, World Bank, World Trade Organization, G-7, and US government—that are seen as pushing an economic agenda that promotes transnational corporations and global markets at the expense of local communities.

What is at stake, ultimately, in all of these developments are the imaginative foundations of global power and authority. Writers in the English language now objectify not only nature on a planetary scale (e.g., global pandemics and the Earth's climate system) but also society (e.g., global markets and global networks) and even politics, routinely attaching the adjective "global" to ideas of civil society, governance, citizenship, politics, obligations, responsibilities, regulation, rules, and order (the scale of this trend can be tracked using Google's ngram viewer). Newspaper articles covering the launch of the Global Health Security Agenda imagined not only global health and global threat but also global effort, global partnership, global networks, global population, global spread, global pandemics, global initiative, global citizens, global community, global systems, and more. We have arrived at a time in which many of the world's imaginings of security differ vastly from a half century ago. Phrases like "global markets," "global networks," and "global systems" may be mere word combinations. Yet, today, these phrases cohere in the thoughts of world elites—and even occasionally among publics—in a way that in prior eras they would not. That coherence is the linguistic trace of a deeper sociotechnical imaginary linking new forms of scientific observation and analysis, new infrastructures of transportation and communication, new patterns of social organization, and new possibilities of politics and governance.

To be sure, globalism remains contested in its encounters with individualism, nationalism, and imperialism. Globalism's rise is neither natural nor totalizing nor hegemonic. The extensive human work of creating globalism as a form of sociotechnical imagination remains contingent on particular arguments and achievements made in particular settings, partial in its successes, and potentially reversible. In recent years, explicit rejections of globalism have grown stronger across key domains of international governance, including both finance and climate change (on the latter, see, e.g., Ostrom 2010). Rebuffing European arguments for new global regulatory powers for the IMF in the wake of the 2008 global financial crisis, US negotiators offered instead to reduce vulnerabilities to financial instabilities by reestablishing buffers between national financial markets. Calls have been made repeatedly to break up the largest and most globally extensive banks to reduce the risks they pose to the global economy. Yet, to date, such calls have largely gone unheeded, and even as US officials rejected new powers for the IMF, they also went to work rewriting rules for bank deposit ratios under the Basel Convention to shore up the foundations of the global banking system. Globalism is thus not a fait accompli. It is nevertheless now a permanent player in humanity's dreams for a better future.

References

Anderson, Benedict. 1991. *Imagined Communities: Reflections on the Origin and Spread of Nationalism.* London: Verso.

Andrews, Edmund, and Michael Grynbaum. 2008. "Fed Weighs Bid to Spur Economy as Markets Plummet Worldwide." *New York Times,* October 7.

Brown, Theodore, Marcus Cueto, and Elizabeth Fee. 2006. "The World Health Organization and the Transition from International to Global Public Health." *American Journal of Public Health* 96: 62–72.

Centers for Disease Control and Prevention. 2014. "Global Health Security." http://www.cdc.gov/globalhealth/security and http://www.cdc.gov/globalhealth/security/why.htm. Accessed March 14, 2014.

Cooper, Helene, and Charlie Savage. 2008. "A Bit of 'I Told You So' Outside of World Bank Talks." *New York Times,* October 11.

Deutsch, Albert. 1958. *The World Health Organization: Its Global Battle against Disease.* New York: Public Affairs Committee.

Edwards, Paul. 2001. "Representing the Global Atmosphere: Climate Models, Data, and Knowledge about Climate Change." Pp. 31–66 in *Changing the Atmosphere: Expert Knowledge and Environmental Governance,* edited by C. A. Miller and P. N. Edwards (Cambridge, MA: MIT Press).

———. 2010. *A Vast Machine: Computer Models, Climate Data, and the Politics of Global Warming.* Cambridge, MA: MIT Press.

Eisenhower, Dwight. 1961. "Farewell Address." January 17.

Ezrahi, Yaron. 1990. *The Descent of Icarus: Science and the Transformation of Contemporary Democracy.* Cambridge: Harvard.

Fenner, F., D. Henderson, I. Arita, Z. Jezek, and I. Ladnyi. 1988. *Smallpox and Its Eradication.* Geneva: World Health Organization.

Foucault, Michel. 1991. "Governmentality." Pp. 87–104 in *The Foucault Effect,* edited by Graham Burchell, Colin Gordon, and Peter Miller. Chicago: University of Chicago Press.

Fuller, R. Buckminster. 1969. *Operating Manual for Spaceship Earth.* Evanston: Southern Illinois University Press.

Grose, Nicholas. 1987. *AIDS: Proposals for Action: The Global Epidemic and the Crisis in Africa.* London: War on Want.

Hacking, Ian. 1990. *The Taming of Chance.* Cambridge: Cambridge University Press.

Hardin, Garrett. 1974. "Living on a Lifeboat." *BioScience* 24: 561–68.

Hays, Samuel. 1959. *Conservation and the Gospel of Efficiency.* Pittsburgh: Pittsburgh University Press.

Hilgartner, Stephen. 2000. *Science on Stage: Expert Advice as Public Drama.* Palo Alto: Stanford University Press).

Hutchinson, G. Evelyn. 1970. "The Biosphere." *Scientific American* 233: 45–53.

Jasanoff, Sheila. 1990. *The Fifth Branch: Science Advisers as Policy Makers.* Cambridge: Harvard.

———. 2001. "Image and Imagination: The Formation of Global Environmental Consciousness." Pp. 309–38 in *Changing the Atmosphere: Expert Knowledge and Environmental Governance,* edited by C.A. Miller and P.N. Edwards. Cambridge, MA: MIT Press.

———. 2004a. "Heaven and Earth: The Politics of Environmental Images." Pp. 31–54 in *Earthly Politics: Local and Global in Environmental Governance,* edited by Sheila Jasanoff and Marybeth Long-Martello. Cambridge, MA: MIT Press.

———. 2004b. *States of Knowledge: The Co-Production of Science and Social Order.* London: Routledge.

———. 2006. *Designs on Nature: Science and Democracy in Europe and the United States.* Princeton: Princeton University Press.

———. 2011. *Science and Public Reason*. Abingdon: Routledge.

Keith, David. 2000. "Geoengineering the Climate: History and Prospect." *Annual Review of Energy and the Environment* 25: 245–84.

Miller, Clark A. 2001. "Scientific Internationalism in American Foreign Policy: The Case of Meteorology (1947–1958)." Pp. 167–218 in *Changing the Atmosphere: Expert Knowledge and Environmental Governance*, edited by C. A. Miller and P. N. Edwards. Cambridge: MIT Press.

———. 2004a. "Climate Science and the Making of Global Political Order." Pp. 44–66 in *States of Knowledge: The Co-Production of Science and Social Order*, edited by Sheila Jasanoff. London: Routledge.

———. 2004b. "Resisting Empire: Globalism, Relocalization, and the Politics of Knowledge." Pp. 81–102 in *Earthly Politics: Local and Global in Environmental Governance*, edited by Sheila Jasanoff and Marybeth Long-Martello. Cambridge: MIT Press.

———. 2006. "'An Effective Instrument of Peace:' Scientific Cooperation as an Instrument of U.S. Foreign Policy, 1938–1950." *Osiris* 21: 133–60.

———. 2007. "Democratization, International Knowledge Institutions, and Global Governance." *Governance* 20: 325–57.

Miller, Clark A., and Paul N. Edwards, eds. 2001. *Changing the Atmosphere: Expert Knowledge and Environmental Governance*. Cambridge: MIT Press.

Monitor's Editorial Board. 2008. "A New Cop for Global Finance." *Christian Science Monitor*, October 17. http://www.csmonitor.com/Commentary/the-monitors-view/2008 /1017/p08s01-comv.html. Accessed May 4, 2013.

Moyn, Samuel. 2010. *The Last Utopia: Human Rights in History*. Cambridge: Harvard University Press.

Nowotny, Helga. 1991. "Knowledge for Certainty: Poverty, Welfare Institutions, and the Institutionalization of Social Science." Pp. 23–41 in *Discourses on Society: The Shaping of Social Science Disciplines*, edited by Peter Wagner, Bjorn Wittrock, and Richard Whitely. Dordrecht: Springer.

Ostrom, Elinore. 2010. "Polycentric Systems for Coping with Collective Action and Global Environmental Change." *Global Environmental Change* 20: 550–7.

Porter, Theodore. 1995. *Trust in Numbers: The Pursuit of Objectivity in Science and Public Life*. Princeton: Princeton University Press.

Rueschemeyer, Dietrich, and Theda Skocpol, eds. 1996. *States, Social Knowledge, and the Origins of Modern Social Policies*. Princeton, NJ: Princeton University Press.

Said, Edward. 1979. *Orientalism*. New York: Random House.

Scheinman, Lawrence. 1993. "Lessons from Post-War Iraq for the International Full-Scope Safeguards Regime." *Arms Control Today* 23: 3–23.

Schneider, Steven. 1997. *Laboratory Earth: The Planetary Gamble We Can't Afford to Lose*. New York: Basic Books.

Scott, James. 1998. *Seeing Like a State*. New Haven: Yale.

Shackley, Simon, and Brian Wynne. 1995. "Integrating Knowledge for Climate Change: Pyramids, Nets, and Uncertainties." *Global Environmental Change* 5: 113–26.

———. 1996. "Representing Uncertainty in Global Climate Change Science and Policy." *Science, Technology and Human Values* 21: 275–302.

Shapin, Steven, and Simon Schaffer. 1985. *Leviathan and the Air Pump: Hobbes, Boyle, and the Experimental Life*. Princeton: Princeton University Press.

Skocpol, Theda. 1992. *Protecting Soldiers and Mothers: The Political Origins of Social Policy in the United States*. Cambridge: Harvard.

Stiglitz, Joseph. 2002. *Globalization and Its Discontents*. New York: Norton.

Stolberg, Sheryl. 2008. "Leaders Move Toward Meetings on Economic Crisis." *New York Times,* October 19.

Takacs, David. 1996. *The Idea of Biodiversity: Philosophies of Paradise.* Baltimore: Johns Hopkins University Press.

Taylor, Charles. 2004. *Modern Social Imaginaries.* Durham: Duke University Press.

Truman, Harry S. 1945. "Address to the United Nations Conference in San Francisco." April 25. http://www.presidency.ucsb.edu/ws/?pid=12391. Accessed May 4, 2013.

White, Aiofe. 2008. "EU, US Call for Global Summit to Reshape Banking." Associated Press, October 15.

Willkie, Wendell. 1943. *One World.* New York: Simon and Schuster.

Wittrock, Bjorn, and Peter Wagner. 1996. "Social Science and the Building of the Early Welfare State." Pp. 90–114 in *States, Social Knowledge, and the Origins of Modern Social Policies,* edited by Dietrich Rueschemeyer and Theda Skocpol. Princeton, NJ: Princeton University Press.

World Bank. 1999. *Confronting AIDS: Public Priorities in a Global Epidemic.* Washington, DC: World Bank.

World Health Organization. 1946. "Official Records of the World Health Organization, No. 2. Proceedings and Final Acts of the International Health Conference Held in New York from 19 June to 22 July, 1946." http://whqlibdoc.who.int/hist/official_records /2e.pdf. Accessed May 4, 2013.

———. 2005. "International Health Regulations." http://www.who.int/ihr/9789241596664 /en/. Accessed May 4, 2013.

FOURTEEN

Global Health Security and the Pathogenic Imaginary

ANDREW LAKOFF

This chapter describes the emergence of "global health security" as a strategic framework designed to prepare for and respond to the threat of catastrophic disease.[1] Global health security focuses on "emerging infectious diseases"—whether naturally occurring or man-made—which are seen to threaten wealthy countries and which typically (although not always) emanate from Asia, sub-Saharan Africa, or Latin America. Such pathogens include severe acute respiratory syndrome (SARS), extremely drug-resistant (XDR) tuberculosis, and humanly transmissible avian influenza; but what is crucial is that this framework is oriented toward outbreaks that have not yet occurred—and may never occur. For this reason, it seeks to implement techniques of preparedness for events whose likelihood is incalculable but whose political, economic, and health consequences could be catastrophic. Its advocates seek to build a real-time, global disease surveillance system that can provide early warning of potential outbreaks and link such early warning to tools of rapid response that will protect against their spread to the rest of the world. To achieve this, global health security initiatives draw together various organizations including multilateral health agencies, national disease control institutes, and collaborative reference laboratories and assemble diverse technical elements such as disease surveillance methods, emergency operations centers, and vaccine distribution systems.

At the heart of the regime of global health security is a sociotechnical imaginary (Jasanoff, this volume) concerning the future problem of infectious disease and its possible solutions: it is a future in which outbreaks of novel diseases continually threaten human life, but catastrophe may be averted if such events are detected and contained in their earliest stages.[2] To justify ongoing investment in pandemic preparedness, an envisioned outbreak must remain on the near-term horizon; it is thus urgent to enact adequate security measures. And importantly, these security measures must be "global" in extent (Miller, this volume), since the perceived threat of emerging disease transcends national boundaries. Despite the force of this imaginary, this global governance regime does not command universal assent. As the chapter will show, its extension is challenged both by existing national agencies that have historically exercised public health responsibilities and by alternative cosmopolitan visions of the problem of global health. To establish its legitimacy, global health security must address a number of questions, including which populations is it responsible for, how does it presume to override national authorities, and what epistemic tools enable it to make authoritative claims?

The chapter begins with a summary of the historical emergence of global health security in relation to the problematic, initially articulated in the late 1980s, of emerging infectious disease. It then looks at the development of a specific governance tool for managing outbreaks of new pathogens on a global scale, the International Health Regulations (IHR), as they were revised in 2005. Protocols for monitoring and responding to outbreaks codified by the revised IHR were at the heart of two recent global health controversies: one around the Indonesian government's refusal to share avian influenza virus strains with a global disease monitoring system and another around the question of whether the World Health Organization (WHO) exaggerated the threat posed by the 2009 swine flu outbreak. The chapter takes up these two controversies in turn as cases for understanding the tensions sparked by the consolidation of global health security as a global governance regime.

The Objects and Aims of Global Health Security

A 2007 report from the WHO articulated the objects and aims of global health security.[3] The report, entitled "A Safer Future: Global Public Health Security in the 21st Century," began by noting the success of traditional public health measures during the twentieth century in dealing with devastating infectious diseases such as cholera and smallpox. But in recent decades, the report continued, there had been an alarming shift in the "delicate balance between humans and microbes" (WHO 2007, 1). A series of factors—including demographic changes, economic development, global travel and commerce, and conflict—had "heightened the risk of disease outbreaks," ranging from new infectious diseases such as HIV/AIDS and drug-resistant tuberculosis to food-borne pathogens and bioterrorist attacks.[4]

The WHO report proposed a strategic framework for responding to this new landscape of threats, which it called "global public health security."

The framework emphasized a terrain of global health that was distinct from the predominantly national organization of classical public health. "In the globalized world of the 21st century," the report began, simply stopping disease at national borders was not adequate. Nor was it sufficient to respond to diseases after they had become established in a population. Rather, it was necessary to prepare for unknown outbreaks in advance, something that could be achieved only "if there is immediate alert and response to disease outbreaks and other incidents that could spark epidemics or spread globally and if there are national systems in place for detection and response should such events occur across international borders" (WHO 2007, 11).

As envisioned by WHO, the framework of global health security was a culmination of two decades of increasing concern over the problematic of "emerging infectious disease." This problematic was initially raised by a group of United States–based infectious disease experts in the late 1980s and early 1990s (King 2002). In 1989, molecular biologist Joshua Lederberg and virologist Stephen Morse hosted a conference on the topic, which led to the landmark volume, *Emerging Viruses* (Morse 1993). Lederberg and Morse shared an ecological vision of disease emergence as the result of environmental transformation combined with increased global migration.[5] Participants in the conference warned of a dangerous intersection. On the one hand, they pointed to a number of new disease threats, including novel viruses such as HIV and Ebola as well as drug-resistant strains of diseases such as tuberculosis and malaria. On the other hand, participants argued, public health infrastructures worldwide had been allowed to decay with the assumption that the problem of infectious disease had been conquered. Moreover, the emergence and spread of new infectious diseases could be expected to continue, owing to a number of processes of global transformation, such as increased travel, urbanization, civil wars and refugee crises, and environmental destruction. For these experts, the AIDS crisis heralded a dangerous future in which more deadly pathogens were likely to appear.

Over the ensuing years, alarm about emerging disease threats came from various quarters, including scientific reports by prominent organizations such as the Institute of Medicine (1992), the reporting of science journalists such as Laurie Garrett (1994), and the dire scenarios of writers such as Richard Preston (1998). For a number of health experts, the emerging disease threat—particularly when combined with weakening national public health systems—marked a troublesome reversal in the history of public health. At just the moment when it seemed that the threat posed by infectious disease had waned and that the critical health problems of the industrialized world now mainly involved chronic disease, these experts warned that we were witnessing a "return of the microbe." It is worth emphasizing the expansive character of the category of "emerging infectious disease." The category made it possible to bring HIV into relation with a range of other microbial threats, including viral hemorrhagic fevers, West Nile virus, dengue, and drug-resistant strains of malaria and tuberculosis. It also pointed toward the imperative to develop means of anticipatory response that could approach a disparate set of disease threats.

In an initial stage of discussion, authorities proposed to address the problem of emerging infectious disease using tools of disease eradication that had been developed as part of Cold War–era international health, such as disease surveillance, outbreak investigation, and containment. For instance, one contributor to *Emerging Viruses* was epidemiologist D.A. Henderson, who had implemented techniques of disease surveillance in the 1960s and 1970s as director of the WHO Smallpox Eradication Program. For Henderson, the problem posed by emerging infections was not one of prevention but rather one of vigilant monitoring. He argued that pathogen emergence was inevitable, that "mutation and change are facts of nature, that the world is increasingly interdependent, and that human health and survival will be challenged, *ad infinitum*, by new and mutant microbes, with unpredictable pathophysiological manifestations" (Henderson 1993, 283). As a result, "we are uncertain as to what we should keep under surveillance, or even what we should look for." What we need, he continued, is a system that can detect novelty: in the case of AIDS, such a detection system could have provided early warning of the new virus and made it possible to put in place measures to limit its spread. Henderson proposed a network of global disease surveillance units to be run by the US Centers for Disease Control and Prevention (CDC), which would be located in periurban areas in major cities in the tropics.

At around this time, the emerging disease problematic entered US national security discussions. Beginning in the mid-1990s, US national security officials began to focus on bioterrorism as one of a number of "asymmetric threats" the nation faced in the wake of the Cold War. They hypothesized an association among rogue states, global terrorist organizations, and the proliferation of weapons of mass destruction (Wright 2006). Reports during the 1990s about secret Soviet and Iraqi bioweapons programs, along with the Aum Shinrikyo subway attack in Tokyo in1995, lent credibility to calls for biodefense measures focused on the threat of bioterrorism. Early advocates of such efforts, including infectious disease experts such as Henderson and national security officials such as Richard Clarke, argued that adequate preparation for a biological attack would require a massive infusion of resources into both biomedical research and public health response capacity.[6] More broadly, they maintained, it would be necessary to incorporate the agencies and institutions of the life sciences and public health into the national security establishment. In the 1990s, Henderson and others connected the interest in emerging diseases among international health specialists with US national security officials' concern about the rise of bioterrorism, suggesting that a global disease surveillance network could serve to address both problems.

Epidemic Intelligence

Henderson's model of disease surveillance was a product of his background at the Epidemic Intelligence Service (EIS) based at the CDC.[7] The EIS approach, introduced in the 1950s by Henderson's mentor Alexander Langmuir, was one of "continued watchfulness over the distribution and trends of incidence through systematic consolidation and evaluation of morbidity and mortality data and other relevant data," as Langmuir (1963, 182–3) put it.[8] Henderson had used this method in tracking the global incidence of smallpox as director of the WHO eradication program. His proposed global network of surveillance centers and reference laboratories extended this approach to as-yet-unknown diseases, providing early warning for response to outbreaks of any kind—whether natural or man-made. Stephen Morse (1992, 29) summarized the justification for "expanding permanent surveillance programs to detect outbreaks of disease" in terms of the shared needs of international health and national security: "A global capability for recognizing and responding to unexpected outbreaks of disease, by allowing the early identification and control of disease outbreaks, would simultaneously buttress defenses against both disease and CBTW [chemical, biological, and toxin warfare]."

This epidemic intelligence approach to emerging infections was institutionalized at a global scale over the course of the 1990s as experts from CDC brought the methods and assumptions of EIS into the WHO. The career of epidemiologist David Heymann is instructive. Heymann began his professional service in EIS, and in the 1970s worked with CDC on disease outbreak containment in Africa and with WHO on the smallpox eradication program (Ashraf 2004). In the early years of the AIDS pandemic, he helped establish a WHO office to track the epidemiology of the disease in developing countries. He then returned to Africa in 1996 to lead the agency's response to a widely publicized Ebola outbreak in Congo. After this he was asked by the director of WHO to set up a program in emerging diseases. "At this time there was an imbalance in participation internationally in the control of emerging and re-emerging infectious diseases," he later recalled, "the burden was falling mainly on the USA" (Ashraf 2004, 787). At WHO, Heymann set up a global funding mechanism that broadened the agency's emerging disease surveillance and response capacities along the CDC model. He and his colleagues soon identified a specific problem to be addressed: how to ensure the compliance of national health agencies with the demands of global health surveillance?

In the wake of the Ebola outbreak, as well as catastrophic outbreaks of cholera in Latin America and plague in India in the early 1990s—now understood as "reemerging diseases"—a "need was identified" for stronger international coordination of response, as Heymann (2004, 1127) later reflected. A major problem for outbreak investigators was that national governments often did not want to report the incidence of a disease that could harm tourism and international trade. The case of the plague outbreak in Surat in 1994, in which Indian officials suppressed international reporting of the event, exemplified the impossibility of forcing countries to publicly report disease emergencies.[9] Although, as Heymann put it, "in our emerging diseases program our idea was to change the culture so that countries could see the advantage of reporting," a practical means of enforcing compliance was needed.

A potential means of such enforcement soon arose from an unexpected source: the creation, during the 1990s, of Internet-based reporting systems such as ProMED in the United States and GPHIN in Canada that scoured international media for stories about possible outbreaks. The development of these digital information networks meant that global public health authorities did not have to rely exclusively on official, nation-state–based epidemiological reporting (Weir and Mykhalovskiy 2007). In 1997, WHO established GOARN (Global Outbreak Alert and Response Network), a network that linked together individual disease surveillance and response systems, which eventually had 120 partners. The resulting potential for the rapid circulation of infectious disease information across national borders undermined national governments' traditional control of public health knowledge, making a global form of disease surveillance possible.

The outbreak of SARS in 2002 in China provided Heymann and his colleagues in WHO's Emerging Infections branch with an opportunity to test the new disease reporting system. As an unknown and unexpected but potentially catastrophic viral disease, SARS fit well into the existing category of emerging infections (Hooker 2007). The Chinese government's initial reluctance to fully report the outbreak to global health authorities led WHO to rely on its new capacity to use nonstate sources of information: SARS was the first time the GOARN network identified and publicized a rapidly spreading epidemic. As opposed to recalcitrant national governments, Heymann later reflected, international scientists "are really willing to share information for the better public good" (Ashraf 2004, 787). GOARN made it possible to electronically link leading laboratory scientists, clinicians, and epidemiologists around the world in a "virtual network" that rapidly generated and circulated knowledge about SARS. WHO tracked the spread of the illness closely and issued a series of recommendations on international travel restrictions. According to Heymann, who led the WHO response, this rapid reaction was key to the containment of the epidemic by July 2003—although he also acknowledged the good fortune that SARS had turned out not to be easily transmissible.

The lesson Heymann drew from the experience of SARS was that, in a closely interconnected and interdependent world, "inadequate surveillance and response capacity in a single country can endanger the public health security of national populations and in the rest of the world" (Heymann 2004, 1128). SARS thus confirmed the pathogenic imaginary of global health security. Processes of globalization, including migration, ecological transformations, and massive international travel, had led to new biological, social, and political risks—risks that transcended national borders and therefore could not be ignored by wealthy countries. Only a global system of rapidly shared epidemiological information could provide adequate warning in order to mitigate such risks. National sovereignty must accede to the demands of global health security. This vision was then applied to the next potential disease emergency, highly pathogenic avian influenza.

The Next Pandemic: H5N1

The space of emerging disease, initially carved out by AIDS and then expanded by SARS, was soon occupied by a new threat: the possibility that a deadly new strain of H5N1 avian influenza would mutate or reassort to become easily transmissible among humans. The risk of such an event could not be calculated using statistical data on historical incidence (since it had not yet occurred), but its occurrence could, experts warned, be catastrophic. As of 2005, when global pandemic flu preparedness efforts intensified, H5N1 had killed nearly 60 percent of those who had contracted it, and was spreading globally among migratory birds and domesticated poultry.

In an article published in *Foreign Affairs*, journalist Laurie Garrett evoked both the dire scenario of an H5N1 pandemic and the uncertainty surrounding it: "In short, doom may loom. But note the 'may.' If the relentlessly evolving virus becomes capable of human-to-human transmission, develops a power of contagion typical of human influenzas, and maintains its extraordinary virulence, humanity could well face a pandemic unlike any ever witnessed. Or nothing at all could happen" (Garrett 2005). Others were less circumspect in their warnings. "It is not a question of if, but when," declared infectious disease expert Michael Osterholm (2005). "I believe an influenza pandemic will be like a 12–18 month global blizzard that will ultimately change the world as we know it today." The prospect of such global catastrophe lent urgency to the enactment of pandemic preparedness measures, including the adoption of major revisions to the venerable IHR.

Governing Global Health Emergencies

According to legal scholar David Fidler (2005, 326), the 2005 IHR revision was "one of the most radical and far-reaching changes in international law on public health since the beginning of international health co-operation in the mid-nineteenth century." For my purposes here, the revised IHR are best understood as a significant element in the emerging framework of global health security. The revised IHR instituted a new set of legal obligations for nation-states to accept global intervention in a world seen as under threat from ominous pathogens circulating ever more rapidly.

The IHR system, dating from the 1851 International Sanitary Law, defines states' mutual obligations in the event of an outbreak of a dangerous communicable disease. Historically, its function has been to guarantee the continued flow of international trade during epidemics, ensuring that countries not take overly restrictive measures in response to the threat of infection. In the context of concern over emerging infectious diseases, the existing IHR had proven ineffectual in forcing disease notification for at least two reasons. For one, its limited list of reportable conditions—cholera, plague, and yellow fever—was of little relevance for the expansive category of emerging infections; second, the existing regulations did not have a mechanism to enforce national compliance with IHR reporting requirements.

The revision of IHR became a vehicle for outbreak investigators to construct the global disease surveillance system that had been proposed by Henderson and others. WHO authorities proposed three key innovations to IHR that would make it possible for the agency to manage a range of potential disease emergencies. The first innovation responded to the problem of the narrow range of disease events covered by the existing IHR. Through the invention of the concept of the "public health emergency of international concern" (PHEIC), the revised regulations vastly expanded the kinds of events to which the regulations might apply. According to the "IHR Decision Instrument" (WHO 2008, 43), naturally occurring infectious diseases such as pandemic influenza and Ebola, intentional releases of deadly pathogens such as smallpox, or environmental catastrophes such as those that occurred at Bhopal in 1984 and Chernobyl in 1986 could all provoke the declaration of a PHEIC. The IHR decision instrument was designed to guide states in determining what would constitute a public health emergency that required the notification of WHO. However, as we will see below, the pathway in the instrument defined as "any event of international public health concern" left considerable room for interpretation of the scope of the regulations.

The second major innovation in the revised IHR responded to the problem of national health agencies' monopoly on epidemiological data. The new regulations expanded the potential sources of authorized reports of outbreaks: whereas the prior IHR had restricted official reporting to national governments, the revised IHR allowed WHO to recognize reports from non-state sources such as digital and print media. In this way, state parties' unwillingness to report outbreaks would not necessarily impede the functioning of the system. The premise was that, with WHO's official recognition of nonstate monitors such as GOARN, reports of outbreaks could no longer be suppressed by national governments and so it would be in states' interest to allow international investigators into the country as soon as possible in order to undertake disease mitigation measures and to assure the public that responsible intervention was underway.

The third innovation of the revised IHR addressed the problem of developing countries' ability to monitor and respond to outbreaks. It required that all WHO member states build national capacity for infectious disease surveillance and response. The construction of "national public health institutes" on the model of the US CDC would make possible a distributed global network that relied on the functioning of nodes in each country. The impetus for such institutes as part of the WHO Global Health Security framework should be distinguished from prior modernizing efforts to build public health systems in the developing world. IHR's reliance on national health systems did not necessarily imply strengthening governmental capacity to manage existing disease; rather, it sought to direct the development of outbreak detection systems according to the needs of global disease surveillance.[10] WHO gave countries until 2016 to fulfill this obligation. However, it was unclear where the resources would come from to implement systems for detecting the onset of emerging diseases in poor countries that already had trouble managing the most common ones.

By 2007, with the official implementation of the revised IHR, accompanied by the release of the WHO Report on "Global Public Health Security in the 21st Century" (described above), the new framework for governing global health emergencies was in place. However, the intensive buildup of preparedness efforts in anticipation of a catastrophic disease outbreak did not go unchallenged. In the remaining portion of the chapter, I focus on two controversies sparked by the consolidation of the global health security framework. The first controversy centered on the political question of whose health was to be protected by the new governance regime; and the second focused on the epistemological question of how to know what constitutes a catastrophic health risk.

Viral Sovereignty

In an opinion piece published in the *Washington Post* in August 2008, diplomat Richard Holbrooke and science journalist Laurie Garrett mounted a sharp attack on what they called "viral sovereignty" (Holbrooke and Garrett 2008). By this term, the authors referred to the "extremely dangerous" idea that sovereign states could exercise ownership rights over samples of viruses found in their territory. Specifically, Holbrooke and Garrett were incensed by the Indonesian government's refusal to share samples of H5N1 avian influenza with the WHO's Global Influenza Surveillance Network (GISN). For over fifty years, this network had collected samples of flu viruses from reference laboratories around the world and used these samples to determine the composition of yearly flu vaccines. More recently, the network had tracked the transformations of avian influenza viruses as a means of assessing the risk of a deadly global pandemic.[11] International health experts feared that the new strain of H5N1, which had already proven highly virulent, would mutate to become easily transmissible among humans—in which case a worldwide calamity could be at hand. GISN thus served as a global "early warning system" enabling experts to track genetic changes in the virus that could potentially lead to a catastrophic disease event.

As the country where the most human cases of avian influenza had been reported, Indonesia was a potential epicenter of such an outbreak. For this reason, the country's decision to withhold samples of the virus undermined GISN's function as a global early warning system. From Holbrooke and Garrett's vantage, which assumed the imaginary of global health security, Indonesia's action posed a significant threat to the world. "In this age of globalization," they wrote, "failure to make viral samples open-source risks allowing the emergence of a new strain of influenza that could go unnoticed until it is capable of exacting the sort of toll taken by the pandemic that killed tens of millions in 1918" (Holbrooke and Garrett 2008). According to Holbrooke and Garrett, Indonesia had not only a moral but also a legal obligation to share its viruses with the WHO. They argued that the country's action was a violation of the newly revised IHR, which held the status of an international treaty for WHO member states.

The opinion piece suggested that the rational and beneficent technocracy of the WHO was faced with antiscientific demagoguery that threatened the world's health. Holbrooke and Garrett painted a picture of the Indonesian Health Minister, Siti Fadilah Supari, as an irrational populist who sought to make domestic political gains through unfounded attacks on the United States and the international health community. Indonesia was apparently withholding these virus samples based on the "dangerous folly" that these materials should be protected through the same legal mechanism that the United Nations Food and Agriculture Organization used to guarantee poor countries' rights of ownership to indigenous agricultural resources—the Convention on Biological Diversity. Further, Holbrooke and Garrett rebuked Supari's "outlandish claims" that the US government was planning to use Indonesia's H5N1 samples to design biological warfare agents—echoing US Secretary of Defense Robert Gates's reaction upon hearing this accusation during a visit to Jakarta: "the nuttiest idea I've ever heard" (*Agence-France Press* 2008).

The controversy over influenza virus sharing was, it turned out, somewhat more complicated than Holbrooke and Garrett allowed. Beginning in late 2006, at Supari's behest, the Indonesian Health Ministry had stopped sharing with the GISN isolates of H5N1 found in patients who had died of avian influenza. The source of Supari's ire was the discovery that an Australian pharmaceutical company had developed a patented vaccine for avian flu using an Indonesian strain of the virus—a vaccine that would not be affordable for most Indonesians in the event of a deadly pandemic. More generally, given the limited number of vaccine doses that could be produced in time to manage such a pandemic—estimates were in the 500 million range—experts acknowledged that developing countries would have little access to such a vaccine. In other words, while Indonesia had been delivering virus samples to WHO as part of a collective early warning mechanism (i.e., GISN), they would not be beneficiaries of the biomedical response apparatus that had been constructed to prepare for a deadly global outbreak. For the Indonesian health minister, this situation indicated a dark "conspiracy between superpower nations and global organizations" (Schnirring 2008).

While less suspicious of US and WHO intentions than Supari, a number of Western journalists and scientists were sympathetic to the Indonesian position, on the grounds of equity in the global distribution of necessary medicines. A *Time* magazine article noted that "they had a point; poor developing nations are often priced out of needed medicines, and they're likely to be last in line for vaccine during a pandemic" (Walsh 2007). An editorial in the *Lancet* (2007, 1763) argued, "to ensure global health security, countries have to protect the wellbeing not only of their own patients but also those of fellow nations." Anxious to secure the operation of its global influenza surveillance apparatus, WHO was willing to strike a bargain: at a World Health Assembly meeting in 2007, members agreed to explore ways of helping poorer countries to build vaccine production capacity. But the financial and technical details of how such a system would function were opaque, and the issue remained unresolved until a settlement was reached in 2011.[12] In October 2008, as Indonesia continued to withhold the vast majority of its virus samples from GISN, *Agence-France Presse* (2008) reported that "Supari does appear to be vindicated by a flood of patents being lodged on the samples of H5N1 that have made it out of Indonesia, with companies in developed countries claiming ownership over viral DNA taken from sick Indonesians." The Australian drug company CSL acknowledged that it had used Indonesian bird flu strains to develop a trial vaccine but insisted that it had no obligation to compensate Indonesia or guarantee access to the vaccine.

A good deal more could be said about this controversy (see, e.g., Lowe 2010), but I want to focus here on just one aspect of Holbrooke and Garrett's attack: their accusation that Indonesia was in violation of the newly revised IHR. International law experts saw the virus-sharing controversy as an early test of how well the revised IHR would function. According to the new regulations, IHR signatories were required to provide WHO with "public health information about events that may constitute a PHEIC" (Fidler 2008). In the case of the virus-sharing controversy, the central legal question was whether biological materials constituted such "public health information." Plausible arguments could be made on both sides. At the May 2007 meeting of the World Health Assembly, WHO Director General Margaret Chan claimed, "countries that did not share avian influenza virus would fail the IHR" (Fidler 2008). The US delegation agreed: "All nations have a responsibility under the revised IHRs to share data and virus samples on a timely basis and without preconditions" (Fidler 2008). Thus, the United States argued, "our view is that withholding influenza viruses from GISN greatly threatens global public health and will violate the legal obligations we have all agreed to undertake through our adherence to IHRs." However, as Fidler noted, the relevance of the revised IHR to the specific issue of virus sharing was ambiguous: the new regulations explicitly referred only to a requirement to share public health information, such as case reports and fatality rates, and the case could be made that biological materials such as virus samples were distinct from such information.

In any case, the Indonesian Health Ministry's response came from outside the legal framework of IHR. Rather, Supari argued that the global virus sharing system was ethically compromised and in need of reform. "We want to change the global virus-sharing mechanism to be fair, transparent and equitable," Supari said in an interview defending the government's decision to withhold the virus (*Agence-France Presse* 2008). "What we mean by fair is that any virus sharing should be accompanied by benefits derived from the shared virus, and these benefits should be coming from the vaccine producing countries." Supari was speaking from within a different sociotechnical imaginary than that of the WHO's framework of "global public health security," which the revised IHR was designed to serve. In speaking of benefits sharing, Supari was invoking a mechanism intended to encourage development—the Convention on Biological Diversity—in order to ground a rhetoric of national sovereignty that ran counter to the transnational authority of the WHO. But her attack on the high price of patented vaccines also resonated with demands for equal access to life-saving medicines coming out of the humanitarian global health movement (Redfield 2013).

A technical and political system designed to prepare for potentially catastrophic disease outbreaks was facing a very different demand: a call for access to essential medicines based on a vision of global equity. The potential for a deadly outbreak of avian influenza had led to an encounter between two different ways of imagining the problem of global health—an encounter that was taking place in the absence of an actual health emergency. At stake was not only the issue of how best to respond to a global outbreak of H5N1, but more broadly, how to define the political obligation to care for the health of populations in a globalizing world in which the capacity of national public health authorities to protect their fellow citizens' well-being was increasingly in question.

Preparing for the Wrong Virus

By 2009, public attention to the threat of an avian influenza pandemic had begun to wane. However, the regime of pandemic preparedness that had been put in place to anticipate its arrival remained in a state of vigilance within national and global public health agencies. And the scenarios that had been generated to manage its potential occurrence took on new life when a different virus emerged: A/H1N1 or swine flu. The strong global response to what turned out to be a relatively weak virus led to a major controversy around the legitimacy of global health security—that is, of the practice of ongoing, intensive anticipation of the onset of a catastrophic disease event.

When the newly reasserted influenza virus A/H1N1 made its appearance among humans in April 2009, it seemed at first to be the pathogen the global health community had been anticipating. Dozens had apparently died in Mexico from a respiratory ailment, and hundreds more were hospitalized. Reports of cases from around the United States indicated rapid transmission of the virus. There was a possibility that the outbreak would become a deadly pandemic, but its key statistical characteristics—in particular, its case fatality ratio—were not yet known. Within weeks an extensive public health apparatus had taken hold of the virus, tracking its global extension through reference laboratories, mapping its genomic sequence, collating data on hospitalization and death rates, working to distribute antiviral medicines and develop a vaccine, and communicating risk to various publics. While some elements of this apparatus were decades old, such as GISN and the egg-based technique of vaccine production—others were quite new, such as Internet-based outbreak reporting systems, molecular surveillance, and national pandemic preparedness plans.

On April 25, based on reports from Mexico and the United States, WHO Director-General Margaret Chan declared a PHEIC under the newly revised IHR. Following IHR protocol, Chan appointed an Emergency Committee consisting of recognized influenza experts, who recommended a Phase Four Pandemic Alert. Given the controversy that ensued, it is important to point out that the definition of "pandemic" from WHO's 2009 preparedness guidance document referred to "sustained community-level outbreaks" in multiple regions but made no reference to the severity of the virus (WHO 2009).

Four days later, on April 29, the Emergency Committee voted to raise the pandemic alert level to Phase Five, indicating that national health authorities should move from "preparedness" to "response" activities. Chan assured members of the public that WHO was tracking the emerging pandemic at multiple registers—clinical, epidemiological, and virological—and advised national health ministers to "immediately activate their pandemic plans" (Chan 2009a). For North American and European governments, this meant among other things triggering advanced purchase agreements with vaccine manufacturers to produce millions of doses in time for anticipated fall immunization campaigns. At this early stage of the pandemic, in the absence of epidemiological data on the severity of the virus, the pandemic alert system alongside national preparedness plans provided government officials with guideposts for action.[13]

On June 11, Chan announced pandemic alert Phase Six, a full global pandemic. In her public statement, she pointed to the agency's ongoing vigilance as the event unfolded: "No previous pandemic has been detected so early or watched so closely, in real-time, right at the very beginning. The world can now reap the benefits of investments, over the past five years, in pandemic preparedness" (Chan 2009b). At the same time, she also warned of ongoing uncertainty: "The virus writes the rules and this one, like all influenza viruses, can change the rules, without rhyme or reason, at any time." Vigilant watchfulness would continue to be necessary.

As of early July, experts were still trying to figure out what H1N1's rules were, in particular its rules of transmissibility and virulence. A critical problem remained the lack of data on the overall incidence, as opposed to the number of fatalities, of H1N1 in the population. This was the well-known "problem of the denominator." A Harvard-based team of epidemiologists argued for immediate investment in serologic surveys so that the case fatality ratio could be calculated: "Without good incidence estimates," they wrote, "estimates of severity will continue to suffer from an unknown denominator. The effectiveness of control measures will be difficult to assess without accurate measures of local incidence" (Lipsitch et al. 2009). The Director of the US Institute of Medicine described such efforts as "epidemic science in real time," through which "scientists can enable policies to be adjusted appropriately as an epidemic scenario unfolds" (Fineberg and Wilson 2009, 987).

Meanwhile, significant political and economic decisions had to be made in the absence of fully elaborated data on risk. The alternative was to invest in preparedness for the worst case. Beginning in the summer 2009, the US government spent $1.6 billion on 229 million doses of vaccine in what the *Washington Post* later called "the most ambitious immunization campaign in US history" (Stein 2010). In the early fall, unanticipated delays in vaccine production combined with high demand led to widespread confusion and criticism, which faded as the anticipated wave of H1N1 arrived without causing a catastrophic number of deaths.

In Europe, when the fall wave of H1N1 arrived, the apparent mildness of the virus led to widespread public skepticism about state-led vaccination campaigns. The French government spent an estimated 500 million euros on a campaign that in the end succeeded in immunizing only 10 percent of the national population. By the winter, the governments of France, Germany, and England all sought to renegotiate their advanced purchase agreements with vaccine manufactures and to unload their excess doses on poor countries in the global south at bargain prices.

A series of political controversies then erupted in Europe over the intensive public health response to H1N1. In *Le Monde*, former French Red Cross president Marc Gentilini admonished the French government for its spending on the immunization campaign, noting that "preparing for the worst wasn't necessarily preparing correctly" (Chaon 2010). A physician and legislator for the governing conservative party decried the misallocation of public health resources, saying "the cost is more than the deficit of all France's hospitals and is three times [the amount spent] on cancer care" (Daneshkhu and Jack 2010). The French government, in turn, defended its actions on the grounds of precaution: "I will always prefer to be too prudent than not enough," said President Sarkozy (Whalen and Gauthier-Villars 2010).

The attention of critics then turned to the warnings from international flu specialists that had led national health ministries to implement mass vaccination campaigns. As Gentilini put it, "I don't blame the health minister, but the medical experts. They created an apocalyptic scenario. There was pressure from the World Health Organization, which began waving the red warning flags too early" (Public Radio International 2010). The head of the French Socialist Party demanded a parliamentary inquiry, calling the vaccination campaign a "fiasco" and arguing that multinational drug companies were "the big winners in this affair" (Daneshkhu and Jack 2010).

The Chair of the Council of Europe's Health Committee, a German physician, convoked public hearings on the matter, charging that the WHO pandemic declaration was "one of the greatest medical scandals of the century" (Macrae 2010). Witnesses before the committee argued that scarce health resources had been squandered on a virus that turned out to be less dangerous than seasonal flu and that such resources should have been spent on "real" killers, whether heart disease in wealthy countries or infant diarrhea in poor ones. A German epidemiologist cited annual mortality statistics to criticize the WHO's emphasis on managing potential outbreaks at the expense of treating the actual "great killers" whose toll was attested to by epidemiological data: "I would like to point out that of the 827,155 deaths in 2007 in Germany about 359,000 come from cardiovascular diseases, about 217,000 from cancer, 4968 from traffic accidents, 461 from HIV/AIDS and zero from SARS or Avian Flu" (Keil 2010). Here, coming from one segment of public health experts, we find the public display of numbers to make the case that rational intervention must be based on risk calculation rather than on precaution against potential catastrophe. This reflected distrust, on the part of certain European health officials, of one of the fundamental premises of global health security: the need for ongoing vigilance against the onset of an event that has never before occurred.

But rather than see the WHO as engaged in a different type of reasoned action, one oriented to preparedness for potential catastrophe, critics decried a lack of objectivity or even corruption, insisting that conflicts of interest among members of the Emergency Committee had led to the pandemic declaration. One source of their suspicion was the removal of the measurement of severity from the WHO preparedness guidance document several months before the appearance of H1N1. In June, an investigative report in the *British Medical Journal* revealed paid consulting relations between leading influenza experts and vaccine manufacturers (Cohen and Carter 2010). The same week, the Council of Europe released its report, concluding that the pandemic declaration had led to "a distortion of priorities of public health services across Europe, waste of huge sums of public money, [and the] provocation of unjustified fears among Europeans," and suggesting that WHO deliberations had been tainted by unstated conflicts of interest between experts and the drug companies that profited from the vaccine campaign (Parliamentary Assembly 2010).

In response to these allegations, Director-General Chan chartered a review of the agency's response under the aegis of IHR. The committee eventually absolved the WHO influenza experts of overstating the seriousness of the pandemic and focused in its final report on the demand for precautionary action in the face of an unfolding and uncertain event. "Reasonable criticism can be based only on what was known at the time and not on what was later learnt," the committee report argued, pointing out that "the degree of severity of the pandemic was very uncertain throughout the middle months of 2009, well past the time, for example, when countries would have needed to place orders for vaccine" (WHO 2011, 17). In its final report, the committee emphasized the problem WHO had faced in adjusting to the unexpected: "Lack of certainty is an inescapable reality when it comes to influenza. One key implication is the importance of flexibility to accommodate unexpected and changing conditions."[14] In other words, in the case of a novel pathogen, the virulence of an encroaching epidemic cannot be determined based on knowledge of the past.

Rather than being the result of a conflict of interest on the part of the WHO experts, the agency's "overreaction" was due to the preparedness plans themselves, which were built upon the scenario of a different kind of outbreak. As an official from the European Center for Disease Prevention and Control later said, explaining the intensive global response to H1N1, "We were all planning for the potential mutation of the avian flu over the next three to five years into a person-to-person transmittable disease."[15] Indeed, it was the use of the avian flu mutation scenario as a guide to authorized action in the absence of statistical data about disease risk that had so exercised the WHO's critics. As epidemiologist Ulrich Keil, a firm advocate of the comparative risk approach, testified, "Governments and public health services are paying only lip service to the prevention of these great killers and are instead wasting huge amounts of money by investing in pandemic scenarios whose evidence base is weak" (Keil 2010, 3). And the Council of Europe report echoed this critique of lack of objective evidence of risk: "It was precisely this lack of watertight evidence about the influenza phenomenon which led to the fears of the pandemic being exaggerated and the subsequent disproportionate response" (Council of Europe 2010, 8).

But it is only from the perspective of traditional public health, which relies on epidemiological risk data to legitimate action, that such an assessment can be made; from the vantage of the framework of global health security, when a sign of the envisioned catastrophic future appears, already existing plans must be put into action. In her September 2010 testimony to the IHR review committee, Director-General Chan revised her earlier statement about the benefits of investments in preparedness: "The world was better prepared for a pandemic than at any time in history. But it was prepared for a different kind of event than what actually occurred" (Chan 2010). Chan admitted that dire scenarios based on H5N1 had structured the agency's response: "Managing the discrepancy between what was expected and what actually happened was problematic."

WHO's pandemic preparedness planners might have responded to the Council of Europe's critical report with this line from philosopher Hans Jonas (1985, 120), writing about the principle of precaution: "the prophecy of doom is made to avert its coming, and it would be the height of injustice to later deride the 'alarmists' because 'it did not turn out to be so bad after all'—to have been wrong may be their merit." Nonetheless, as we have seen, in the absence of the anticipated catastrophic event, the sociotechnical imaginary of global health security finds itself challenged to defend its tools of anticipatory vigilance and precautionary response.

Notes

1. I would like to thank Sheila Jasanoff and Sang-Hyun Kim for their detailed and helpful comments on an earlier draft of this chapter.

2. Jasanoff (this volume) defines sociotechnical imaginaries as "widely accepted and actively pursued visions of social futures underpinned by expectations of what is possible, attainable, and worth securing through science and technology." She urges analysts to remain "attentive to the ways in which imaginaries frame and represent alternative futures, relate past and future times, enable or restrict actions, and naturalize ways of thinking about possible worlds."

3. This section draws on portions of Collier and Lakoff (2008).

4. Charles Rosenberg (1998) contrasts this new form of "civilizational risk" with those that sparked early public health efforts, noting that anxieties about the risk of modern ways of life are here explained not in terms of the city as a pathogenic environment but in terms of evolutionary and global ecological realities.

5. Warwick Anderson (2004) describes this vision as follows: "Evolutionary processes operating on a global scale were responsible for the emergence of 'new' diseases. As environments changed, as urbanization, deforestation, and human mobility increased, so, too, did disease patterns alter, with natural selection promoting the proliferation of microbes in new niches" (Anderson 2004, 60). He traces its history to the mid-twentieth-century work of disease ecologists such as Theobald Smith, Macfarlane Burnet, and René Dubos.

6. As Wright (2006) argues, the very use of the term "weapons of mass destruction" to link nuclear weapons to biological weapons was a strategic act on the part of biodefense advocates.

7. The Epidemic Intelligence Service was founded in 1951 by Alexander Langmuir. For Henderson's recollections, see Henderson (2009).

8. Langmuir pioneered a method of epidemiological surveillance designed to track each instance of a disease within a given territory—one that would serve the needs of both public health and biodefense. See Fearnley (2010) for an insightful historical analysis.

9. See Garrett, *The Coming Plague* (1994).

10. As one document suggested, "It is proposed that the revised IHR define the capacities that a national disease surveillance system will require in order for such emergencies to be detected, evaluated and responded to in a timely manner" (Fidler 2005, 353).

11. The Director of WHO's communicable diseases cluster leader, David Heymann, described the network as follows: GISN "identifies and tracks antigenic drift and shifts of influenza viruses to guide the annual composition of vaccines, and provides an early alert to variants that might signal the start of a pandemic" (Heymann 2004).

12. This was known as the "Pandemic Influenza Preparedness" (PIP) Framework. See Fidler and Gostin (2011).

13. The WHO preparedness guidance explained the function of the alert system as follows: "This phased approach is intended to help countries and other stakeholders to anticipate when certain situations will require decisions and decide at which point main actions should be implemented" (WHO 2009).

14. Intriguingly, this was the same conclusion that the Review Committee chair, Harvey Fineberg, had reached in his coauthored book evaluating the much-criticized US CDC response to swine flu in 1976. See Neustadt and Fineberg (1983).

15. The IHR Review Committee agreed with this assessment: "The response to the emergence of pandemic influenza A (H1N1) was the result of a decade of pandemic planning, largely centered on the threat of an influenza A (H5N1) pandemic" (WHO 2011).

References

Agence-France Presse. 2008. "Indonesia's Bird Flu Warrior Takes on the World," October 12.

Anderson, Warwick. 2004. "Natural Histories of Infectious Disease: Ecological Vision in Twentieth Century Bioscience." *Osiris* 19: 39–61.

Ashraf, Haroon. 2004. "David Heymann—WHO's Public Health Guru." *Lancet Infectious Diseases* 4 (12): 785–8.

Chan, Margaret. 2009a. "Statement by WHO Director-General, Dr Margaret Chan." April 29. Available at: http://www.who.int/mediacentre/news/statements/2009/h1n1_20090429/en/ index.html.

———. 2009b. "Statement to the Press by WHO Director-General Dr Margaret Chan." June 11. http://www.who.int/mediacentre/news/statements/2009/h1n1_pandemic_phase6_20090611/en/index.html.

———. 2010. "External review of WHO's response to the H1N1 influenza pandemic." September 28. Available at: http://www.who.int/dg/speeches/2010/ihr_review_20100928/en/.

Chaon, Anne. 2010. "France Joins Europe Flu Vaccine Sell-off," *Agence France-Presse*. January 3.

Cohen, D., and P. Carter. 2010. "WHO and the Pandemic Flu 'Conspiracies." *British Medical Journal* 340, June, p. c2912.

Collier, Stephen J., and Andrew Lakoff. 2008. "The Problem of Securing Health." In *Biosecurity Interventions: Global Health and Security in Question*, edited by Stephen J. Collier and Andrew Lakoff. New York: Columbia University Press.

Daneshkhu, Scheherazade, and Andrew Jack. 2010. "Sarkozy under Fire on Flu Vaccine 'Fiasco.'" *Financial Times*, January 5.

Fearnley, Lyle. 2010. "Langmuir and the Birth of Disease Surveillance." *Behemoth: A Journal on Civilisation* 3 (3): 36–56.

Fidler, David P. 2005. "From International Sanitary Conventions to Global Health Security: The New International Health Regulations." *Chinese Journal of International Law* 4: 2.

———. 2008. "Influenza Virus Samples, International Law, and Global Health Diplomacy." *Emerging Infectious Diseases* 14 (1): 88–94.

Fidler, David P., and Lawrence O. Gostin. 2011. "The Pandemic Influenza Preparedness Framework: A Milestone in Global Governance for Health." *Journal of the American Medical Association* 306 (2): 200–201.

Fineberg, Harvey V., and Mary Elizabeth Wilson. 2009. "Epidemic Science in Real Time." *Science* 324, May 22, p. 987.

Garrett, Laurie. 1994. *The Coming Plague: Newly Emerging Diseases in a World out of Balance.* New York: Farrar, Straus and Giroux.

———. 2005. "The Next Pandemic?" *Foreign Affairs* (July/August).

Henderson, D.A. 1993. "Surveillance Systems and Intergovernmental Cooperation." In *Emerging Viruses*, edited by Stephen S., Morse. New York: Oxford University Press.

———. 2009. *Smallpox: The Death of a Disease.* New York: Prometheus.

Heymann, David L. 2004. "The International Response to the Outbreak of SARS in 2003." *Philosophical Transactions of the Royal Society of London* B 359.

Holbrooke, Richard, and Laurie Garrett. 2008. "'Sovereignty' That Risks Global Health." *Washington Post*, August 8.

Hooker, Claire. 2007. "Drawing the Lines: Danger and Risk in the Age of SARS." In *Medicine at the Border: Disease, Globalization, and Security, 1850–Present*, edited by Alison Bashford. Basingstoke and New York: Palgrave MacMillan.

Institute of Medicine. 1992. *Emerging Infections: Microbial Threats to Health in the United States.* Washington, DC: National Academy Press.

Jonas, Hans. 1985. *The Imperative of Responsibility: In Search of an Ethics for the Technological Age.* Chicago: University of Chicago Press.

Keil, Ulrich. 2010. "Introductory statement by Prof. Dr. Ulrich Keil." Social, Health and Family Affairs Committee of the Parliamentary Assembly of the Council of Europe. Hearing on "The Handling of the H1N1 Pandemic: More Transparency Needed?" Strasbourg, January 26.

King, Nicholas. 2002. "Security, Disease, Commerce: Ideologies of Post-Colonial Global Health." *Social Studies of Science* 32 (5/6): 763–89.

Lancet. 2007. Editorial: "International Health Regulations: The Challenges Ahead. " *Lancet* 369, May 26.

Langmuir, Alexander. 1963. "The Surveillance of Communicable Diseases of National Importance." *New England Journal of Medicine* 268 (4): 182–92.

Lipsitch, M., S. Riley, S. Cauchemez, et al. 2009. "Managing and Reducing Uncertainty in an Emerging Influenza Pandemic." *New England Journal of Medicine* 361 (2): 112–5.

Lowe, Celia. 2010. "Viral Clouds: Becoming H5N1 in Indonesia." *Cultural Anthropology* 25 (4): 625–649.

Macrae, Fiona. 2010. "The 'False' Pandemic: Drug Firms Cashed in on Scare over Swine Flu, Claims Euro Health Chief." *Daily Mail*, January 17.

Morse, Stephen S. 1992. "Epidemiologic Surveillance for Investigating Chemical or Biological Warfare and for Improving Human Health." *Politics and the Life Sciences* 11: 1.

———, ed. 1993. *Emerging Viruses*. New York: Oxford University Press.

Neustadt, Richard, and Harvey Fineberg. 1983. *The Epidemic That Never Was: Policy Making and the Swine Flu Scare*. New York: Vintage.

Osterholm, Michael. 2005. "Avian Flu: Addressing the Global Threat." Testimony before the House Committee On International Relations. http://www.flutrackers.com/forum /showthread.php?t=62. Accessed on May 16, 2013.

Parliamentary Assembly of the Council of Europe. 2010. "The Handling of the H1N1 Pandemic: More Transparency Needed."

Preston, Richard. 1998. *The Cobra Event*. New York: Ballantine.

Public Radio International. 2010. The World: "Flu Vaccine Overstock." January 11. Transcript available at http://www.theworld.org/2010/01/flu-vaccine-overstock/.

Redfield, Peter. 2013. *Life in Crisis: The Ethical Journey of Doctors without Borders*. Berkeley and Los Angeles: University of California Press.

Rosenberg, Charles. 1998. "Pathologies of Progress: the Idea of Civilization as Risk." *Bulletin of the History of Medicine* 72: 4: 714–30.

Schnirring, Lisa. 2008. "Supari Accuses Rich Nations of Creating Viruses for Profit."

Stein, Rob. 2010. "Millions of H1N1 Vaccine Doses May Have to Be Discarded." *Washington Post*, April 1.

Walsh, Bryan. 2007. "Indonesia's Bird Flu Showdown." *Time*, May 10.

Weir, Lorna, and Eric Mykhalovskiy. 2007. "The Geopolitics of Global Public Health Surveillance in the 21st Century." Pp. 240–63 in *Medicine at the Border: Disease, Globalization, and Security, 1850—Present*, edited by Alison Bashford. Basingstoke and New York: Palgrave MacMillan.

Whalen, Jeanne, and David Gauthier-Villars. 2010. "European Governments Cancel Vaccine Orders." *Wall Street Journal*, January 10.

World Health Organization. 2007. *The World Health Report 2007: A Safer Future. Global Public Health Security in the 21st Century*. Geneva: WHO.

———. 2008. International Health Regulations (2005). 2d ed. Geneva: WHO.

———. 2009. "Pandemic Influenza Preparedness and Response: A WHO Guidance Document." Geneva: WHO.

———. 2011. *Implementation of the International Health Regulations (2005)*. Report of the Review Committee on the Functioning of the International Health Regulations (2005) in Relation to Pandemic (H1N1) 2009, May 5.

Wright, Susan. 2006. "Terrorists and Biological Weapons." *Politics and the Life Sciences* 25 (1): 57–115.

FIFTEEN

Imagined and Invented Worlds

SHEILA JASANOFF

Science and technology have been involved in efforts to reimagine and reinvent human societies for close to two hundred years. Social theory, however, has yet to embrace this key dimension of modernity and to acknowledge the centrality of these two institutions in constructing the futures toward which we direct our presents. The concept of sociotechnical imaginaries aims to do just that. It takes as its starting point the resurgence of theoretical interest in the nature of collective self-understandings but offers a more comprehensive framework for thinking critically about why societies follow the paths they do and why some formations endure while others weaken and wither. It incorporates pervasive elements of modern life that normally are not included in the analysis of power and organization, such as expertise, intellectual property, bioethics, nuclear power, computers, the fear of pandemics, or genetically modified (GM) seeds. In this respect, the sociotechnical imaginary as laid out in this book is a voyaging concept: it facilitates theorizing across disciplinary boundaries by taking in ordinarily neglected dimensions of social thought and practice. And in so doing, it offers as much analytic mileage to traditional social sciences such as anthropology, history, sociology, legal studies, and political theory as it does to science and technology studies understood narrowly.[1]

Just as imagination liberates the mind to rise beyond the constraints of the possible, so too the lens of sociotechnical imaginaries enables us, as analysts, to look for patterns and juxtapositions that cut across the conventional grid lines of disciplines. Imaginaries, to some extent, belong to the disciplinary common property of anthropology, but they have found their way into political theory and thereby into new projects of analysis and explanation, through such anthropologically minded scholars as Benedict Anderson and Charles Taylor. Imaginaries have also migrated into science and technology studies through writers concerned with the interpenetration of knowledge, materiality, and power. How, for example, do our coproduced sociotechnical formations bear on the potential for concerted action, stability, resistance, or conflict (Jasanoff 1996, 2001, 2004; Mitchell 2002)? As noted in the introduction, sociotechnical imaginaries offer enlightening perspectives on questions of difference, time, space, and identity that form the classic subject matter of social thought. The introductory chapter mapped the major theoretical and methodological issues raised in working with sociotechnical imaginaries: when does it make sense to invoke the concept; with what methods can we study them; and what counts as evidence of their existence or effects? In this concluding chapter, I offer a more comprehensive narrative of social change, accounting for both lock-ins and transcendence, that emerges from empirical work on the nature and operation of sociotechnical imaginaries. This is a reflection on what we should be looking for when we study sociotechnical imaginaries and what lessons we can glean from such undertakings about the social fabrication of power and meaning.

The essays in this collection serve at one level as case studies, illustrating how sociotechnical imaginaries take shape in varied social and cultural contexts and how they in turn help reorient the evolution of those contexts. These studies exemplify but also elaborate on the definition offered in the introduction. Sociotechnical imaginaries, in our view, are "collectively held, institutionally stabilized, and publicly performed visions of desirable futures, animated by shared understandings of forms of social life and social order attainable through, and supportive of, advances in science and technology." Standing on their own, and marrying theory and observation, the chapters play rich variations on this definition. Taken together, however, they also provide a powerful new angle on world making: one that rejects linear causality and excessively actor-centered histories while at the same time retaining an empirical focus on where transformative ideas come from, how they acquire mass and solidity, and how imagination, objects, and social norms—including accepted modes of public reasoning and new technological regimes—become fused in practice.

Reading across the cases, one finds a number of recurrent emphases that add up in effect to an account of collective belief formation in scientifically and technologically engaged societies. First, work in the imaginaries framework necessarily invites us to examine the origin of new scientific ideas and technologies and the social arrangements or rearrangements they help sustain. This is a fundamentally humanistic inquiry that recognizes the capacity of individuals and groups to see and think things differently from what was previously seen or thought. Second, by inquiring into imagination as a social practice, we follow the embedding of ideas into cultures, institutions, and materialities, whereby the merely imagined is converted into the solidity of identities and the durability of routines and things. Third, several of the cases illustrate moments of resistance, when new conceptions of how to change the world bump up against the old, or when powerful competing imaginations struggle to establish themselves on the same social terrain. These periods of emergence and conflict shed light not only on the components of successful imaginaries but also on the social apparatus that supports their continued potency. Resistant imaginaries function as a political force that can either enable or disable thoroughgoing social transformations. Last, in tracing sociotechnical projects from conception to realization, we explore the phenomenon of extension, the complex of processes by which unconventional ideas gain traction, acquire strength, and cross scales, for example, by persisting through time or by overcoming geopolitical boundaries.

These four phases in the development of sociotechnical imaginaries bridge several perennially problematic boundaries that have long troubled the social sciences: descriptive and normative, structure and agency, material and mental, local and translocal. Imaginaries reveal a dynamic interplay between binaries that are too often kept analytically distinct; they build on the world as it is, but they also project futures as they ought to be. Articulated and championed by agents of change, working within and beyond localized sites of action, imaginaries get built into the hard edifices of matter and praxis. Once situated in the specifics of time, place, and social worlds, they still have power to move minds and actions at a distance; and, as constructs in part of human thought, they remain continually open ended and subject to revision. In analyzing how sociotechnical imaginaries have played out in varied concrete settings, we thus position the concept to do more general work within the analytic and explanatory repertoires of the social sciences.

This chapter follows sociotechnical imaginaries from origin to extension, noting how the four phases of turning imagination into social practice are exemplified in each preceding chapter. In revisiting these chapters, I roughly track the order in which they appear in the table of contents, and yet this is an imperfect organizational strategy, since to a greater or lesser extent each chapter illustrates all four phases of the dynamic described above—a point we try to make clear through cross referencing. Nonetheless, origin stories figure more prominently in some early chapters, especially those focused on individuals, while material or institutional embeddings and resistance feature more in others, as does extension in still a third group. To repeat a metaphor borrowed from the introduction, the chapters form a braided whole in the study of sociotechnical imaginaries. Each tells its own distinct saga of an imagined and invented world, originating in dreams and ambitions but substantiated into people, objects, and practices. Yet each chapter also relates to one or more others with respect to sites and modes of action and the mobilization of specific aspects of imagination and of material technologies that provide instruments for drawing together a collective social or political imagination.

Origins

It takes daring individuals to dream up new worlds and give personal embodiment to loosely circulating cultural aspirations, and individuals figure prominently in many of the stories told in the preceding chapters. Thus, in William Storey's account, Cecil Rhodes arrives in South Africa in 1870, a seventeen-year-old vicar's son from Hertfordshire, to try his hand at farming. The land overwhelms him, much as the Hudson Valley painters of a slightly earlier day were overcome by the sweeping vistas of their grand new world. Later, as capitalist and industrialist, dreamer and technocrat, Rhodes learns to master the land. He exploits South Africa's natural resources, especially its mines, as well as its native-born people. But the farming instinct remains alive and well in this transplanted Englishman, who in time develops vineyards as well as mines and railroads. His infamous political imaginary of racial segregation, precursor to the scourge of apartheid, comes to fruition in tandem with a less well documented imaginary of segmented sociotechnical development. In Rhodes's vision, Storey argues, overlapping dichotomies of black and white, labor and leisure, city and country map onto a nation segregated not only by race but by the parallel economies of its dirty, extractive, mining industries and its sometimes idyllic agricultural enclaves.

A continent and a century away, Vannevar Bush, President Franklin D. Roosevelt's unofficial wartime science adviser, confronted a future in which the problem was no longer how to meld technology and politics into governing institutions but how to save science from its own instrumental successes. Bush envisioned a grave threat to science from its close partnership with the state at the end of the Second World War. This, as Michael Dennis argues, was a specter of state control so like the Soviet Union's disastrous embrace of Lysenkoism that it figured as a "monster" in Bush's imagination. Dennis traces the production of a Cold War US imaginary of science and the state through the tension between Bush's attempts to carve out a space of "basic science," wholly free from politics, and the political scientist Don K. Price's alternative plan to integrate politics with science in his 1954 book *Government and Science*. Price, who had the enviable opportunity to work his ideas into pedagogical practice at Harvard's newly minted School of Government, in effect won that contest. He rightly foresaw that there was no turning back on the union between science and politics forged during the war, if not long before (Forman 2007). In Price's imagination, however, the potential power imbalance could be managed by a networked state, in which universities and other independent entities would deliver wise, confidential advice on how to direct research in the public (read *military*) interest. A deep irony of Price's sanitized vision was that it ignored the plight of J. Robert Oppenheimer, the physicist and national hero whose suspect loyalty to the state became the centerpiece of the infamous 1954 security hearings. Caught in the monstrous jaws of the Cold War imaginary that Bush so feared, Oppenheimer—a human symbol of science's subservience to politics—illustrates in his fall from grace the pathologies of the networked state that Price sought to render tractable through a professionalized science policy service.

The figure of the monster that energized Vannevar Bush's postwar science policy surfaces, in one form or another, in any attempt to imagine worlds radically different from those that look threatening in the present. As such, the fear of monsters may be integral in some sense to imaginaries of resistance or revolution such as those discussed in the chapters by Felt, Moon, and Barker. A point worth noting here, however, is that monsters can fire up institutional as well as individual imaginations. In my comparative work on biotechnology policy, for example, I suggested that divergent national regulatory choices in the late twentieth century responded, in part, to culturally specific notions of what is morally repugnant, or monstrous, in projects to manipulate nature (Jasanoff 2005a; see also Burri; Hurlbut, this volume). Such potent ideas of what has to be avoided—the discordant harmonics of risk and disorder that trouble the uplifting strains of technological progress and liberation—are part and parcel of the construction of imaginaries.

Cecil Rhodes's African nation building, and eventually his imperial project, drew heavily on the model of the British Empire as he saw it. He imagined bringing "the whole uncivilized world under British rule" (Storey, this volume), although his efforts ultimately led to a form of rule entirely particular to South African conditions and circumstances. Similarly, late twentieth-century nation builders found models to appropriate, rhetorically if not in practice, from postcolonial states that elbowed their way into modernity during the postwar period. Rwanda's Paul Kagame belongs to this cadre, as does to some extent Onno Purbo, the "father of the Indonesian Internet," described by Joshua Barker. Kagame, in Warigia Bowman's narrative, confronts problems that are almost the opposite of those faced by Rhodes: a nation already divided by blood and the backwash of genocide, a shattered economy, elites in disarray, and ruling institutions in ruins. With so much destruction on the ground, it is poetically just that this seasoned warrior seized on an aerial metaphor, the imaginary of Southeast Asia's "flying geese" (Akamatsu 1962), and of Singapore in particular, to will a new Rwanda into being. Bowman's chapter describes the inevitable disjunctions between the envisioned model and its mundane contexts of application. Kagame's is a dream of technology in the abstract, divorced from the institutional and cultural supports—such as English language proficiency and relative income equality—that made the Singaporean information technology revolution succeed. Kagame's Rwanda progressed toward healing its deadly ethnic rifts, but big gaps remained between the leader's soaring rhetoric of progress and on-the-ground social and technological realities.

Embedding

Ideas matter in the origin stories of imaginaries, whether they begin in the minds of single individuals, in projects of like-minded activists (Barker, Kim, Moon, this volume), in corporate boardrooms (Smith, this volume), or among professionals such as bioethicists trained to think together for the common good (Hurlbut, this volume). But ideas about scientific and technological futures need to gain assent outside such bounded communities in order to become full-fledged imaginaries. Often, they must latch onto tangible things that circulate and generate economic or social value: commodities like wine or diamonds; artifacts such as defensive weaponry or GM crops; legal instruments such as licenses allocating intellectual property rights; or, as in the cases of bottom-up innovation described by Moon and Barker in their stories about Indonesia, the relative hardness of long-entrenched cultural expectations and interpersonal relations. This hybridization, or coproduction (Jasanoff 2004) of ideas, materiality, values, and sociality happens through processes that we call embedding.

Sociotechnical imaginaries are similar in many respects to large technological systems, although they are made up in part of individual visions, dreams, and ambitions. The idea of inspiration still permeates much talk and thinking about the imagination. Imaginative faculties are imputed to specially gifted people who have not lost the childlike capacity for wonder and can call new worlds into being through sheer creative brilliance. The "technoscientific imaginaries" that George Marcus (1995) and other anthropologists have written of are largely the figments of individual envisioning. Reward structures in science, moreover, recognize and reinforce the idea of genius, which creates powerful role models for scientists; more recently, parallel structures have arisen to reward not only discovery but also invention. But one person's vision does not make an imaginary any more than one swallow calls a summer into being. It may take considerable mental effort to rethink the process of creative projection not as one person's "intellectual scheme" but rather as a collective reflection on a group's "social existence" (Taylor 2004, 23); not as "mere contemplation" but as an "organized field of social practices" (Appadurai 2002, 50). Nonetheless, historians of technology have long insisted that creating novel material objects involves more than individual insight; indeed, this may be the reason why the "charismatic inventor" is a rarer figure in popular mythology than the "genius scientist." And work in science and technology studies has extensively documented that even Nobel Prize–winning discoveries rely on social, cultural, economic, and normative structures in order to achieve their effects in the world (see, for example, Forman 1971, 2007; Rabinow 1996).

As many of the foregoing chapters illustrate, a foretaste of change, even when it originates with a sole progenitor, needs to be laid down upon economic, material, and social infrastructures in order to take hold at population-wide or nationwide levels. Cecil Rhodes had to move mountains, not merely in his mind but literally, and to harness new machines in order to get at the gems that propelled his stratospheric rise along South Africa's economic and political axes. His early reflections were prescient: "Some day I expect to see the kopje one big basin where once there was a large hill" (Storey, this volume). His was a grounded, physical vision of progress, one that could not have come into being without skills, labor, connections, money, and machines—and the subjugation of bodies that worked to minds that planned the digging of pits, the laying of rails, and the ruling of continents.

Through deployments of labor and capital, including nature's capital, imaginaries get embedded in the concrete artifacts of industrial civilization, be they massive in size like engineered landscapes, medium like nuclear power plants (Felt, this volume; Jasanoff and Kim 2009), or tiny and mobile like genetically modified organisms (GMOs; Chen, Smith, this volume). Analogies and histories matter. Kagame in Rwanda buys into a model of development sanctioned by assumed success in Singapore. The Chinese government, according to Chen, draws on its own long history of state-sponsored rice cultivation when it buys into a genomic vision of plant breeding. To be sure, that move gains strength from a newer imaginary of scientific modernization; it also displaces farmers in the innovation chain, allowing genomic scientists to take the lead in solving the nation's food security problem.

Embedding also occurs through group reflection by publics and other

328 / Sheila Jasanoff

nonstate actors on remembered pasts and desired futures. Ulrike Felt provides a fascinating account of the mobilization of citizens' memories to produce a new imaginary of Austria as free from technologies seen as incompatible with national identity. Focus group research by Felt and her colleagues documents that, by the first decade of the twenty-first century, Austrian citizens took pride in being a small nation that had stood up against technologies deemed both harmful and non-Austrian. Ordinary Austrians congratulated themselves on having innovated policy instruments, such as warning labels, to enable a regime of nuclear and GMO-free coexistence within the European Union. This assertion of "epistemic subsidiarity" (Jasanoff 2012) rested on an assembling of factual and counterfactual events that together constituted a new narrative of Austrian autonomy. A powerfully imagined "balcony scene" in which the words "Austria is free" were allegedly spoken, although there is no record of that utterance being made there, an inconclusive antinuclear referendum later seen as dispositive, and a successful though possibly ineffectual referendum against agricultural biotechnology—all these were pieced together by citizens and the media into a compelling picture of a small state going its own way in choosing and rejecting technologies.

Memory work and imaginaries similarly loop together in J. Benjamin Hurlbut's chapter on the legacy of Asilomar in US deliberations on the ethics of biotechnology. Here, in contrast to the Austrian case, it is not citizens or nongovernmental organizations who play upon collective memory, but rather scientists and ethicists who jointly affirm the potent myths of value-free and self-regulating science that are so central to American political culture (Jasanoff 2005b). These expert communities recollect Asilomar as a golden moment in which scientists took responsibility for the risks and rewards of their novel creations. Bearing little resemblance to the actualities of the 1975 Asilomar meeting, which was notable mainly for ratifying a reassuring discourse of containment around GMOs, Hurlbut's "Asilomar-in-memory" shores up broader political and policy settlements consistent with the overarching American national imaginary of progress through scientific and technological innovation. Most significant here is the idea of a linear progression from innovation at the bench to technological applications to eventual release into the market and the world. That linear model remains a powerful resource for policy makers, even though it has been repeatedly, and thoroughly, discredited through empirical investigation (see, e.g., Pielke 2007; Stokes 1997). Hurlbut's analysis shows how the adoption of that convenient model by leading bioethicists in effect disciplines democracy itself, by permitting scientists to define when, where, and even in what terms debates about the value of technological advances should take place. Science's right to self-governance, legitimated through the construct of Asilomar-in-memory, thus produces as its significant other a constrained understanding of democracy, in which scientists in effect define important parameters of political representation while purporting merely to declare the state of the world as it is.

The embedding of sociotechnical imaginaries takes place, then, through many partially overlapping pathways—from the production of things such as GM rice or nanomaterials that have hoped-for futures designed into them to the subtle, unacknowledged processes of collective "remembering" of events, which possibly never took place, in order to construct meaningful translations from pasts that were, to presents that are, to futures as people would like them to be. It is through embedding, whether material as in objects or psychosocial as in memories and habits of social interaction, that imaginaries are effectively translated into new contexts. Embedding thus performs an important part of the work of extension that allows imaginaries to spread across cultures, time, and space; but it is not friction free.

Resistance

Imaginaries, as we have argued throughout this volume, occupy a hybrid zone between the mental and the material, between individual free will and group habitus (Bourdieu 1990), between the fertility of ideas and the fixity of things. Most importantly, however, sociotechnical imaginaries can become integrated into the discourses and practices of governance, and thereby structure the life worlds of larger groups, including entire nations and even transnational communities. That integration of imagination with rulership is perhaps easiest to discern in the early phases of a technology's introduction, when evolving regulatory systems are grappling with alternative framings of risks and benefits. Moments of resistance, which threaten the disintegration of older settlements, offer additional insights into the underlying structures and assumptions of power.

Imaginaries move through the realm of resistance in double guise, sometimes raising impediments to the spread of new ideas and at other times crystallizing the dissatisfactions of the present into possibilities for other futures that people would sooner inhabit. Revolutions, whether in science (Kuhn 1962) or in social order (Taylor 2004), can be seen as the overthrow of one no longer sufficient imaginary by another that looks more promising.

Key to such complete and radical transformations is a widespread resistance to the status quo that makes the projected alternative appealing, believable, and worth attaining, even through immense struggle and sacrifice.

Heterodox imaginations are by no means guaranteed to succeed, especially when the dominant imaginary itself is strongly rooted in culture and history. Sang-Hyun Kim describes a situation in South Korea where repeated, forceful attempts to assert visions counter to those of state-supported, technocratic, developmental nationalism failed to win wide popular support. Kim shows that opposing imaginaries were at play in the development of nuclear power, the regulation of biotechnology, and the demonstrations against beef potentially contaminated with bovine spongiform encephalopathy imported from the United States. It may matter that two of Kim's cases—nuclear power and biotechnology—involve technologies through which states in the postwar period have particularly sought to affirm their high standing among nations. Given nuclearity's wartime history and its ties to both military and energy self-sufficiency, nuclear power resonates quite differently in Korea from more recent new and emerging technologies. Nonetheless, it is worth noting that, unlike in Felt's account of the Austrian case, the repertoires of resistance in South Korea seemed unable to coopt the discourses of nationhood and national autonomy.

In each South Korean controversy, activist efforts foundered against the national fear (a spectral monster) of losing an all-important competitive edge in relation to other countries. Put differently, the top-down sociotechnical imaginary of industrially driven development, repeatedly expressed in South Korea's desire to be among the world's technology leaders (Kim 2014), proved resistant to the countervailing demands of groups wishing to democratize the national politics and ethics of technology. There was no other equally compelling model of what the nation might stand for to provide the foundations for a radically new sociotechnical imaginary.

By contrast, the Indonesian father-son activists Hasan Poerbo and Onno Purbo relied less on physical vehicles to carry their imaginations and more on long-standing cultural notions of what connects people to people to form robust networks of concerted action. Muhammad Hatta and Hasan Poerbo, as described by Suzanne Moon, imagined Indonesian development in active opposition to the imaginary of New Order developmentalism espoused by President Suharto. Overthrowing President Sukarno, modern Indonesia's first leader, Suharto grabbed power in a bloodbath of anticommunist violence in which half a million people died. He then enacted a vision of economic development that was patronage dependent, dominated by large conglomerates, and preferring centralized industrial policy over decentralized agricultural production. Against this controlling imaginary, the anticolonial activist and revolutionary thinker Hatta stood for a model of Indonesian development that centered on cooperatives rather than on conglomerates.

Hatta's vision was both communitarian and utopian: an understanding of the nation as a family, realizing its potential through participatory self-governance and initiative (*swakarya*). The contrast between Hatta and Indonesia's first two presidents underscores a point not sufficiently obvious from Anderson's theorization of nationalism: the imagination that binds nations together rests in turn on normative, culturally conditioned ideals of solidarity, permitting multiple "nationalisms" to coexist within a single nation. Which nationalism governs is then itself a prime matter of contestation, as borne out in early twenty-first-century electoral conflicts all over the world. While industrialization was not central to Hatta's project, Moon suggests that his imaginary was sociotechnical in its emphasis on infrastructures that would advance social cohesiveness as well as economic production. Hatta advocated development from the bottom up, through self-help and concern for others in linked, small-scale projects.

It fell to the architect Hasan Poerbo to realize Hatta's imaginary in a project for mass housing. Although specific connections between the two men are hard to document, Moon finds persuasive convergences in their thinking. Poerbo saw mass housing as a sociotechnical project that was also an alternative mode of building Indonesia's future, relying, as Hatta had favored, on cooperatives and substituting communitarian "people mindedness" for an industrialized "efficiency mindedness." Instead of drawing on networks of patronage radiating out from Jakarta, Poerbo's project aimed to capitalize on widely distributed artisanal skills and traditional knowledge of residential construction throughout the nation. In the lineage from Hatta to Poerbo, then, Moon finds a resilient counterarticulation of an Indonesian imagined community, one that is both social and technological, yet one that relies on an utterly different networking of knowledge, skills, and communal values from the New Order put in place by Suharto.

Joshua Barker's story of the founding of the Indonesian Internet likewise illustrates the force of an imaginary shaped outside the structures of state power and integrating technological capacities with a self-conscious "politics of freedom." The originator of this vision, Onno Purbo (Hasan Poerbo's son), conceives of the Internet, while studying in Canada, as a counterpoise to the top-down, state-controlled development of satellite communications by the Suharto regime. *That* system, based on the Javanese oath of kingship (*palapa*), after which it was named, corresponds more closely to Anderson's

analysis of nationalism as a communal bond enforced through centralized control of thought and communication by capital and the state. Key
to Purbo's thinking, by contrast, is the dispersed character of the Internet,
enabling communication between one and many and uniting people across
geographical space as well as across time. The network within Indonesia,
Purbo wrote, should resemble the network that already connected Indonesian students living abroad, but in a community imagined as both national
and antidespotic. This oppositional vision of an emerging technological
system illustrates at once the phenomena of extension, in its borrowing of
experiences from other points of origin, and of resistance, in the local reart-
iculation of those distant visions to fight realities on the ground.

Notably, the tactics Purbo uses to bring his model of the Indonesian
Internet into being are those of the political organizer rather than the engineer of physical interfaces. He and his associates see themselves (in their
own terms) as guerilla warriors, building a freedom movement from within.
Minds must be persuaded and hearts won over, in addition to expertise
and infrastructure being built. Much effort is therefore expended on teaching and training a widening network (work never undertaken in Kagame's
Rwanda with its fractured civil society) and on speeches and rallies to form a
counterpublic that in time grows strong enough to influence state policy. On
display here are the classic dynamics of coproduction, with talented actors
using social levers to reconfigure the technological: an ethos (freedom), a
set of social identities (guerillas, liberationists), an understanding of what
the Internet is (a commons), and aspects of the Internet's design (accessibility) are produced in a single transformative process of enacting a new
sociotechnical imaginary.

Extension

In an early article, Bruno Latour (1990) offered an influential account of
how scientific knowledge becomes universal. His explanation centered on
the moves by which localized observations are converted into readable representations, or inscriptions, that in turn are disseminated by centers of calculation, perhaps better conceptualized as centers of power, discipline, and
control (see introduction). Many of today's welter of portable knowledge
claims, especially those rendered in graphic, quantified, or cartographic
forms, seem on the surface to conform well to Latour's analysis—although,
as two decades of climate change controversy illustrate, even the most carefully constructed representations may encounter unexpected friction and
resistance when they travel (Jasanoff and Wynne 1996; Mahony 2014).

Work on imaginaries, however, suggests a fundamentally different explanation for how scientific and technological ideas acquire dominion over time and territory. This dynamic works not through the frictionless mobility and erasures of context that black-box scientific facts or technological systems but rather through the global circulation of already powerful sociotechnical imaginaries—such as those of the endless scientific frontier, innovation-driven progress, or global health crisis—which are then re-embedded into local constellations of production and practice. Imaginaries, unlike Latour's notion of circulation, are symmetrical with respect to the production and reception of ideas and artifacts; scientific and technological ideas, in short, are produced together with ideas about science and technology. Sociotechnical imaginaries belong in this respect to the framework of coproduction developed in *States of Knowledge* (Jasanoff 2004).

This more complex circulation depends, as many of this volume's contributions demonstrate, on translation agents who are capable of moving imaginaries from one sociopolitical setting to another. Cecil Rhodes imports a remembrance of rural England into South Africa and couples that with ideas of British rule, eventually inscribing lines of division between industry and agriculture, and rulers and ruled, on a foreign landscape. Paul Kagame launches his information technology policies in Rwanda with an eye toward Singapore, hoping that his East African nation will take off like the flying geese of another continent. Onno Purbo, in Joshua Barker's telling, accentuates with specifically Indonesian overtones the discourse of liberation associated with the early years of cyberspace. The imaginary constructed by Purbo and his colleagues, Barker says, "was an extension of a globalizing rhetoric about the Internet commons, but it was also inflected by the specific history of technology and nationalism in Indonesia." Extension, in other words, calls for a situated re-embedding in order for translated imaginaries to take root and flourish in new soil.

Though individuals matter (as also in classic accounts of actor-network theory), unsurprisingly it is institutions of governance that operate as some of the most effective agents of extension. Institutions already have jurisdiction, that is, they control well-demarcated tracts of physical or virtual territory in which they exercise authority and implement the rules of the game. In the modern world, moreover, few institutions fulfill their jurisdictional functions without recourse to science and technology; institutional imaginaries often come with technological promises and perils built right in (see Burri, Hurlbut, Kim, this volume).

Nancy Chen dissects one such mode of extension through governance, the incorporation of biotechnology into China's long national engagement with culturing rice. Operating through the private, but publicly funded, Beijing Genomics Institute (BGI), China muscled its way into rice genome sequencing with a "China first" strategy that led to the selection of the nation's preferred *indica* strain as the target for sequencing. At BGI, work on the rice genome played out alongside other projects of distinctively Chinese interest, such as research on the panda and the silkworm. BGI emerged in short order as a site for hybridizing international genomic science and technology with Chinese priorities so as to underwrite a biotechnological imaginary of, by, and for the nation.

The process of integrating biotechnology into the Chinese government's promise to safeguard the nation's food security required strengthening some networks at the expense of others, Chen argues. The genome sequencing effort forged strong links between BGI scientists and the state but diminished the role of farmers, who were relegated to the position of technicians while scientists became de facto rice breeders through their specialist knowledge of plant genetics. While giving agricultural biotechnology a national face, then, the Chinese story also paradoxically extends and internationalizes a transnational imaginary of science-led development that puts China on the same page as the United States with respect to naturalizing—indeed imperializing (Jasanoff 2006)—a particular vision of GM technology. BGI also participated in demoting the relatively more complex understandings of ecologists and farmers in relation to the techniques of molecular biology. These are moves that Austria, along with other European states, successfully resisted (Felt, this volume), showing that extension can never be taken for granted or free from contestation.

Extension, moreover, does not imply an abandonment of political particularities. Regula Burri's chapter uses cross-national comparison to illustrate variations between US and German approaches to defining nanotechnology as a regulatory object. She shows how benefits and risks were differently weighted in these two policy environments, although both were extremely hospitable to and supportive of this promising technological sector. The US sociotechnical imaginary emerged as less symmetric. American policy documents assessed the presumed benefits of nanotechnology more highly than possible negative consequences and put greater emphasis on achieving economic and political leadership than on environmental hazards. US governmental authorities also imagined risks to be manageable, whereas German authorities stressed the minimization or avoidance of risk. In keeping with its more positive outlook on the future, US policy placed greater emphasis on producing the ideal citizen-consumer who would help ensure a successful market for nanoproducts across a wide range of applications.

Accordingly, while German policy envisioned a need for two-way dialogue with publics, the US National Nanotechnology Initiative (NNI) invested in research centers to educate citizens so as to foster what NNI planners saw as more informed decision making. Burri suggests that these differences in framing and policy focus speak to the institutionalized ways of public knowledge making, reasoning, and uptake that I have called civic epistemologies (Jasanoff 2005b, 2012). In other words, sociotechnical imaginaries are embedded in the political cultures of nations. Projects of world making succeed best when they are well synchronized with ongoing projects of nation building and the reaffirmation, or reperformance, of dominant national identities.

Burri directs her attention to the public sphere, where institutionalized discourses of reason are always and already in play. Elta Smith offers one of the clearest examples of a private sector institutional imaginary in her analysis of Syngenta Corporation's "humanitarian contract" for Golden Rice. In her analysis, the process of licensing this vitamin-enriched rice strain for use in developing countries becomes a locus for sophisticated forms of boundary drawing. Smith focuses particularly on the tacit accommodations between Syngenta's private economic interests and its interpretation of its "corporate social responsibility," itself shorthand for justifying self-governance by multinational corporations in the twenty-first century. This too is at bottom a story of coproduction, showing how a new biological entity, Golden Rice, operates simultaneously as an engineered device promising to deliver needed nutrients to at-risk bodies and as a bearer of corporate ideologies of property and ownership. As a material-legal package, the rice and the license that enables its distribution divide the world's consumers into two distinct classes—imagined as either capable or incapable of innovating for themselves. Indeterminate contractual terms such as "humanitarian research and use" in effect delineate, and potentially hold in place, distinctions between passive consumers (the "developing"), who may derive present benefit from "donations" but are not allowed to profit from their use, and active producers (the "developed"), who are regarded as capable of deriving future value by adding to Syngenta's inventions.

International organizations are prominently involved in creating, institutionalizing, and extending sociotechnical imaginaries. One salient example of such extension is the emergence of globalism itself as a newly imagined space of governance in the latter part of the twentieth century. Clark Miller traces the growth of the global sociotechnical imaginary through the interaction of three frames: global security, global systems, and global governance. Humanity as a whole confronted threats perceived to be on a worldwide scale during and soon after the Second World War, largely as a result of the atomic bombing of Hiroshima and Nagasaki. But as Miller demonstrates through a close reading of statements by President Harry Truman and other officials, response capability was still thought to lodge in nation-states, whose job was to create the "one world" in which all could work together to solve those big problems.

The crafting of a new imaginary of governance, with the scale of problem definition and response moved up and away from the nation-state, was a product, Miller argues, of science-driven thinking in the specialized agencies of the United Nations system, most notably the World Health Organization (WHO) and the World Meteorological Organization (WMO). Increased modeling capabilities played their part, along with other scientific techniques of visualization and calculation that allowed populations to be constructed independently of national boundaries, such as people infected by HIV-AIDS or at risk from sea level rise or pandemics. The result is not a simple superseding of national authority by global institutions but rather the emergence of new, expert-driven imaginations of how threats to the human condition should be framed and which institutions are best positioned to offer relief.

Andrew Lakoff's chapter on global health offers an interesting counterpoint to Miller's story. Like Miller, Lakoff focuses on the emergence of the global as a space of problem framing and agenda setting, with the WHO again taking center stage by developing the imaginary of "global health security." This potentially apocalyptic scenario feeds on the confluence of novel pathogens, rapid circulation of people, and poorly distributed monitoring and response capabilities around the world. Outbreaks of infectious diseases such as Ebola, SARS or severe acute respiratory syndrome, and humanly transmissible avian flu (e.g., H5N1) have prompted a discourse of preparedness for public health disasters, which, in the ideal case, will never be realized. To implement this imaginary, WHO has attempted to hold together a vast network of global surveillance with the aid of a new governance tool, the International Health Regulations, whose protocols and provisions remain open to contestation.

Lakoff's study of a global sociotechnical imaginary in the making illustrates much that is problematic about extension. WHO's project of preparedness for global health emergencies depends on the acquiescence, indeed submission, of older centers of calculation (Latour 1990), the national public health authorities that came into being with the rise of the modern state. As so often happens in national political disputes, disagreements between levels of governance, fundamentally a matter of jurisdictional struggle, translate into expert controversies. Lakoff offers two illustrations: the Indonesian Health Ministry's refusal, starting in 2006, to share influenza virus with the international community under a doctrine of "viral sovereignty"; and the debate over whether WHO experts had overreacted in advising expensive and unwarranted precautionary measures against swine flu in 2009. Charges against the WHO experts ranged from faulty modeling to conflicts of interest, the latter rejected by a review committee appointed to study the episode. For our purposes, however, the key point here is that the WHO's imaginary of global governance, underwritten by new knowledge and enabled by new technologies of monitoring and surveillance, remains as yet imperfectly realized. Even in an era of intensified globalization, nation-states remain alive and well as sovereign centers of imagination and governance.

Miller's and Lakoff's chapters chronicle not only the birth of an imaginary of supranational governance but also the frictions and conflicts that accompany any effort to extend imagination's rule over new subjects and territories. Resistance in these cases operates almost as a physical force impeding the supposedly free flow of globalization. But oppositional movements could also stimulate the creation of new sociotechnical imaginaries that might lay the groundwork for far-reaching reforms in current understandings of sovereignty, constitutionalism, and democracy.

Conclusion: Fabricating the Future

Science fiction, I suggested in the introduction, is a repository of sociotechnical imaginaries, visions that integrate futures of growing knowledge and technological mastery with normative assessments of what such futures could and should mean for present-day societies. Utopic or dystopic, these fictions underscore the self-evident truth that technologically enabled futures are also value-laden futures. Science fiction stories express fears and yearnings that are rooted in current discontents, either signaling possible escape routes or painting in morbid colors the horrific consequences of heedlessness in the present. They thus offer a deeper look into—possibly even predictions of—what harms societies are most desperate to avoid and what good they may achieve through foresight and imagination.

Marking its fiftieth anniversary in 2013, the celebrated British television series Doctor Who, the longest running science fiction show ever, and like James Bond a marker of Britain's high cultural standing, gave New Statesman columnist Laurie Penney a chance to comment on fiction's power to shape the future. The time-traveling Doctor, Penney noted, had gone through numerous incarnations in the show's half-century of colonizing the national psyche, with a third of the British population tuning in to the series. Yet, though change was built into the plotline from early on, the eleven Doctors since the show's inception had all been white men. Arguing that it was time to shift basic assumptions about the Doctor's color and gender, Penney (2013, 21) observed, "Sweeping social change usually happens in stories first, and science fiction often has an agenda. What could be more political, after all, than imagining the future?" Indeed! The twelfth Doctor, Peter Capaldi, however, was not to be the harbinger of that particular revolution.

This volume attests to the basic rightness of Penney's intuition—that imagining the future is political—but it carries that intuition one symmetrical step forward, showing that political action is also profoundly imaginative. People, in Penney's view, "do the work of changing the world—but stories give us permission to reimagine it" (Penney 2013, 21). That statement overlooks the degree to which the political life of societies is itself a form of collective storytelling, a joint and several imagining of the purposes and the potential of living and working together on an Earth at once malleable and constraining. Political imaginaries shape the future as they reinscribe or reconfigure the past (see Ezrahi 2012). Politics, in other words, continually enacts, and therefore also engenders, the dreamscapes of modernity. This volume has sought to establish the centrality of science and technology in those acts of imagining, not only through the material productions of technoscience, but through the very ideas and practices of "science" and "technology" as formative, and normative, forces in the world.

Politics, in the terminology of this volume, is a space in which sociotechnical imaginaries originate and flourish. Those imaginaries help explain why societies differ (Burri, Felt), how they evolve through time (Dennis, Kim, Moon, Storey), how powerful visions spread through space (Bowman, Chen, Lakoff, Miller, Smith), and how they in turn burrow into human identity and subjectivity (Barker, Felt). The framework of imaginaries allows analysts to gain purchase on the dynamics of social change, asking how reality comes about at any given moment rather than taking the plainly visible structures of society for granted. In this respect, sociotechnical imaginaries are part of the repertoire of the constructivist and interpretive social sciences. They consistently direct our attention toward the practices of collective sense making and the tacit assumptions that allow collectives to hold together in understandable, sustainable, livable modes of being.

Reading across the preceding chapters, we have identified four phases in the construction of sociotechnical imaginaries: origins, embedding, resistance, and extension. In each phase, there is a tension between stability and change that the lens of imaginaries permits us to interrogate more closely. That push and pull is perhaps clearest in the phase of resistance, which can block change as well as facilitate it. The Indonesian stories told by Moon and Barker suggest, however, that resistance is most likely to lead to reorderings of technology and society when an imagined new order draws on deeper notions of how societies ought to fit together and how they ought to be. Memory work serves a similar function in the phases of embedding and extension, allowing novel technoscientific constructs to be most readily naturalized when they fall in line with the way things are remembered as being. Those recovered memories, as Felt and Hurlbut show in their chapters, become internalized in human subjects, giving new meaning to what it means to be an Austrian, for example, or a member of a biosciences community.

A persistent risk of working with imaginaries is that they may come to be taken as unproblematic, as mere descriptions of things as they are in the collective consciousness of various sorts of groups. Ascribing a fixed ontological status to sociotechnical imaginaries, however, would rob them of their analytic value. Such an approach could easily become both overwhelming and superficial, elevating any and every act of projection or prediction to the status of an imaginary. Instead, what makes the careful study of imaginaries rewarding is that one is forced to look at the stylized moves through which collective mindfulness is trained, moves that may, as we have repeatedly seen, endow social actors with some forms of prescient vision, even with the capacity to move mountains, while rendering them oblivious to alternate forms of organization, order, and justice. An inquiry into sociotechnical imaginaries allows at its best a deep meditation on the basis of a technological society's particular forms of sightedness and blindness, and the trade-offs that inevitably accompany attempts to build a shared normative order.

Above all, the turn toward the imagination, together with an emphasis on the creative potential of science and technology, makes possible a study of alternative futures. That sociotechnical orders are not natural, that they do not reflect any intrinsic properties of humans or things, is by now too well established to need belaboring—at least for interpretive analysts. But the corollary that other worlds are always there for the making is less well understood and still less acted upon. At times, the juggernaut of global capital, driven by the furious whip of technological innovation and tuned to a univocal discourse of progress, seems unstoppable. Yet, as almost all of the studies in this book plainly demonstrate, multiple imaginaries can be spun from the same raw materials of invention and will. By situating their stories in specific contexts of struggle and achievement, the authors in this volume continually point to roads less seen and less traveled but that are there nonetheless for the critical imagination to map and explore. Analyzing sociotechnical imaginaries emerges, then, as a form of intensely political narration, reminding both observers and observed that the seen reality is not the only one about which we can dream.

Notes

1. Many outside the field of science and technology studies understand science and technology studies (STS) to be simply about characterizing how science works and how technological objects and systems are produced. This is STS in a narrow sense. STS, properly understood, includes the full-blown investigation of science and technology in society, hence not only how truth claims are established or machines are made, but also how the social, political, and cultural authority of science interacts with that of other powerful institutions.

References

Akamatsu Kaname. 1962. "A Historical Pattern of Economic Growth in Developing Countries." *Journal of Developing Economies* 1 (1): 3–25.

Appadurai, Arjun. 2002. "Disjuncture and Difference in the Global Cultural Economy." Pp. 46–64 in *The Anthropology of Globalization: A Reader*, edited by Jonathan Xavier Inda and Renato Rosaldo. Oxford: Blackwell; earlier version published in *Public Culture* 1990;2 (2): 1–24.

Bourdieu, Pierre. 1990. *The Logic of Practice*. Cambridge: Polity Press.

Ezrahi, Yaron. 2012. *Imagined Democracies: Necessary Political Fictions*. New York: Cambridge University Press.

Forman, Paul. 1971. "Weimar Culture, Causality, and Quantum Theory: Adaptation by German Physicists and Mathematicians to a Hostile Environment." *Historical Studies in the Physical Sciences* 3: 1–115.

———. 2007. "The Primacy of Science in Modernity, of Technology in Postmodernity, and of Ideology in the History of Technology." *History and Technology* 23 (1/2): 1–152.

Jasanoff, Sheila. 1996. "Science and Norms in International Environmental Regimes." Pp. 173–97 in *Earthly Goods: Environmental Change and Social Justice*, edited by Fen O. Hampson and Judith Reppy. Ithaca, NY: Cornell University Press.

———. 2001. "Image and Imagination: The Formation of Global Environmental Consciousness." Pp. 309–337 in *Changing the Atmosphere: Expert Knowledge and Environmental Governance*, edited by Clark Miller and Paul Edwards. Cambridge, MA: MIT Press.

———, ed. 2004. *States of Knowledge: The Co-Production of Science and Social Order*. London: Routledge.

———. 2005a. "In the Democracies of DNA: Ontological Uncertainty and Political Order in Three States." *New Genetics and Society* 24 (2): 139–55.

———. 2005b. *Designs on Nature: Science and Democracy in Europe and the United States*. Princeton, NJ: Princeton University Press.

———. 2006. "Biotechnology and Empire: The Global Power of Seeds and Science." *Osiris* 21 (1): 273–92.

———. 2012. *Science and Public Reason*. Abingdon, Oxon: Routledge-Earthscan.

Jasanoff, Sheila, and Brian Wynne. 1998. "Science and Decisionmaking." Pp. 1–87 in *Human Choice and Climate Change*, edited Steve Rayner and Elizabeth L. Malone. Washington, DC: Battelle Press.

Kim, Sang-Hyun. 2014. "Contesting National Sociotechnical Imaginaries: The Politics of Human Embryonic Stem Cell Research in South Korea." *Science as Culture* 23(3): 293–319.

Kuhn, Thomas S. 1962. *The Structure of Scientific Revolutions.* Chicago: University of Chicago Press.

Latour, Bruno. 1990. "Drawing Things Together." Pp. 19–68 in *Representation in Scientific Practice,* edited by Michael Lynch and Steve Woolgar. Cambridge, MA: MIT Press.

Mahony, Martin. 2014. "The Predictive State: Science, Territory and the Future of the Indian Climate. *Social Studies of Science* 44 (1): 109–33.

Marcus, George E., ed. 1995. *Technoscientific Imaginaries: Conversations, Profiles, and Memoirs.* Chicago: University of Chicago Press.

Mitchell, Timothy. 2002. *Rule of Experts: Egypt, Techno-Politics, Modernity.* Berkeley: University of California Press.

Penney, Laurie. 2012. "In the Red." *New Statesman,* June 21–27.

Pielke, Roger A., Jr. 2007. *The Honest Broker: Making Sense of Science in Policy and Politics.* Cambridge: Cambridge University Press.

Rabinow, Paul. 1996. *Making PCR: A Story of Biotechnology.* Chicago: University of Chicago Press.

Stokes, Donald. 1997. *Pasteur's Quadrant: Basic Science and Technological Innovation.* Washington, DC: Brookings Institution Press.

Taylor, Charles. 2004. *Modern Social Imaginaries.* Durham, NC: Duke University Press.

ACKNOWLEDGMENTS

This book grows in large part from research funded by a grant from the National Science Foundation (NSF), "Sociotechnical Imaginaries and Science and Technology Policy: A Cross-National Comparison," award no. 0724133. Sheila Jasanoff was the Principal Investigator and Sang-Hyun Kim and J. Benjamin Hurlbut were funded as Postdoctoral Fellows. We gratefully acknowledge the NSF's support, which was indispensable for developing our theoretical framework as well as for carrying out the research on South Korea and the United States that was reported in the chapters by Kim and Hurlbut in this volume.

Any long, collective, intellectual project needs additional support along the way to foster exchanges that allow early ideas to mature. We were very fortunate to benefit from a further NSF grant, "Life in the Gray Zone: Governance of New Biology in Europe, South Korea, and the United States," award no. 1058762, and an award for a project entitled "Biology and Law" from the Faraday Institute's program on "Uses and Abuses of Biology." Sheila Jasanoff was Principal Investigator on both grants; Ben Hurlbut and Krishanu Saha were coinvestigators on the Faraday Institute award.

Further, we would like thank the National Research Foundation of Korea for supporting Sang-Hyun Kim's research and scholarly activities through the HK Transnational Humanities Project (NRF-2008-361-A00005).

We are extremely grateful to Harvard University's Weatherhead Center for International Affairs for a conference grant that brought many of the contributing authors to Cambridge in November 2008 to discuss early versions of their chapters. That conference did much to deepen our understanding of the comparative and cross-national dimensions of sociotechnical imaginaries. It also provided an invaluable opportunity to apply this concept to nations and time periods not covered by our grants and to include additional disciplinary perspectives besides science and technology studies (STS), such as anthropology, history, and history of science.

Colleagues from many countries contributed in varied ways to helping us refine the concept of sociotechnical imaginaries, most notably the 2012 and 2013 Fellows of the STS Program at Harvard, who reviewed and commented on the introduction, and the international scholars who attended a double session on the topic at the 2010 meeting of the Society for Social Studies of Science in Tokyo.

We are indebted to a number of colleagues, including Aziza Ahmed, Stève Bernardin, Yaron Ezrahi, and Rob Hagendijk, who offered sage advice and discerning suggestions for improving the introduction and conclusion. Two reviewers for the University of Chicago Press read the final manuscript with great generosity and critical insight. Their comments helped us tighten the connections among chapters and further highlight the book's central themes. We owe special thanks to Shana Rabinowich for her indispensable technical and administrative support in preparing the manuscript for the Press. We are deeply grateful to Gili Vidan for her heroic assistance in compiling the index.

Finally, and above all, we would like to thank the contributors to the volume for demonstrating through their efforts how an expansive but nebulous concept such as sociotechnical imaginaries can be put to work to illuminate a wide range of problems and topics in the politics of science and technology. Several authors offered crucially important comments on the two chapters that bookend the collection. All responded to our multiple requests for revisions with patience and good humor, lightening our task as editors and making the production of this book an exercise in the best kind of academic give and take.

CONTRIBUTOR BIOGRAPHIES

JOSHUA BARKER is associate professor of anthropology at the University of Toronto. His research interests include political anthropology and the anthropology of new media. He is a contributing editor of the journal *Indonesia* and coedited *State of Authority: State in Society in Indonesia* (Cornell Southeast Asia Program, 2009).

WARIGIA BOWMAN is assistant professor at the University of Arkansas Clinton School of Public Service. Previously, she taught at the University of Mississippi and at the American University in Cairo, Egypt. Bowman holds a PhD in Public Policy from the Harvard Kennedy School and a JD from the University of Texas.

REGULA VALÉRIE BURRI is professor of science and technology studies at HafenCity University in Hamburg, Germany. Her research at the intersections of science, technology, and society centers on topics such as visual knowledge, science and art, and the governance of science and technology. She is the coeditor of *Biomedicine as Culture* (Routledge, 2007).

NANCY N. CHEN is professor of anthropology at University of California Santa Cruz. Her current research focuses on genetically engineered foods and nutriceuticals to explore new boundaries of taste, consumption, and health. She is the author of *Food, Medicine, and the Quest for Good Health* (Columbia, 2009) and *Breathing Spaces: Qigong, Psychiatry, and Healing in China* (Columbia, 2003).

MICHAEL AARON DENNIS is an adjunct in Georgetown University's Security Studies Program and the BioDefense Program at George Mason University. His research interests center on the history, historiography, and the politics of American science and technology. He is completing a book manuscript entitled *A Change of State: Political Culture, Technical Practice and the Making of Cold War America.*

ULRIKE FELT is professor of science and technology studies and head of the STS Department at the University of Vienna. Her research interests focus on issues of governance, democracy, and public participation in technoscience and changing research cultures, especially in the life sciences, biomedicine, and nanotechnologies. From 2002 to 2007 she was editor-in-chief of *Science, Technology, and Human Values.*

J. BENJAMIN HURLBUT is assistant professor in the School of Life Sciences at Arizona State University. He studies the history of governance of bioscience research in the United States. He holds an AB in classics from Stanford and a PhD in the History of Science from Harvard. He was a postdoctoral fellow in STS at the Harvard Kennedy School.

SHEILA JASANOFF is Pforzheimer Professor of Science and Technology Studies at Harvard University's John F. Kennedy School of Government. Her research centers on the production and use of science in legal and political decision making. Her books include *The Fifth Branch* (Harvard, 1990), *Science at the Bar* (Harvard, 1995), *Designs on Nature* (Princeton, 2005), and *Science and Public Reason* (Routledge, 2011).

SANG-HYUN KIM is associate professor at the Research Institute of Comparative History and Culture, Hanyang University, Korea. He holds a DPhil in chemistry from the University of Oxford and a PhD in the history and sociology of science from the University of Edinburgh and is currently involved in the HK Transnational Humanities Project funded by the National Research Foundation of Korea.

ANDREW LAKOFF is associate professor of sociology, anthropology, and communication at the University of Southern California, where he directs the Research Cluster in Science, Technology and Society. He is the author of *Pharmaceutical Reason: Knowledge and Value in Global Psychiatry* (Cambridge, 2006) and coeditor of *Biosecurity Interventions: Global Health and Security in Question* (Columbia, 2008).

CLARK MILLER is associate director of the Consortium for Science, Policy, and Outcomes and Chair of the PhD Program in Human and Social Dimensions of Science and Technology at Arizona State University. His research explores the global politics of science and technology. He is the coeditor of *Changing the Atmosphere* (MIT, 2001).

SUZANNE MOON is associate professor in the history of science at the University of Oklahoma and editor-in-chief of *Technology and Culture*. Her recent research explores sociotechnical imaginaries and narratives about social justice in the history of Indonesia's postcolonial industrialization. She is the author of *Technology and Ethical Idealism: A History of Development in the Netherlands East Indies* (CNWS, 2007).

ELTA SMITH is a managing consultant with ICF International. Her work covers food and environmental policy from "farm to fork," including biotechnology and novel foods, plant variety rights, seed market development, food labeling, organic foods, "better regulation" approaches to policy making, and voluntary agreements in the food sector. She holds a PhD in Public Policy from the Harvard Kennedy School.

WILLIAM KELLEHER STOREY is professor of history at Millsaps College. He is the author of *The First World War: A Concise Global History* (Rowman and Littlefield, 2009); *Guns, Race, and Power in Colonial South Africa* (Cambridge, 2008); *Writing History: A Guide for Students* (Oxford, 1999; 4th ed. 2012); and *Science and Power in Colonial Mauritius* (Rochester, 1997). He is currently writing a biography of Cecil Rhodes.

INDEX